Macmillan

KENYA SECONDARY SCHOOL ATLAS

Editors

Dr Dunstan A Obara
Senior Lecturer in Geography
Department of Geography
University of Nairobi

Dr Richard T Ogonda
Senior Lecturer in Geography
Department of Geography
Kenyatta University

PREFACE

The **Macmillan Kenya Secondary School Atlas** has been designed specifically to meet the requirements of the Kenya Certificate of Secondary Education Syllabus for the teaching of geography. This unique resource atlas has been developed over a period of three years in close consultation with the editors. In addition university departments, geography teachers, ministry officials and individuals have provided most valuable information for the completion of the maps within the atlas. Thus its suitability to the current syllabus, and the latest teaching methods, has been ensured.

The atlas is the result of exhaustive research combined with careful consideration of the educational level, content and presentation of map data. The aim being to enable the student to derive the maximum benefit from the information provided.

In addition to covering the more traditional topics of physical and human geography, the atlas also emphasises current issues of importance such as economic development, population growth, environmental problems and conservation. Concentrating initially on maps of Kenya, the atlas broadens its scope to encompass the neighbouring countries, the East African region and the continent of Africa. Statistics, graphs, diagrams, colour photographs, satellite images, case studies and explanatory text accompany the maps wherever applicable.

The **Macmillan Kenya Secondary School Atlas** is the most appropriate and up to date atlas available for use by students in Kenya.

Macmillan Kenya (Publishers) Ltd
Nairobi
Kenya

CONTENTS

ACKNOWLEDGEMENTS

The publishers would like to acknowledge, with thanks, the following individuals, companies and ministries who have provided information and assistance in the production of this atlas.

KENYA
Central Bureau of Statistics
Department of Geography, Kenyatta University
Department of Geography, Nairobi University
J. Kiguru
M. Kinyanjui
Meteorological Office
Ministry of Tourism
Ministry of Transport and Communications
National Irrigation Board
J. Odero
Population Institute, University of Nairobi
Tana River Development Authority
UNEP (United Nations Environment Programme)

INTERNATIONAL
Department of Geology, Reading University, UK
Department of Soil Science, Reading University, UK
Ewbank Preece Ltd, UK
FAO (Food and Agriculture Organisation of The United Nations)
Flag Institute, UK
Meteorological Office, UK
Nigel Press Associates, UK
UN (United Nations)
UNESCO (United Nations Educational, Scientific and Cultural Organisation)
WLPU Engineering Consultants, UK

The publishers wish to acknowledge, with thanks, the following photographic sources: Compix, Robert Harding Picture Library, Hutchison Library, J R Parry, Picturepoint, Spectrum Colour Library, TROPIX Photo Library, WLPU Consultants
Front cover photographs: Daily Telegraph Colour Library — ESA Meteosat, J R Parry, J Scofield
Map extracts and aerial photographs: Survey of Kenya
Satellite images: Nigel Press Associates Ltd
Back cover artwork (flags): Courtesy of GEOprojects (UK) Ltd

Cartography by Lovell Johns Ltd., Oxford

First published 1990. Reprinted 1990

Macmillan Kenya (Publishers) Ltd
Kijabe Street
PO Box 30797
NAIROBI
Kenya

MACMILLAN
KENYA

ISBN 0-333-45628-9

The Structure of the Earth

The earth is an oblate spheroid in shape, slightly flattened at the poles and bulging at the equator. Its circumference around the equator is approximately 40 000 kilometres; the equatorial diameter being 12 756 kilometres, and the polar diameter 12 714 kilometres.

The Earth's Interior

The earth consists of a series of uniform concentric shells. Its internal structure can be divided into three parts: the crust, a relatively thin surface layer; the mantle, extending down to 2 900 kilometres; and the core, which makes up its centre.

The Crust

The crust is the earth's outer layer, consisting of comparatively low density material. There is a fundamental difference between the composition of the continental crust and the crust found beneath the oceans.

A Section through the Earth's Crust

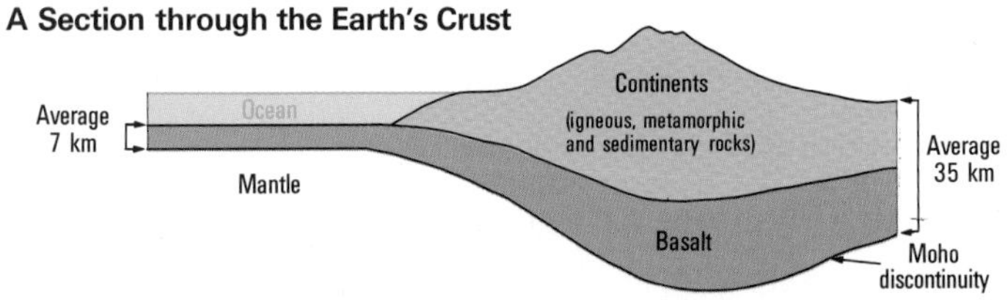

The ocean crust averages about 7 kilometres in thickness. The upper layer is made up of sediments and volcanic lava and is approximately 2 kilometres thick. The principal layer beneath this is made up of basalt and is about 5 kilometres thick. The main constituents of the oceanic crust rocks are silica and magnesium, thus this heavier part of the earth's crust is known as the 'sima' layer.

The continental crust is thicker, averaging about 35 kilometres and its structure is more complex. In most places it can be divided into an upper and lower part; the lower part being similar to the basaltic layer under the oceans. The upper part is composed of a complicated mixture of igneous, metamorphic and sedimentary rocks – all comparatively light rocks. The most abundant elements of these rocks are silicon and aluminium; hence the continental crust is known as the 'sial'.

The base of these crustal rocks is marked by a sharp change in rock density to that of the mantle rocks which are much heavier. This junction is called the Mohorovičić discontinuity (Moho for short) after the Yugoslav scientist who detected it in 1909.

The Mantle

The mantle differs from the crust in chemical composition. Its principal rock is peridotite which consists mainly of iron and magnesium. The upper, rigid part of the mantle can extend to about 100 kilometres below the Moho, and together with the earth's crust, forms the lithosphere.

The lower mantle, extending to 2 900 metres below the earth's surface, is less rigid and hotter. Known as the asthenosphere, this section of the mantle is capable of being deformed over long periods of time. Thus the concept of plate tectonics (see page 110) or the movements of the earth's crust, result from movement of the lithosphere over the asthenosphere.

At a depth of 2 900 kilometres below the earth's surface lies another important boundary where the rock density almost doubles. Discovered in 1914, this is known as the Gutenberg discontinuity, and below this level lies the earth's core.

A Cross-section through the Earth

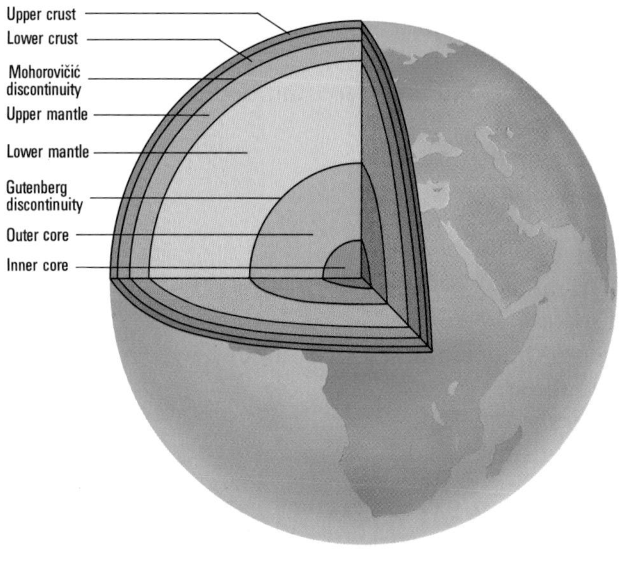

The Core

The core has a radius of about 3 500 kilometres and forms about 33% of the earth's mass. Seismic evidence suggests that the outer core is liquid whilst the inner core, with a radius of 1 220 kilometres is believed to be a solid nickel-iron alloy.

Temperature

There has been much speculation about the temperature of the earth's interior. It is known that temperatures increase with depth, probably at between 20°C-40°C per kilometre to begin with, but this rate is not believed to continue for long. The diagram below shows the estimated temperature at various levels.

The Main Subdivisions of the Earth

	Approximate depth in kilometres	State	Estimated temperature in °C
Upper crust			
Lower crust	50	Solid	400
Upper mantle	100		1 000
Lower mantle	2 900	Semi-solid	3 500
Outer core	5 180	Liquid	3 700
Inner core	6 400	Solid	4 000–4 500

A Section through the Atmosphere

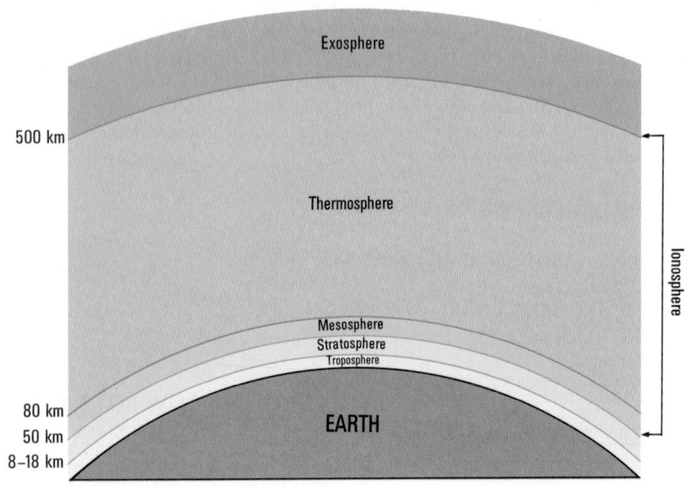

The Atmosphere

The earth is surrounded by a thin layer of gases which form the atmosphere. The atmosphere protects the earth from temperatures which would otherwise reach such extremes between day and night that life on earth would be destroyed.

The lowest part of the atmosphere, known as the troposphere, extends to about 8 kilometres over the poles and to about 18 kilometres over the equator. The temperature falls by 6°C for every kilometre in height until it stabilises at the tropopause, or the top of the troposphere.

The next section of the atmosphere is the stratosphere, which extends to about 50 kilometres. Within the stratosphere temperatures rise again from -55°C to 10°C until the ozone layer is reached, which filters out the sun's harmful ultra-violet rays.

Above the stratosphere is the mesosphere where cooling takes place again. Temperatures of about -100°C are reached at the top of the mesosphere at a height of 80 to 100 kilometres.

Beyond the mesosphere are the thermosphere and exosphere, where temperatures increase again. The exosphere eventually merges into space.

The region of the atmosphere between 50 and 500 kilometres is known as the ionosphere. It contains layers of ions, or electrically charged particles, which reflect radio waves around the earth, making long distance communications possible.

The Solar System

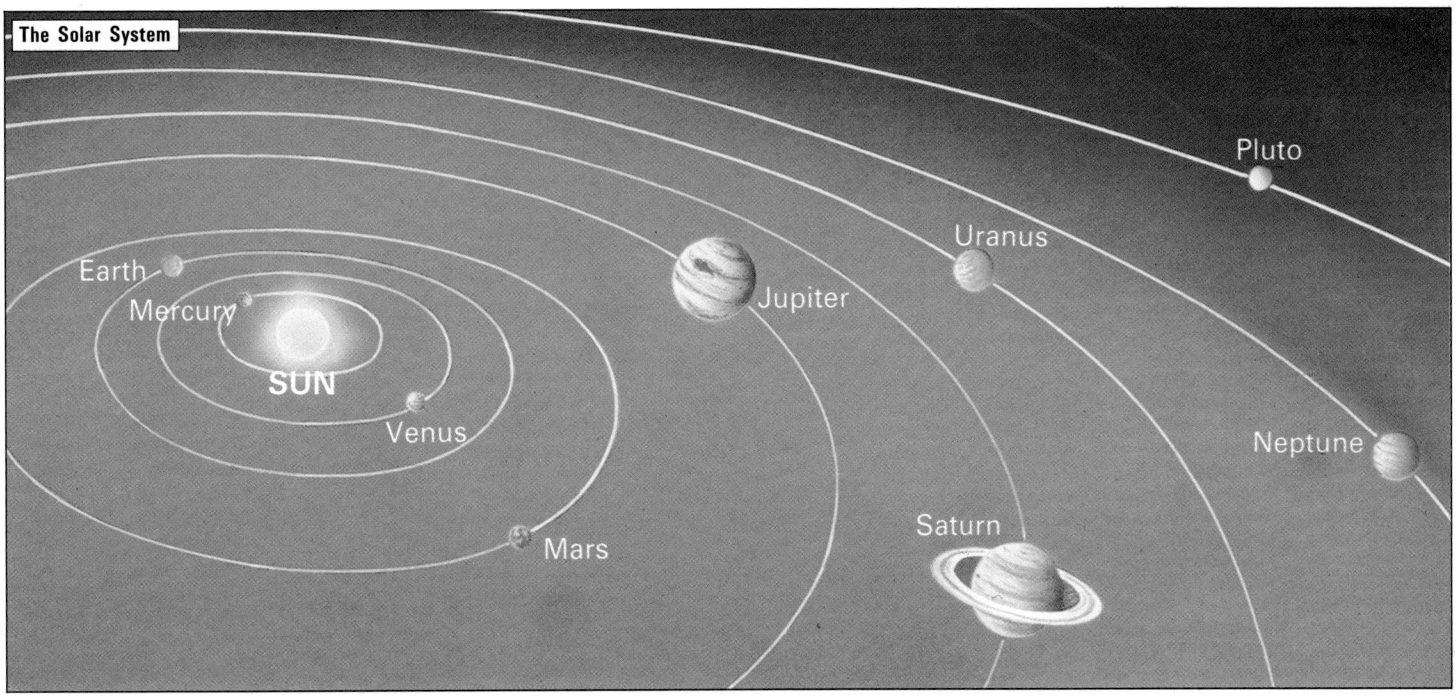

Relative Sizes of the Planets

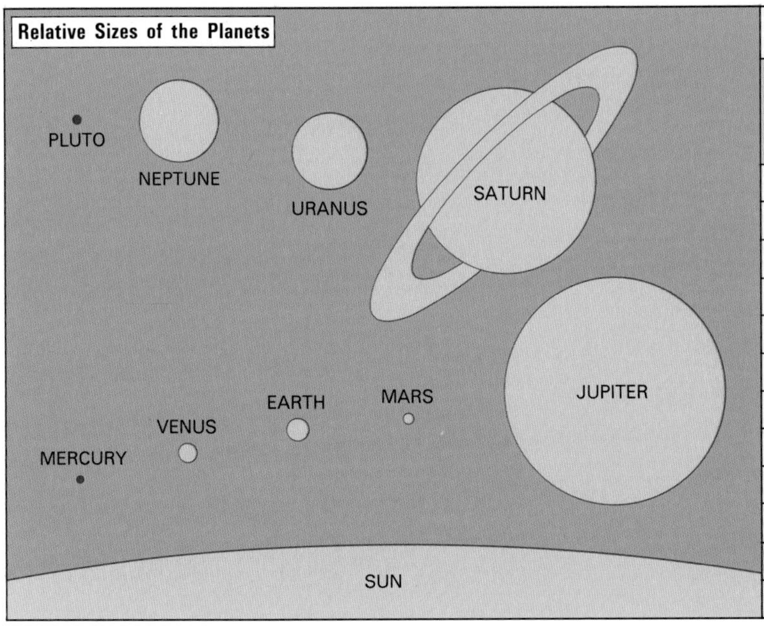

The Solar System (1)

Planet	Distance from Sun (millions of km.)	Period of revolution round the Sun (earth years and earth days)	Diameter (km.)	Number of Satellites
PLUTO	5900	248years 292days	5500?	none
NEPTUNE	4495	164years 292days	48400	2
URANUS	2868·9	84years	47100	5
SATURN	1426	29years 168days	119300	10
JUPITER	777·9	11years 314days	142200	13
MARS	228	1year 322days	6790	2
EARTH	149·5	365·26days	12756	1*
VENUS	107·9	225days	12100	none
MERCURY	57·9	88days	4880	none
SUN	–	–	1390634	none
	(from earth)	(round the earth)		
MOON*	0·38	27·3days	3476	–

The Seasons

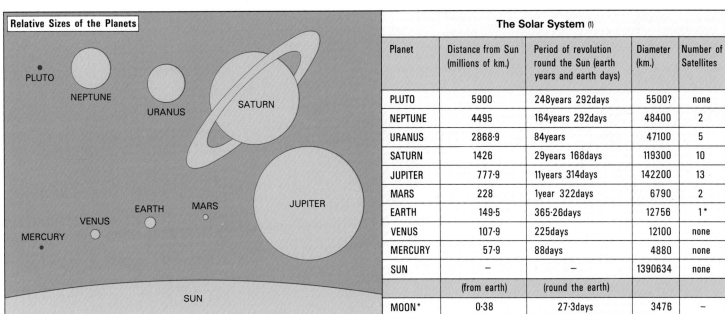

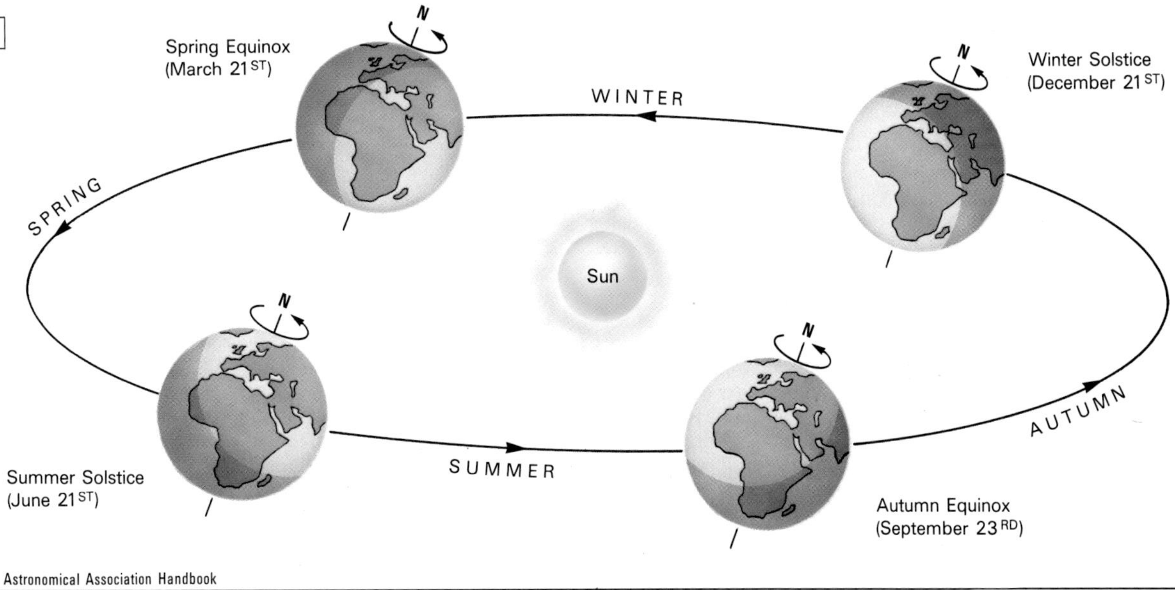

(1) Figures from the British Astronomical Association Handbook

LATITUDE, LONGITUDE AND MAP PROJECTIONS

Latitude and Longitude

Features on the earth' s surface can be located by using a geographical grid consisting of a network of imaginary intersecting lines. The lines run from east to west and north to south.
Lines running from east to west are lines of latitude. The line around the centre of the earth is the equator, and is numbered 0°. All lines of latitude run parallel to the equator, and are numbered in degrees north or south of the equator. The north pole is 90°N and the south pole is 90°S. Lines of latitude are also referred to as parallels.
Lines running from north to south connecting the poles are called lines of longitude or meridians. The prime meridian, numbered 0°, passes through Greenwich, London. All lines of longitude are numbered in degrees east or west of the Greenwich meridian. The line directly opposite the Greenwich meridian is numbered 180°.
Any place can therefore be accurately located by referring to the point where its line of latitude intersects its line of longitude.

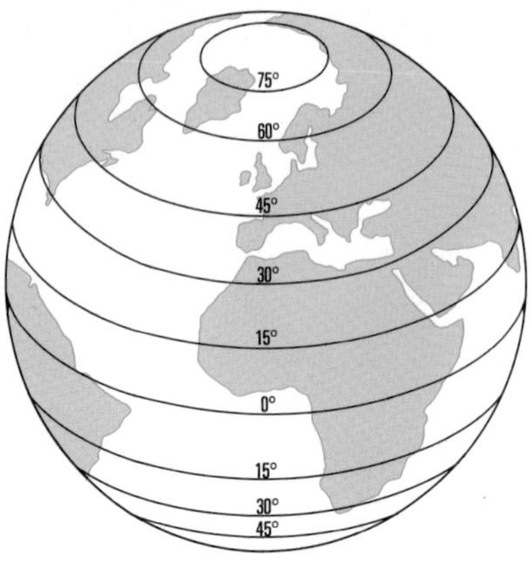

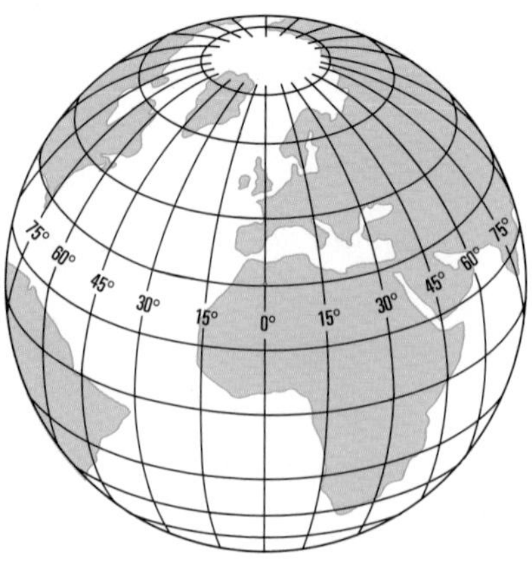

Map Projections

The most accurate method of representing the spherical shape of the earth is by using a scale model called a globe. However it often becomes necessary to represent this three dimensional shape on a flat two dimensional surface such as a piece of paper. To do this a map projection must be used.
The parallels of latitude and the meridians of longitude are projected from the globe onto a flat surface as a network, or graticule, of intersecting lines. To achieve a perfect map we should expect a representation which gives correct shape and direction to features on the earth' s surface whilst keeping areas in proportion. However, because features are being transferred from a spherical surface to a flat surface there will always be some distortion of these properties. The choice of map projection will therefore depend largely on which of the three properties of correct shape, direction or area is the most important, bearing in mind the subject of the map and the geographical region to be covered.
Map projections can be classified into three major groups – cylindrical, conical and azimuthal, but many different forms of projection are derived from these major groups.

Conical Projection

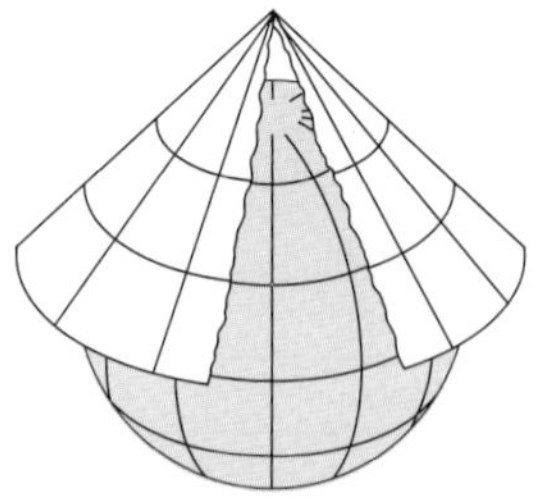

Azimuthal Projection

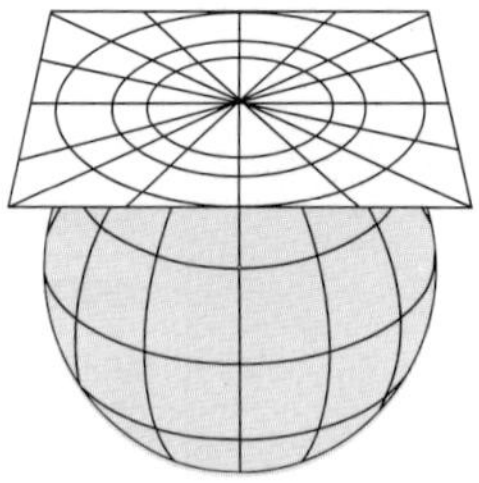

Cylindrical Projection

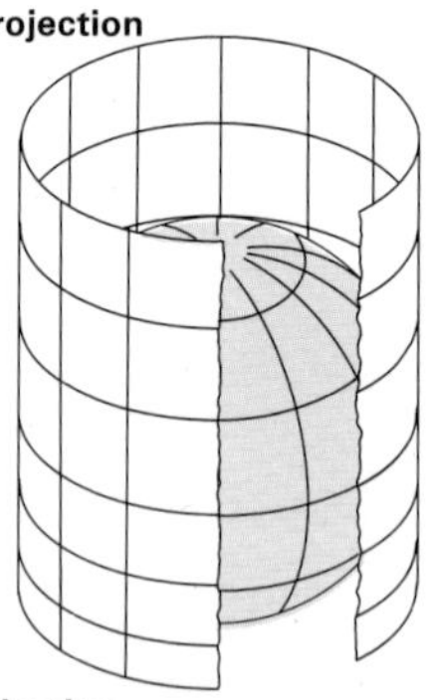

Bonne's Projection

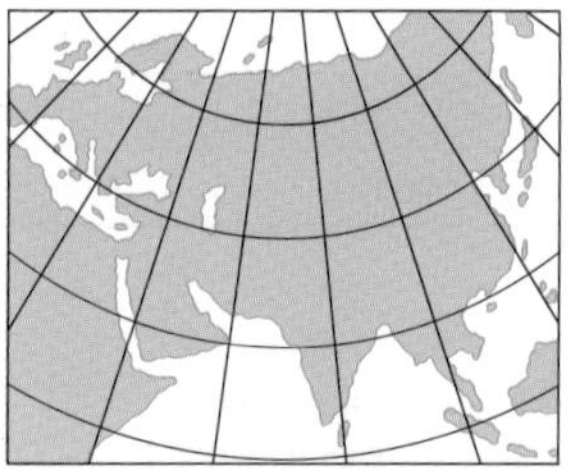

Lambert Equal Area Projection

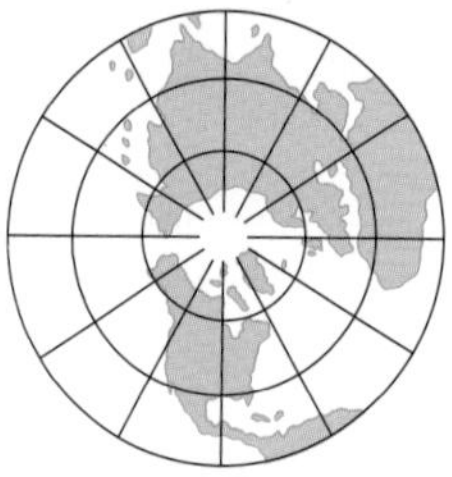

Mercator Projection

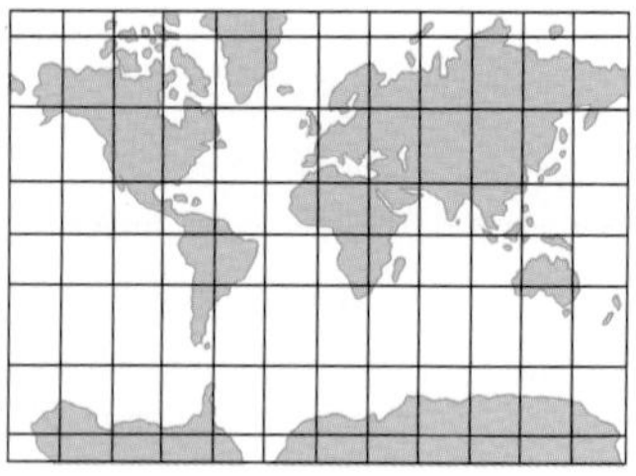

Conical Projections

Conical projections are produced by transferring lines of latitude and longitude (parallels and meridians) from a globe onto a cone. The cone is then developed into a flat map.
Scale is correct at the point where the cone touches the globe – the standard parallel. All other parallels are concentric circles; all meridians are straight lines which converge at either the north or south pole. Exaggeration of scale along both meridians and parallels increases with the distance from the standard parallel.
Bonne' s projection, above, is a simple conic projection which has been modified to remove exaggeration of scale by spacing all meridians to their true distance along each parallel.

Azimuthal Projections

Azimuthal projections are constructed by projecting part of the globe onto a flat surface, or plane, placed directly on a particular point of the globe, such as a pole or the equator.
All meridians radiate as straight lines from the central point, and the parallels of latitude are concentric circles. Direction is true outwards from the centre of the projection, but distortion of scale and shape occurs uniformly away from the centre.
The above diagram illustrates how an azimuthal projection depicts the globe as a flattened disc, in this case with the north pole as its central point.
Azimuthal projections are most frequently used for mapping polar regions.

Cylindrical Projections

Cylindrical projections are constructed by projecting lines of latitude and longitude onto a cylinder wrapped around the globe. The cylinder is then cut along a convenient line and spread out.
The cylinder may touch the globe at the equator, in which case the equator is the only line true to scale. Meridians are equally spaced vertical lines and are at right angles to the parallels. Cylindrical projections are essentially rectangular in shape and the whole globe can be shown.
One of the best cylindrical projections is Mercator' s. In this case distortion of shape is avoided by increasing scale along the parallels, but consequently area is greatly distorted away from the equator.

TIME ZONES AND DAY AND NIGHT

Time Zones

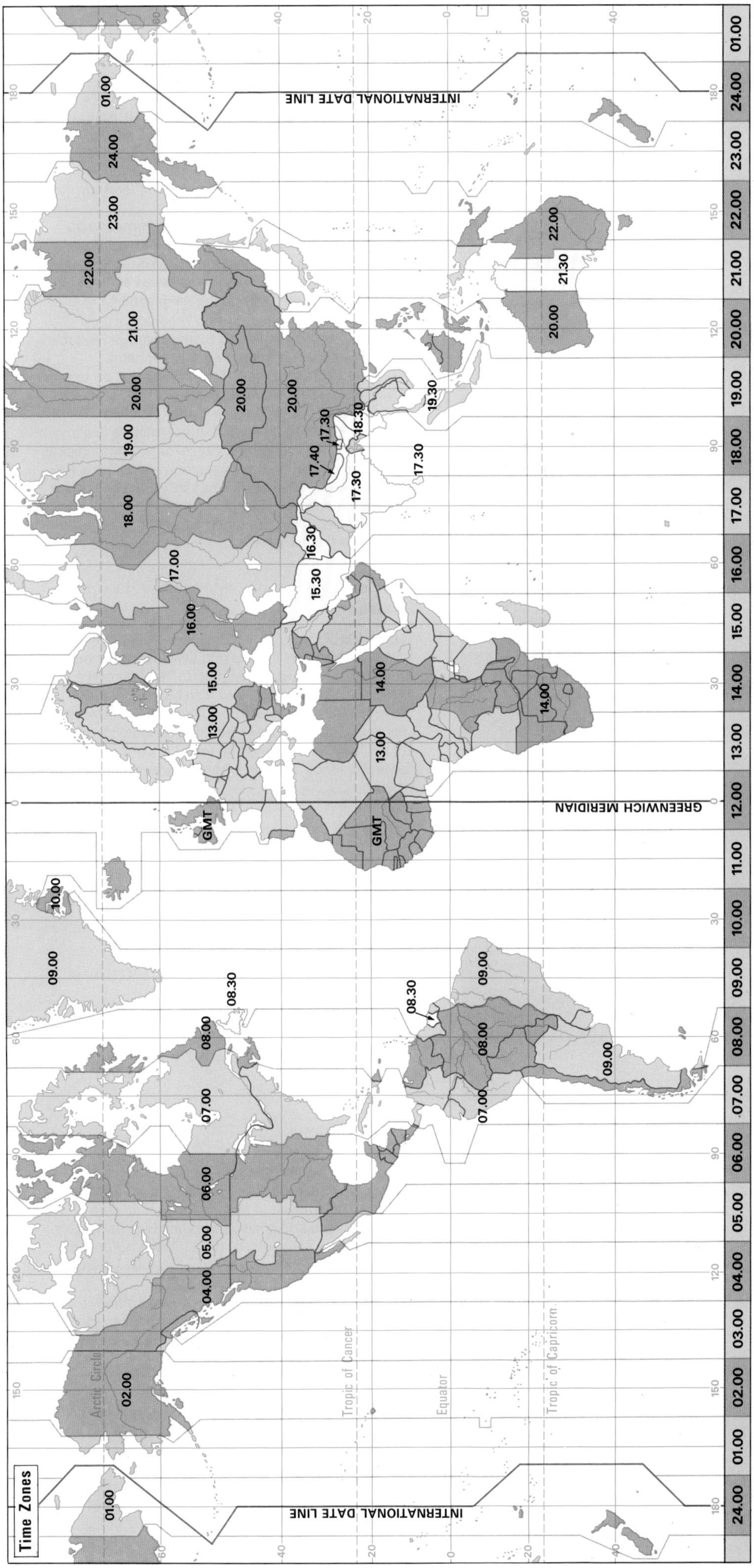

Time Zones

Time varies from standard time zones

Solar Time

GMT Greenwich Mean Time

17.40 Local time

Scale at the Equator
0 1000 2000 3000 4000 5000 kilometres
1:160 000 000

Day and Night

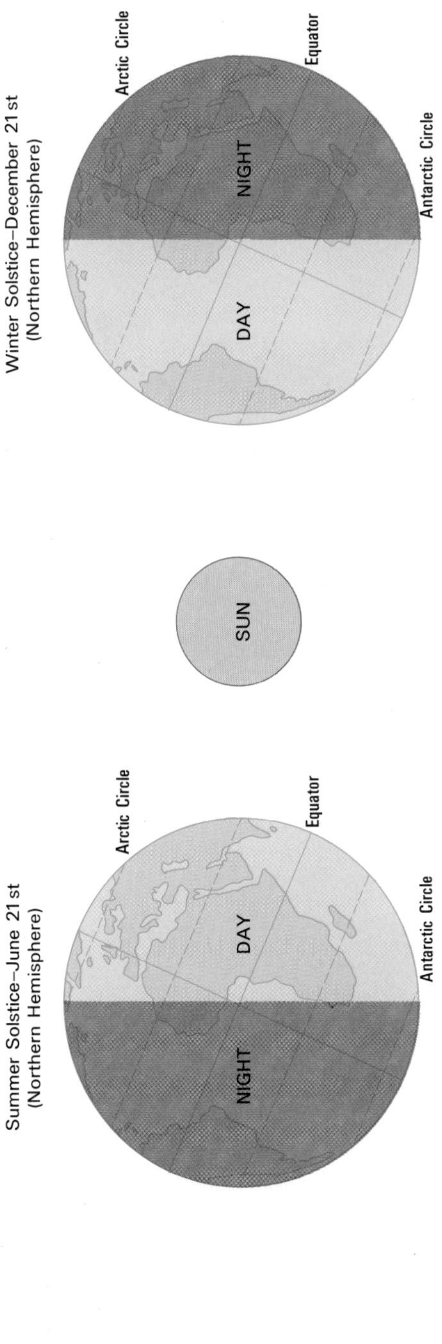

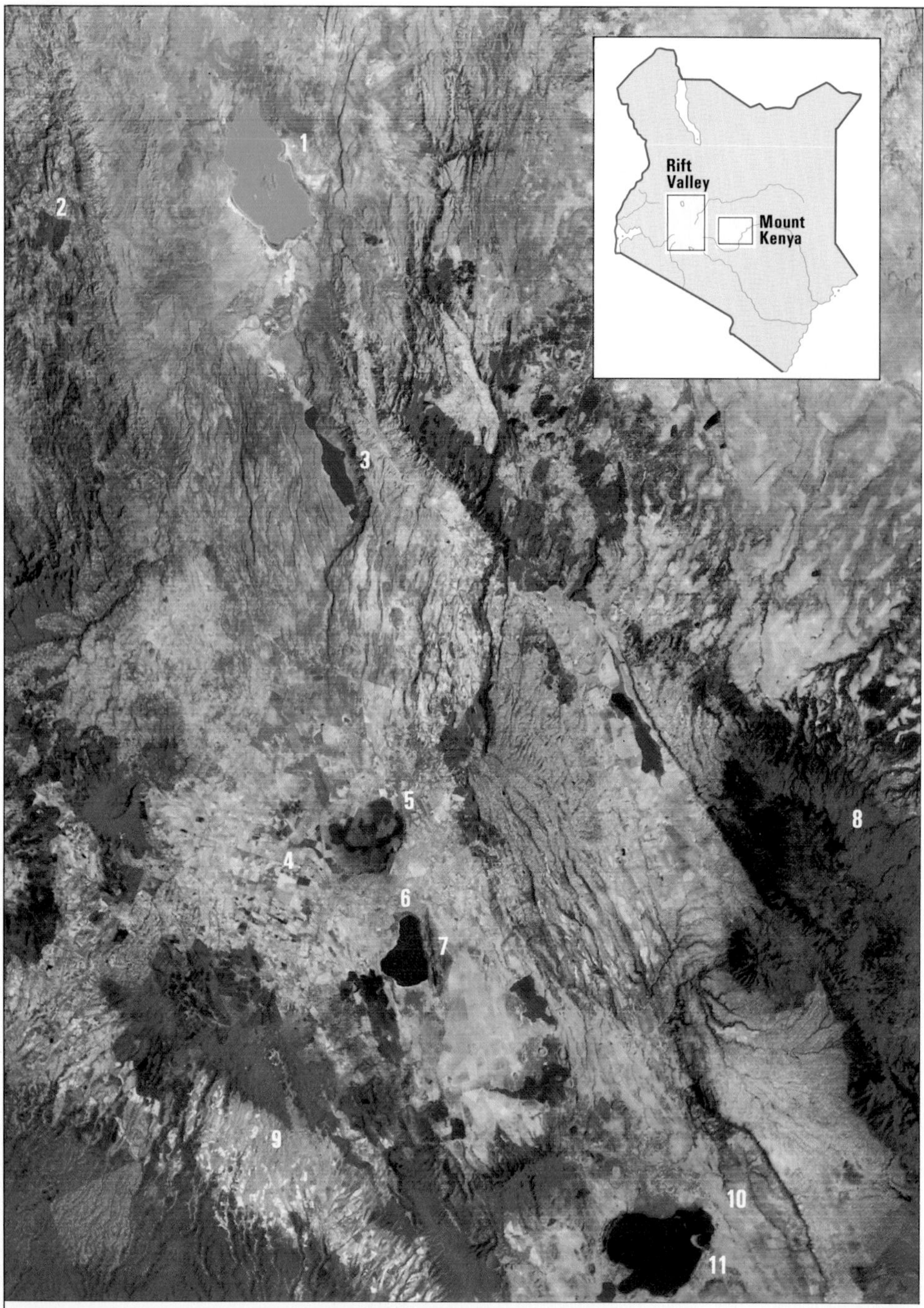

Satellite Imagery

Data gathered by scanning systems aboard satellites can be processed into false colour images which reveal detailed information about the earth' s surface. The images on the left are from the U.S. satellite 'Landsat 2' which orbits the earth at an altitude of 930 kilometres and repeats its orbital pattern every 18 days. Images of the earth' s surface are therefore regularly updated.

Satellite data is a valuable source of information on natural resources and can be used extensively to monitor environmental deterioration. The images assist in the location of mineral resources and ground water, and are invaluable during the planning stages of development projects. Progress of crop growth can also be revealed, thereby contributing to accurate estimates of agricultural production.

The two satellite images of Kenya illustrated on this page are at a scale of approximately 1:1 000 000. The false colour of the image reveals the following information: the red/brownish tones indicate healthy, chlorophyll-rich vegetation. Green/blue tones indicate areas of dry vegetation or vegetation which has a very small leaf area. Solid black areas are open water more than 5 metres deep. The white areas with corresponding black areas immediately to their left are clouds and their shadows.

Rift Valley

This satellite image shows a section of the Kenya Rift Valley. The parallel faults along the sides of the main rift valley can be clearly identified. The floor of the valley, with its series of lakes, can be recognised by the green/blue tones which indicate areas of relatively sparse vegetation. The higher, surrounding slopes are richly vegetated by either forest or cultivated land. Several distinctive features can be indentified on the image:

1. Lake Baringo
2. Tugen Hills
3. Lake Bogoria
4. Fields and pasture land
5. Menengai Crater
6. Nakuru
7. Lake Nakuru
8. Aberdare Range
9. Wheat and barley fields on the Mau Escarpment
10. Naivasha
11. Lake Naivasha

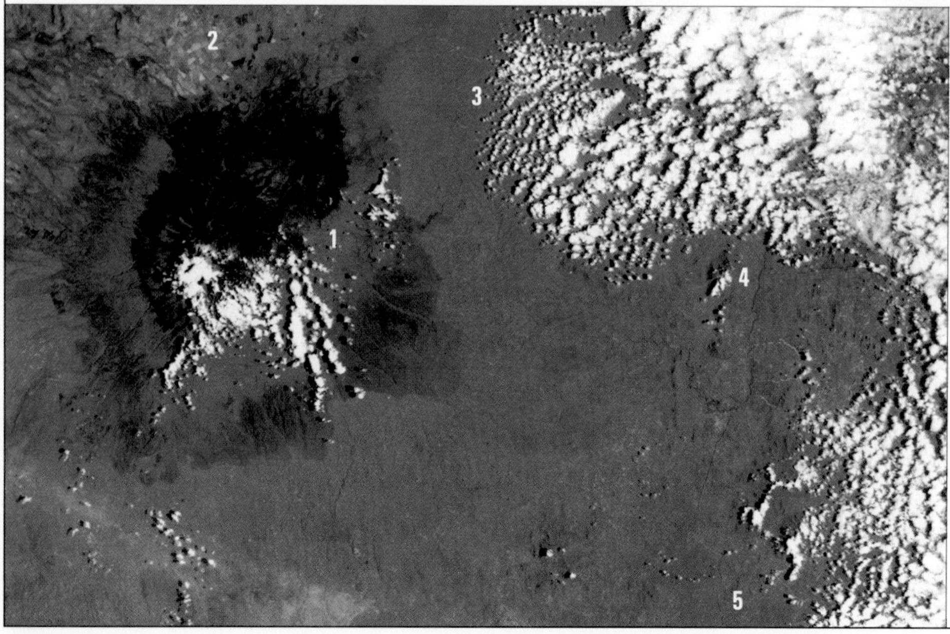

Mount Kenya

The forested slopes of Mount Kenya are the predominant feature of this satellite image. To the north and north-west of Mount Kenya is an area of mixed farmland and grazing; the shapes of fields with different crops can easily be seen. To the north-east of Mount Kenya, the slopes of the Nyambeni Range are an important tea growing area.

The course of the Tana River can be indentified to the right of the image. Since this image was taken the Kiambere Reservoir has been constructed. Page 22 of the atlas shows how the area has benefitted from the development of the Tana River hydro-electric power scheme in recent years.

1. Mount Kenya
2. Mixed farmland and grazing
3. Tea growing area
4. Tana River
5. Site of the Kiambere Reservoir

This page explains the relief representation and most important symbols and lettering styles used in this atlas. The scale of the map is stated on each individual page. Abbreviations used appear at the beginning of the index.

Relief Representation

Diagram A

Diagram B

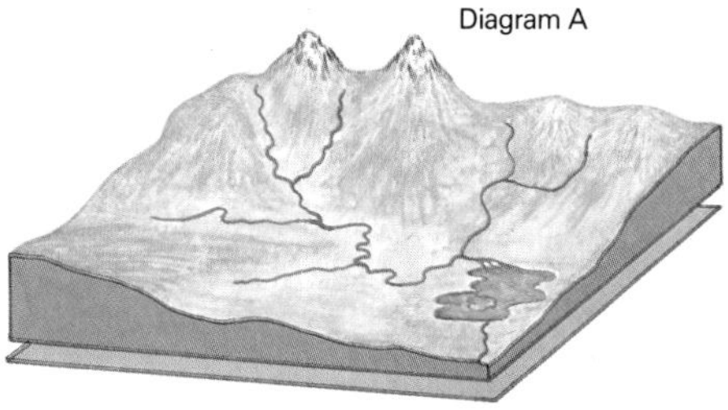

Diagram A illustrates an imaginary landscape. It shows an area of mountain peaks, hills, valleys and a river system that flows into a lake in the lowlands.

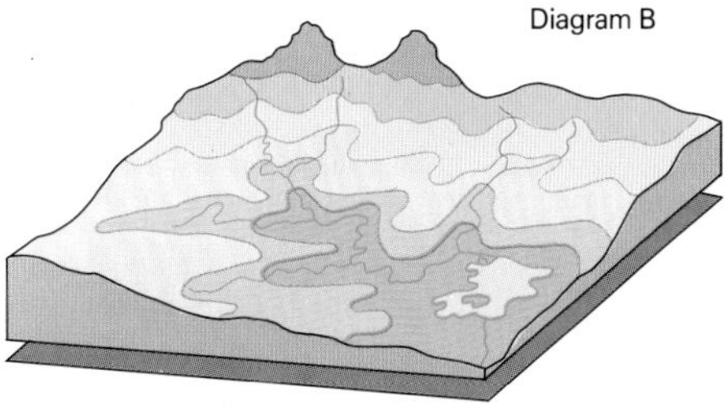

Diagram B shows the same landscape but the relief is shown by a series of coloured layers known as hypsometric layers.

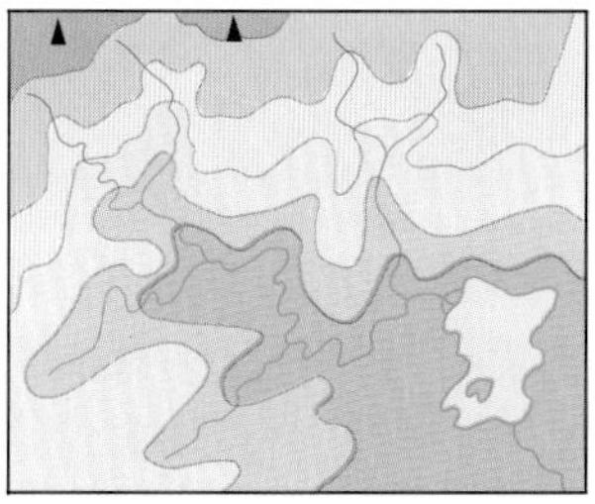

This map is derived from Diagram B, it shows a map of exactly the same area. The lines separating the hypsometric layers are called contour lines. These are lines that join all the places on a map which are at the same height above sea level.

Altitude above sea level in metres

6000
5000
4000
3000
2000
1000
500
200
0

Land below sea level

Depth below sea level in metres

0
200
2000
4000
6000
8000

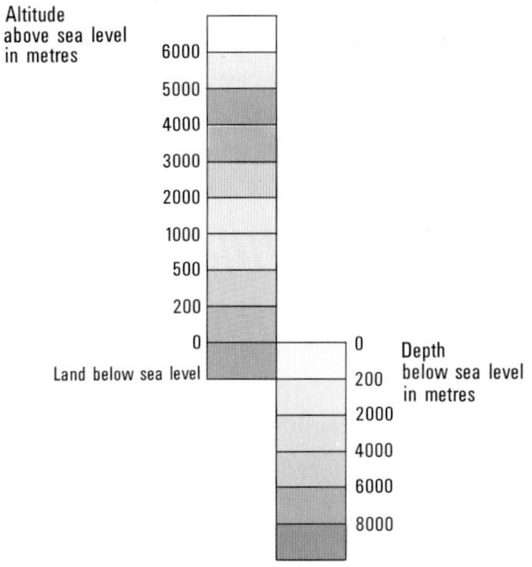

Physical Features

∼	Land contour
▲	Spot height (elevation in metres)
∼	Sea contour
+	Ocean depth (in metres)
◯	Permanent lake, reservoir
◌	Intermittent, seasonal lake
◌	Salt pan
⁎	Marsh, swamp

∼	River
- - ∼	Intermittent, seasonal river
∼	Waterfall, cataract
∼	Dam
⊥⊥⊥	Canal
⊨	Pass
◌	Permanent ice cap
ⵔ	Oasis

Political and Administrative

═══	International boundary
- - -	Undefined or disputed international boundary
━━━	Internal administrative boundary: First order
────	Internal administrative boundary: Second order

Settlement

Each settlement is allocated a town symbol according to its population. The size of lettering used for each settlement is graded to correspond with the appropriate town symbol. Symbols contained within squares denote capital cities

▪	■	over 1 000 000 inhabitants
◉	●	500 000–1 000 000 inhabitants
◎	●	100 000–500 000 inhabitants
▣	●	25 000–100 000 inhabitants
▢	○	less than 25 000 inhabitants

Note: For population of settlements on large scale maps refer to individual key information.

Other Features

×	Prehistoric site
∴	Historic site
⚔	Fort
★	Place of interest
◯	National park, reserve

Communications

Road representation varies with scale

───	Principal road
━━━	Principal road
───	Other road
- - -	Track

large scale maps only

───	Railway
- - -	Railway under construction
✈	International airport

Lettering

KENYA	Country name
NAROK	Administrative area
Réunion (Fr.)	Sovereignty of dependent territory
Cairo	Capital city
Boston Tanga	Settlement
Sahara *Rift Valley*	Physical feature
C. Horn	Cape, point, peninsula
Pemba	Island
INDIAN OCEAN	Ocean, sea
G. OF ADEN	Gulf
Lake Turkana	Bay, lake
Nile	River

Thematic Map Symbols

These are a selection of symbols used on the thematic maps; additional symbols are explained in a key accompanying each map.

Energy

●	Coal	⊗	Nuclear power station
⚒	Hydro-electric power station	⋀	Oil
▱	Natural gas	▢	Uranium

Iron ore, ferro-alloys and steel

⊠	Chrome	▽	Nickel
●	Cobalt	▬	Steel
I	Iron ore	H	Tungsten
◀	Manganese	▽	Vanadium
▶◀	Molybdenum		

Other minerals

▲	Antimony	✣	Mica
Ω	Asbestos	⍉	Nitrates
△	Bauxite	▪	Phosphates
▐	Beryllium	◀	Platinum
▼	Copper	P	Potash
◇	Diamonds	●	Pyrites
⏀	Gold	○	Silver
⊖	Lead and zinc	▱	Sulphur
▬	Lithium	×	Tin
Ⅲ	Mercury	T	Titanium

Forested slopes of the Aberdare Range

Mt. Kenya, the second highest mountain in Africa.

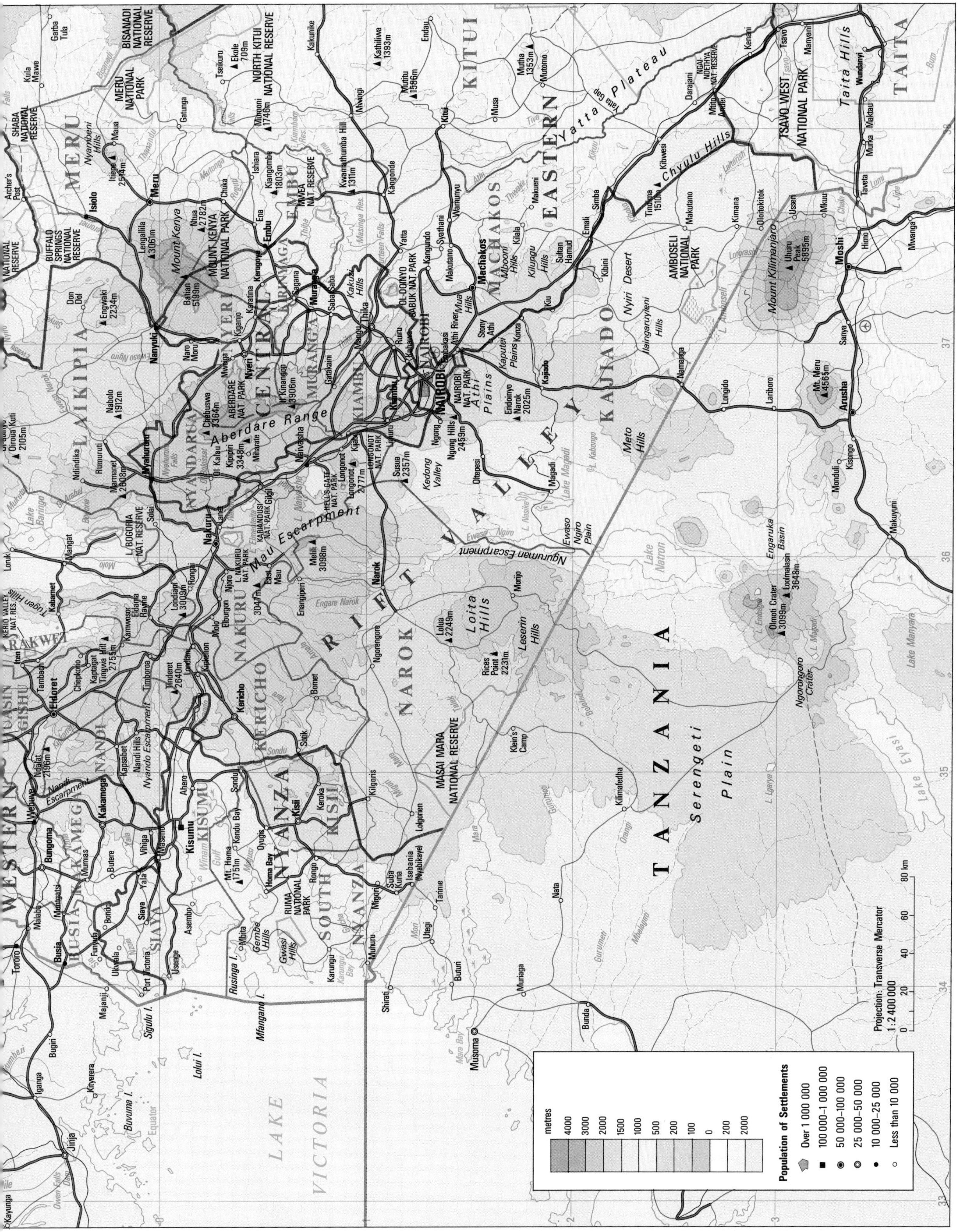
TANZANIA
LAKE VICTORIA
RIFT VALLEY
EASTERN
CENTRAL
NAIROBI
NYANZA
WESTERN
KAJIADO
NAROK
MACHAKOS
KITUI
MERU
LAIKIPIA
TAITA
Serengeti Plain
Aberdare Range
Mau Escarpment
Nguruman Escarpment
Yatta Plateau
Chyulu Hills
Taita Hills
Mount Kenya
Mount Kilimanjaro
Uhuru Peak 5895m
Batian 5199m
Nakuru
Naivasha
Kisumu
Eldoret
Kericho
Narok
Machakos
Kitui
Nyeri
Embu
Meru
Isiolo
Nanyuki
Thika
Kajiado
Moshi
Arusha
Musoma
Kakamega
Bungoma
Busia
Kisii
Jinja
AMBOSELI NATIONAL PARK
MASAI MARA NATIONAL RESERVE
TSAVO WEST NATIONAL PARK
MOUNT KENYA NATIONAL PARK
ABERDARE NAT. PARK
NAIROBI NAT. PARK
metres
4000
3000
2000
1500
1000
500
200
100
0
200
2000
Population of Settlements
Over 1 000 000
100 000–1 000 000
50 000–100 000
25 000–50 000
10 000–25 000
Less than 10 000
Projection: Transverse Mercator
1:2 400 000
0 20 40 60 80 km

KENYA – NORTH EASTERN, EASTERN AND COAST PROVINCES

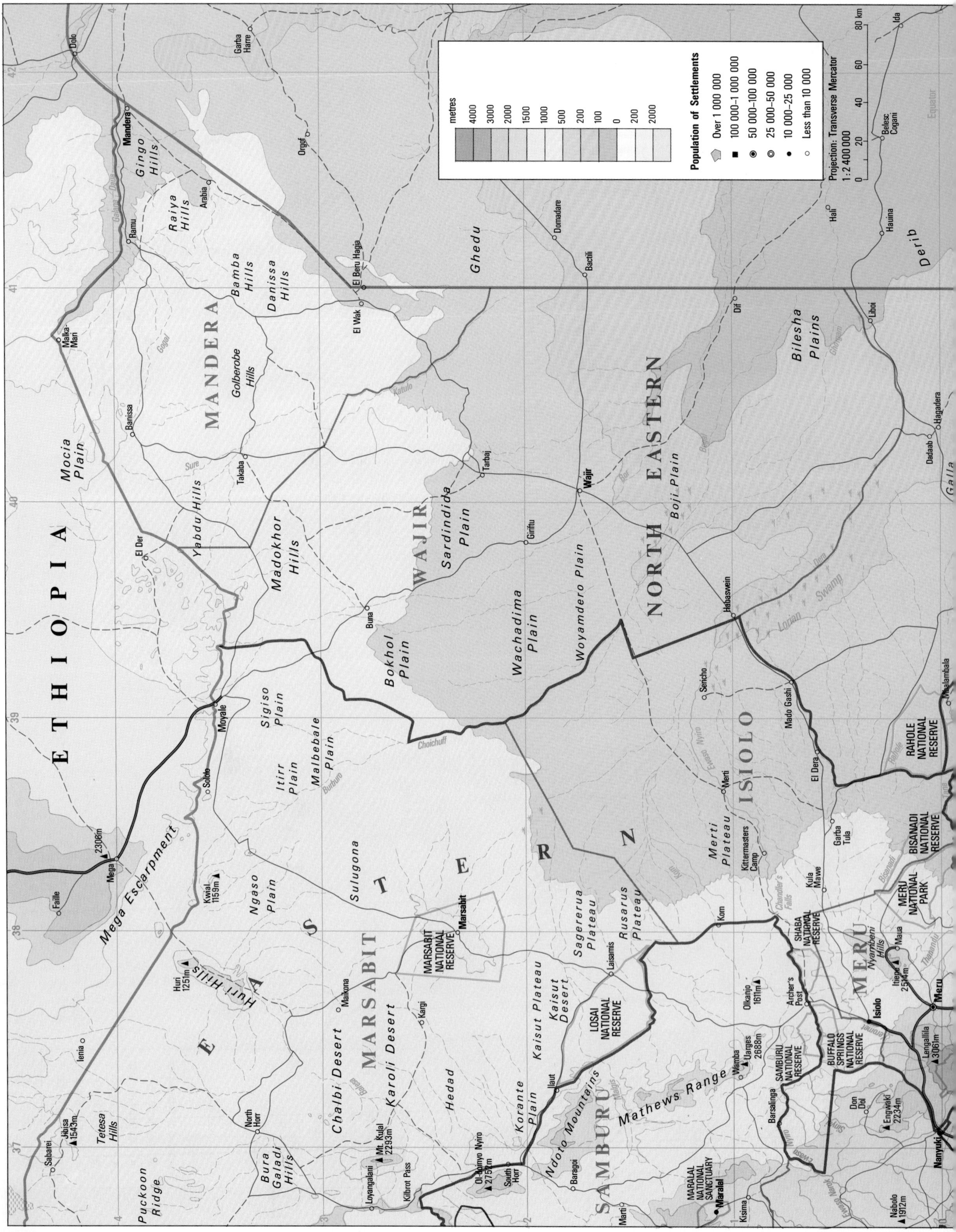

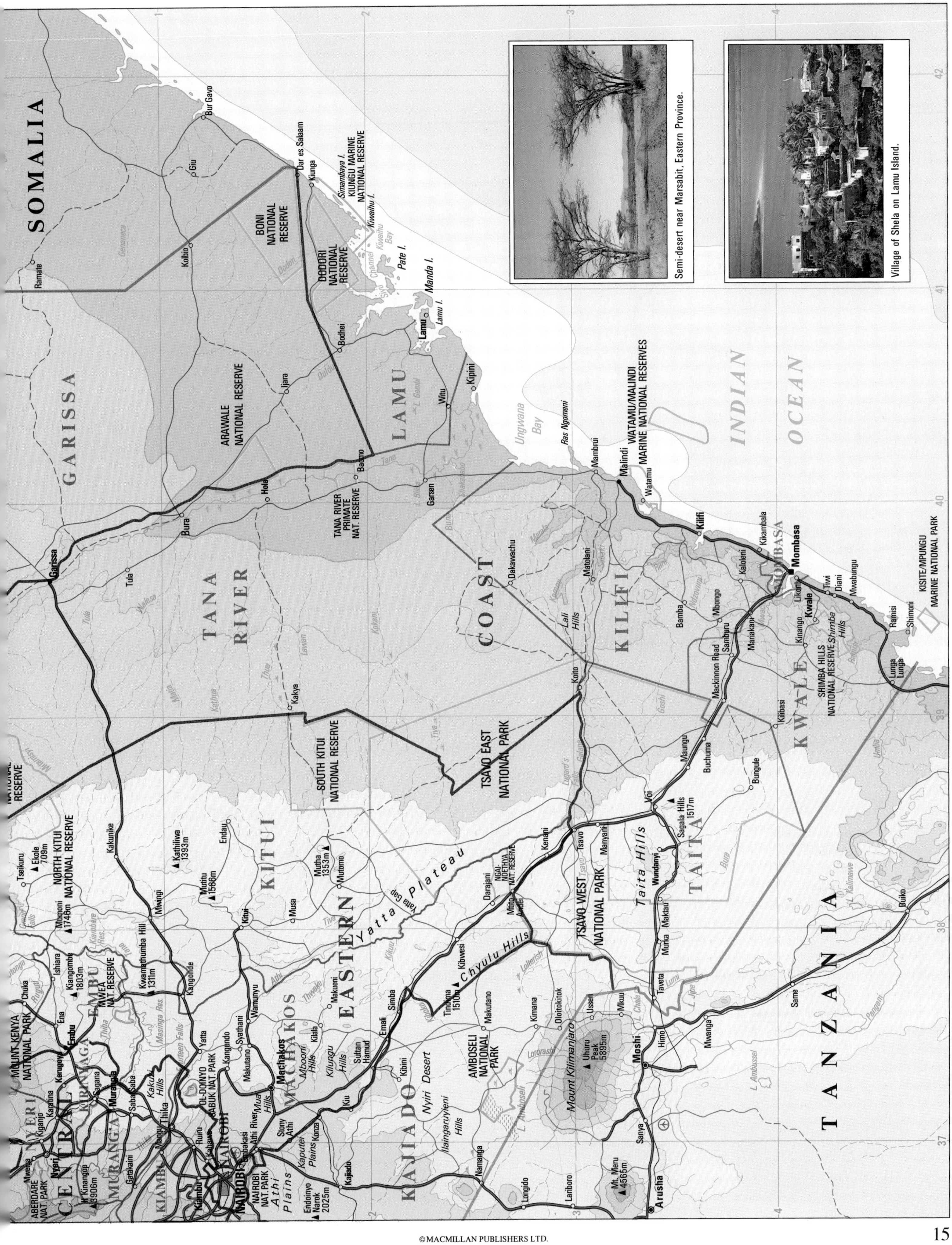

Semi-desert near Marsabit, Eastern Province.
Village of Shela on Lamu Island.

KENYA – ADMINISTRATIVE

Province and District	Population*
Kiambu District	686 290
Kirinyaga District	291 431
Murang'a District	648 333
Nyandarua District	233 302
Nyeri District	486 477
Central Province	**2 345 833**
Kilifi District	430 986
Kwale District	288 363
Lamu District	42 299
Mombasa District	341 148
Taita District	147 597
Tana River District	92 401
Coast Province	**1 342 794**

Embu District	263 173
Isiolo District	43 478
Kitui District	464 283
Machakos District	1 022 522
Marsabit District	96 216
Meru District	830 179
Eastern Province	**2 719 851**
Nairobi	**827 775**
Garissa District	128 867
Mandera District	105 609
Wajir District	139 319
North Eastern Prov.	**373 787**

Kisii District	869 512
Kisumu District	482 327
Siaya District	474 516
South Nyanza District	817 601
Nyanza Province	**2 643 956**

Baringo District	203 793
Elgeyo Marakwet Dist.	148 868
Kajiado District	149 005
Kericho District	633 348
Laikipia District	134 524
Nakuru District	522 709
Nandi District	299 319
Narok District	210 306
Samburu District	76 908
Trans Nzoia District	259 503
Turkana District	142 702
Uasin Gishu District	300 766
West Pokot District	158 652
Rift Valley Province	**3 240 402**
Bungoma District	503 935
Busia District	297 841
Kakamega District	1 030 887
Western Province	**1 832 663**

* Source: Population census 1979 See page 26 for 1990 population figures by province (official estimates)

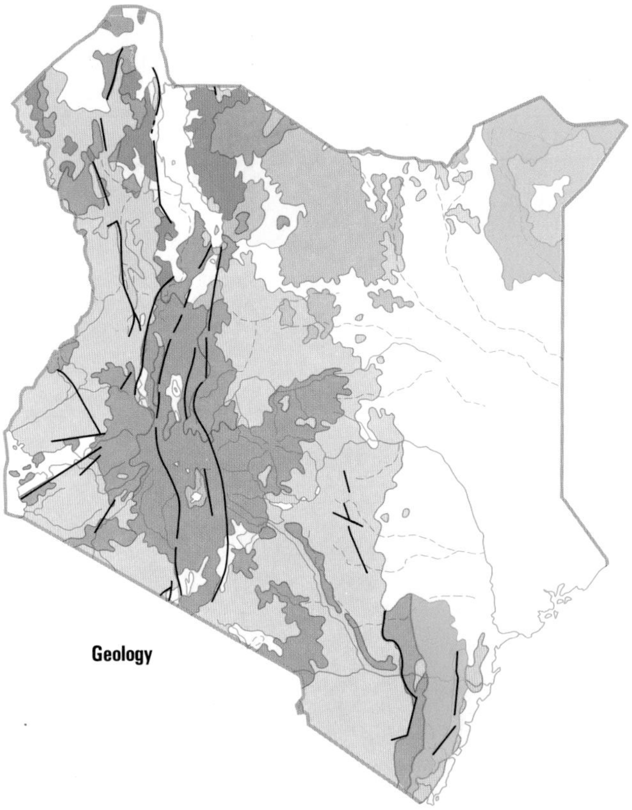

Geology

Geology

- QUATERNARY
- Pliocene-Pleistocene
- TERTIARY
- Neogene

MESOZOIC

- Cretaceous
- Jurassic
- Triassic

PALAEOZOIC

- Carboniferous-Permian
- PRECAMBRIAN

EXTRUSIVE IGNEOUS ROCKS

- Quaternary
- Tertiary
- Neogene
- Pleistocene
- Fault line

Note: for the Geological Time Chart see page 58

Source: UNESCO Geological Map of the World

1:10 500 000

0 100 200 300 400 km

Natural Vegetation

- Heath and moorland
- Mountain forest, mainly bamboo
- Rain forest- tall hardwood trees, dense undergrowth, some grassland
- Evergreen bushland with acacia wooded grassland
- Equatorial forest patches
- Coconut woodland, bushland and tall savanna grasses
- Deciduous bushland and thicket with acacia
- Semi-desert grassland with dry stunted thorny shrubs
- Grassland with semi-aquatic vegetation
- Mangrove swamp

Source: UNESCO Vegetation Map of Africa

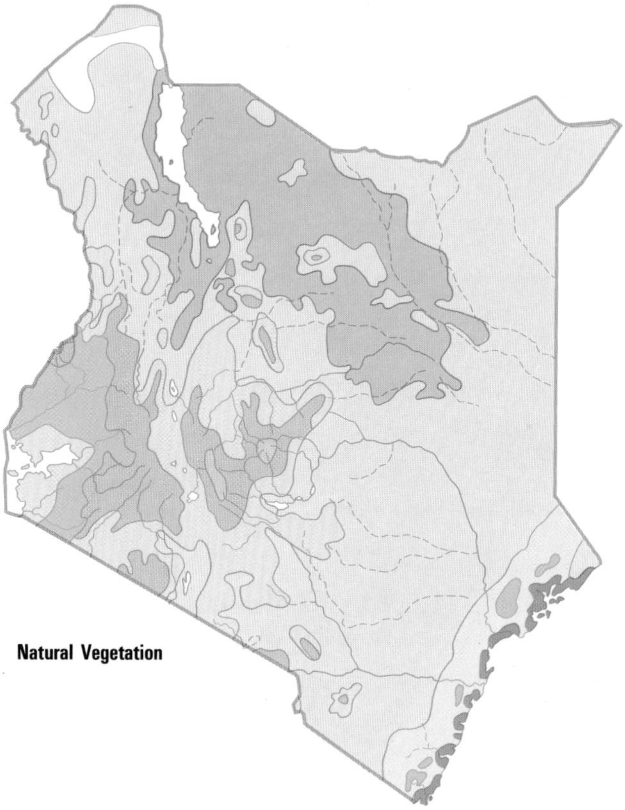

Natural Vegetation

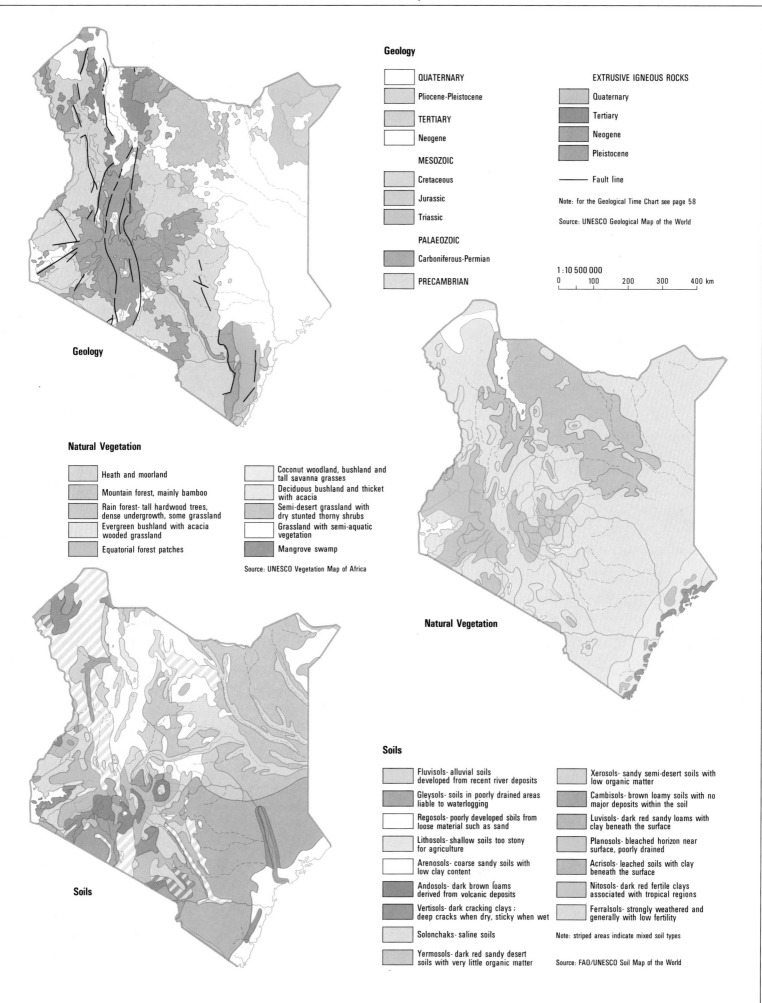

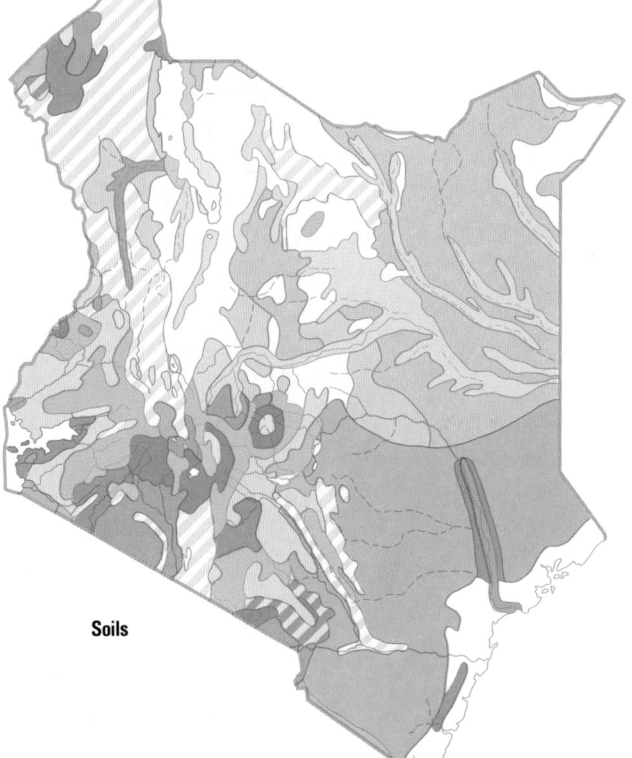

Soils

Soils

- Fluvisols- alluvial soils developed from recent river deposits
- Gleysols- soils in poorly drained areas liable to waterlogging
- Regosols- poorly developed soils from loose material such as sand
- Lithosols- shallow soils too stony for agriculture
- Arenosols- coarse sandy soils with low clay content
- Andosols- dark brown loams derived from volcanic deposits
- Vertisols- dark cracking clays; deep cracks when dry, sticky when wet
- Solonchaks- saline soils
- Yermosols- dark red sandy desert soils with very little organic matter
- Xerosols- sandy semi-desert soils with low organic matter
- Cambisols- brown loamy soils with no major deposits within the soil
- Luvisols- dark red sandy loams with clay beneath the surface
- Planosols- bleached horizon near surface, poorly drained
- Acrisols- leached soils with clay beneath the surface
- Nitosols- dark red fertile clays associated with tropical regions
- Ferralsols- strongly weathered and generally with low fertility

Note: striped areas indicate mixed soil types

Source: FAO/UNESCO Soil Map of the World

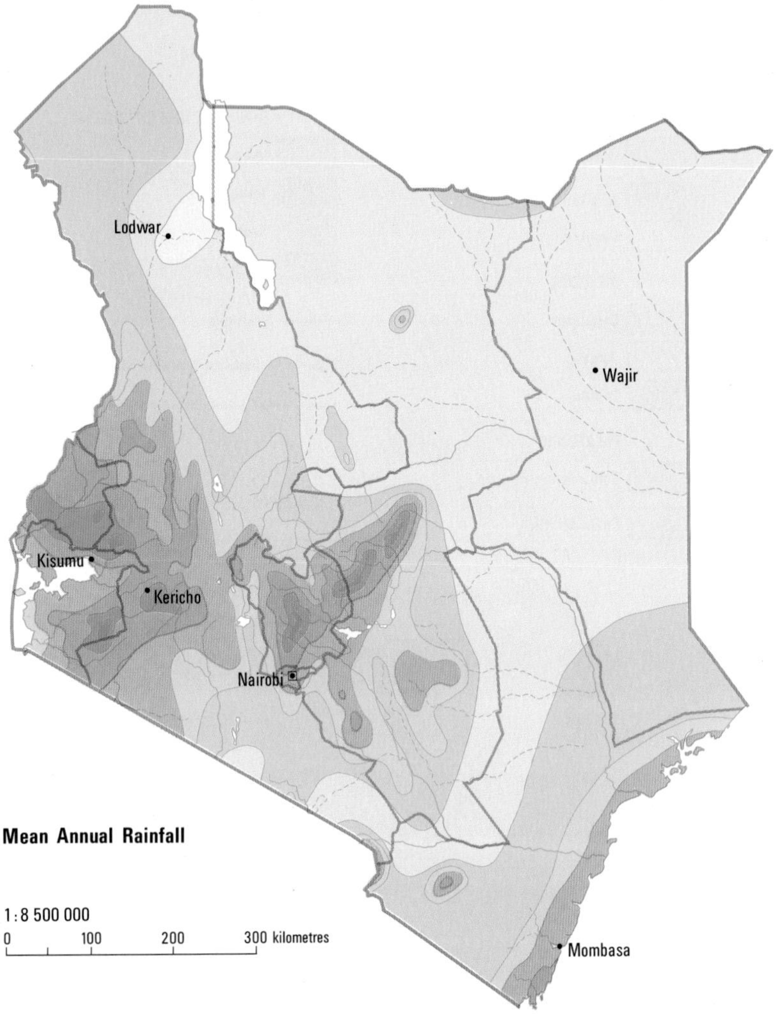

Mean Annual Rainfall

1:8 500 000

0 100 200 300 kilometres

Mean Annual Rainfall

Rainfall in millimetres

- Over 2000
- 1600–2000
- 1200–1600
- 800–1200
- 600–800
- 400–600
- 200–400
- 0–200

Rainfall Reliability

The amount and distribution of rainfall varies considerably from year to year. This variability of rainfall can result in difficult conditions for agriculture.

The likelihood of receiving 750mm of rain a year

- Good
- Quite good
- Unlikely
- Very unlikely

750 mm of rainfall a year represents an adequate amount for most types of agriculture

1:20 000 000

0 200 400 600 kilometres

Mean Monthly Rainfall and Winds

Rainfall in millimetres

- Over 400
- 200–400
- 150–200
- 100–150
- 50–100
- 25–50
- 10–25
- 0–10

→ Prevailing winds

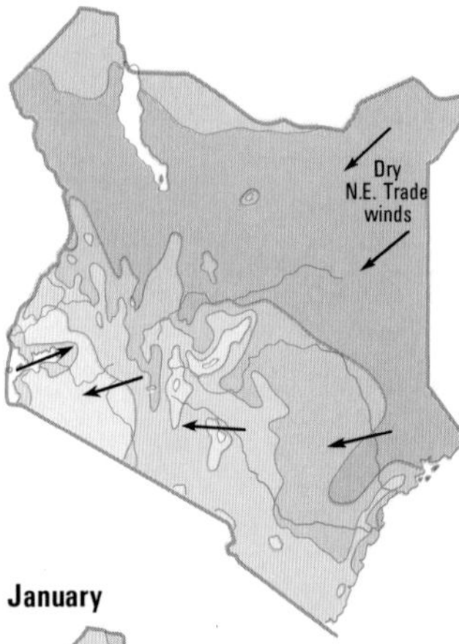

January

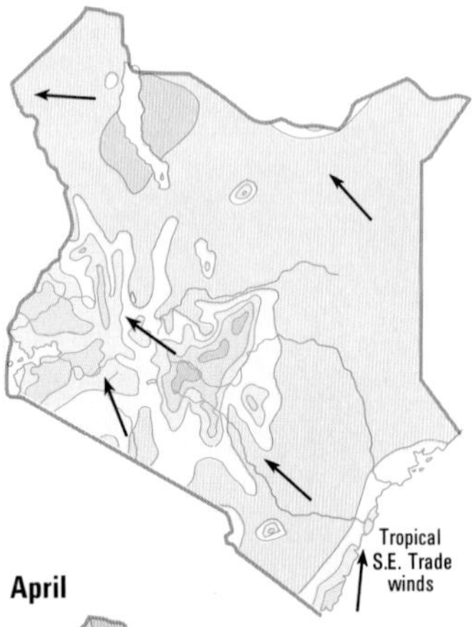

April

Climate station	Number of days when rainfall over 1 mm	Number of days of thunder	Maximum rainfall in 24 hours (millimetres)
Kericho	195	194	121.4
Kisumu	121	198	154.9
Lodwar	19	21	101.6
Mombasa	98	26	152.4
Nairobi	93	27	165.2
Wajir	32	24	215.9

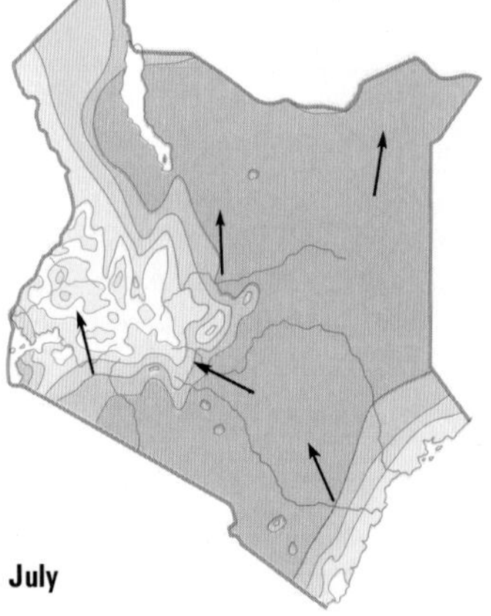

July

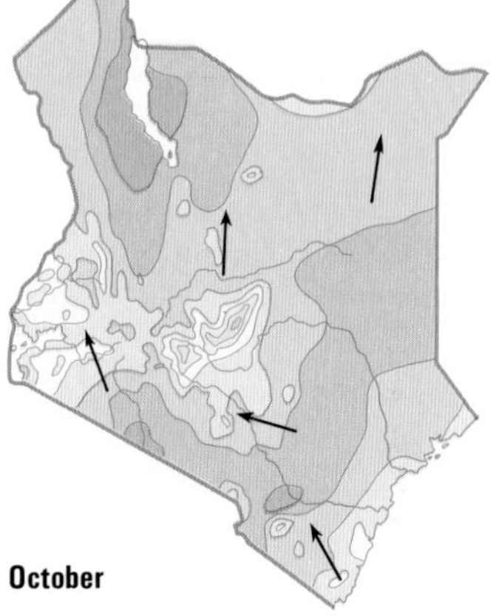

October

Rainfall in Kenya is relatively low and unreliable in nature. Apart from the Kenya highlands, Lake Victoria basin and the coastal strip, the majority of the country can be classified as semi-arid or arid. The inter-tropical convergence zone passes over Kenya twice in the year resulting in two rainy seasons (March-May and November-December) with intervening dry seasons.

Two main wind systems also affect the climate. The dry north-east trade winds and the tropical south-east trade winds which bring heavy rains in April. Higher rainfall in western Kenya is a result of south-westerly winds passing over Lake Victoria.

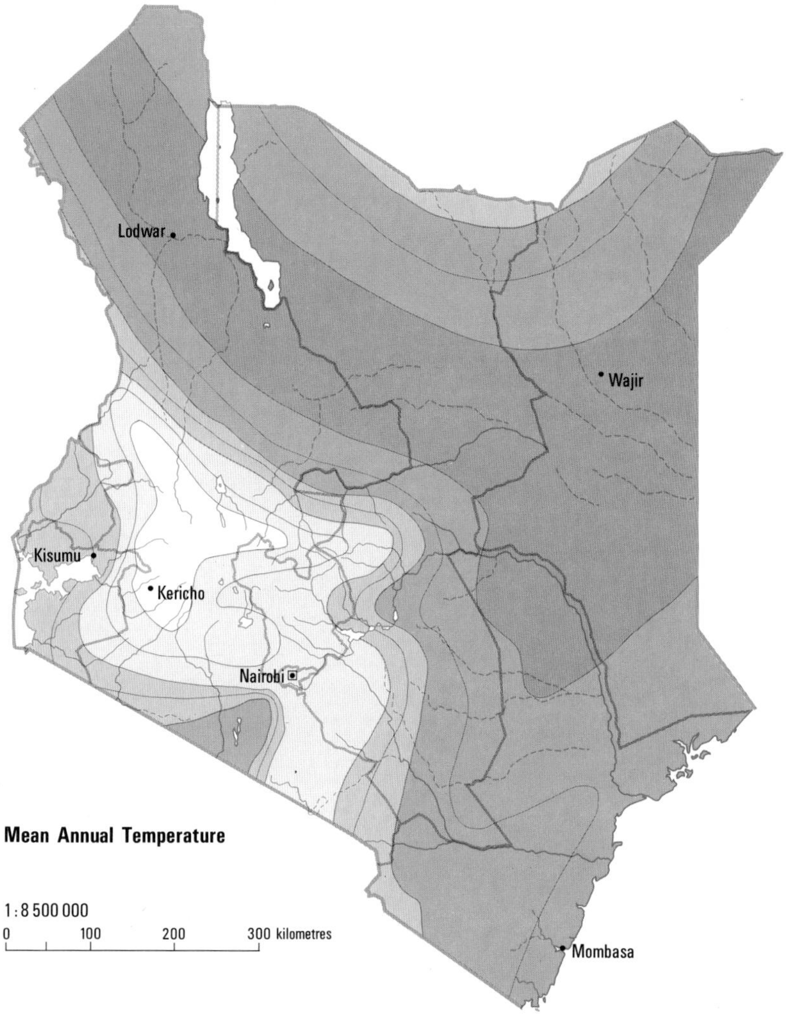

Mean Annual Temperature

1:8 500 000

0 100 200 300 kilometres

Mean Annual Temperature

Surface Temperature in °C

- Over 28
- 26–28
- 24–26
- 22–24
- 20–22
- 18–20
- 16–18
- 14–16

Temperatures are affected by both altitude and distance from the sea. The coastal region has relatively high temperatures and humidity throughout the year.

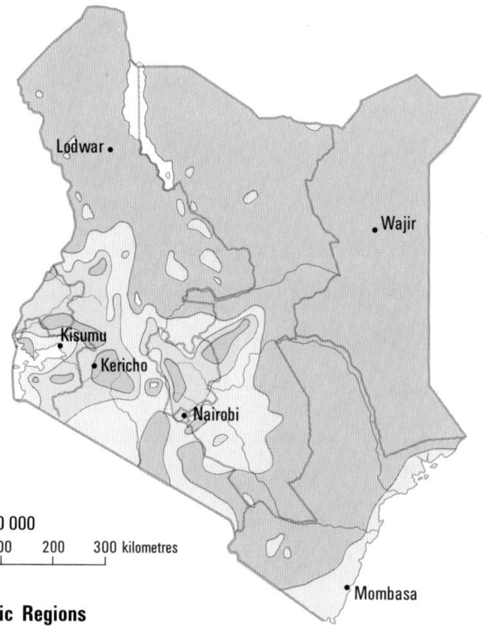

1:13 500 000

0 100 200 300 kilometres

Climatic Regions

- Moist, sub-humid and humid
- Dry sub-humid
- Semi-arid
- Arid

Note:
This climatic classification is based on the Thornthwaite system with an adjustment for evaporation.

Climate station	Sunshine hours per day	Relative humidity 0300 hrs	0600 hrs	1200 hrs
Kericho	6.1	—	70%	63%
Kisumu	8.3	85%	68%	47%
Lodwar	9.8	65%	56%	35%
Mombasa	8.1	93%	82%	65%
Nairobi	6.9	94%	83%	50%
Wajir	8.0	79%	73%	44%

Climate Graphs

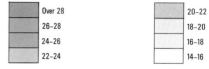

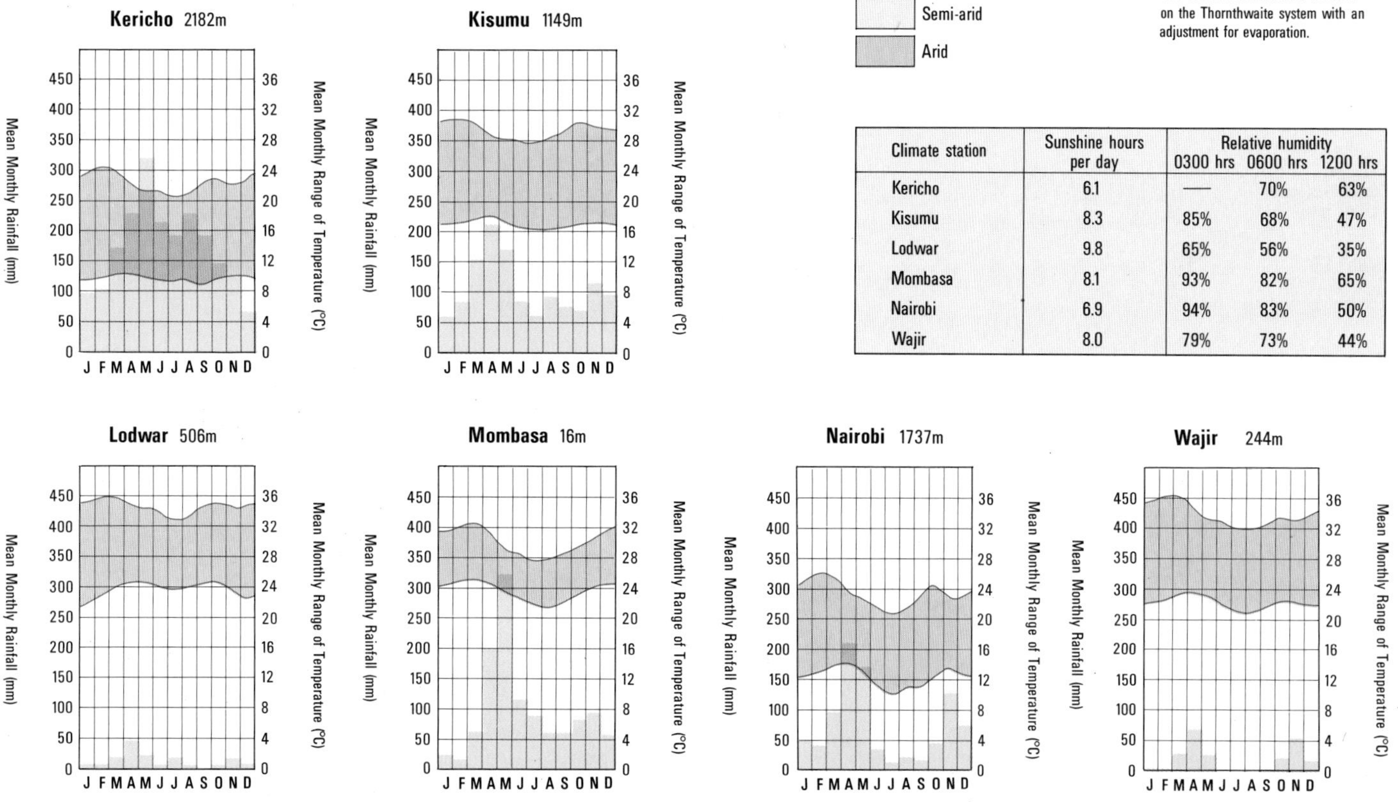

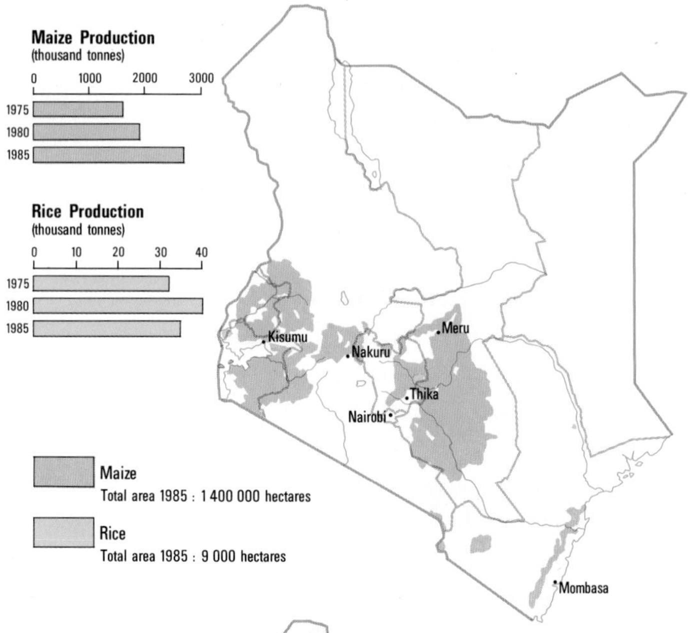

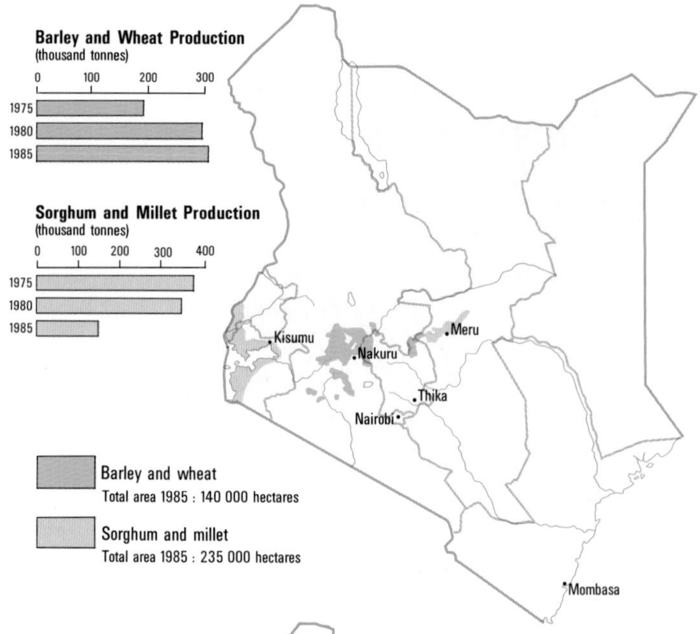

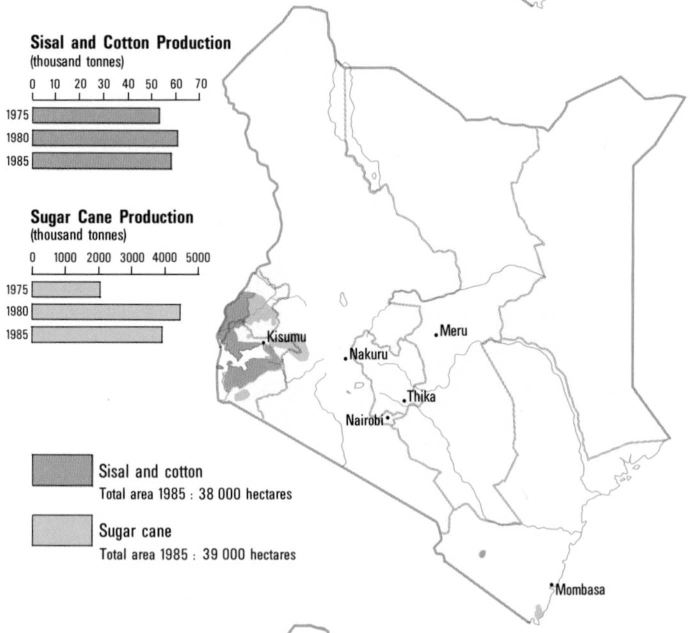

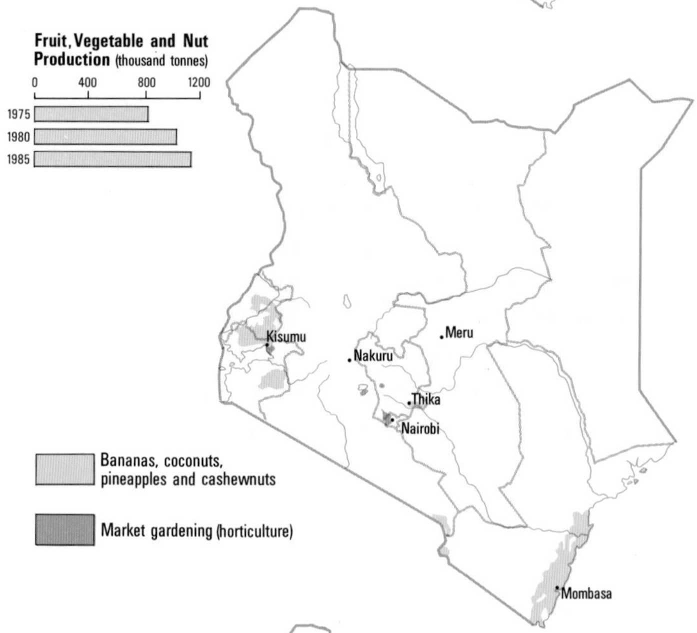

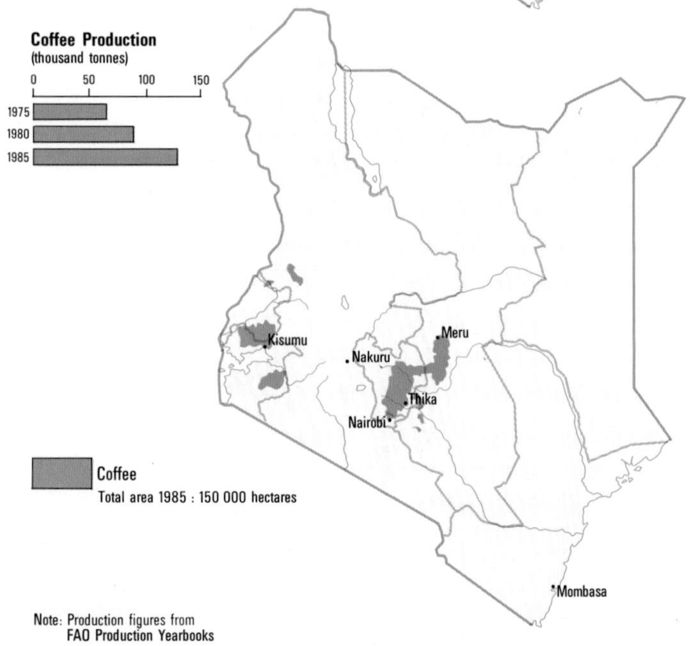

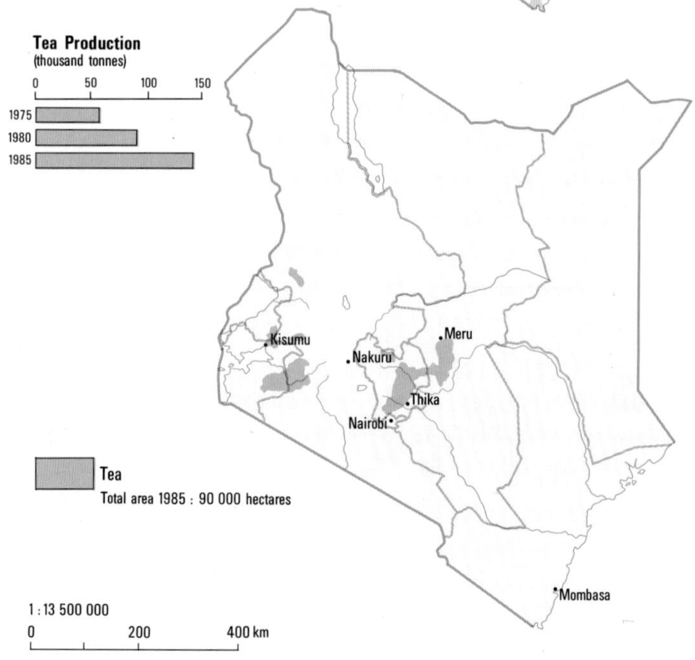

Maize Production
(thousand tonnes)
0 1000 2000 3000
1975
1980
1985
Rice Production
(thousand tonnes)
0 10 20 30 40
1975
1980
1985
Kisumu
Nakuru
Meru
Thika
Nairobi
Mombasa
Maize
Total area 1985 : 1 400 000 hectares
Rice
Total area 1985 : 9 000 hectares
Barley and Wheat Production
(thousand tonnes)
0 100 200 300
1975
1980
1985
Sorghum and Millet Production
(thousand tonnes)
0 100 200 300 400
1975
1980
1985
Barley and wheat
Total area 1985 : 140 000 hectares
Sorghum and millet
Total area 1985 : 235 000 hectares
Sisal and Cotton Production
(thousand tonnes)
0 10 20 30 40 50 60 70
1975
1980
1985
Sugar Cane Production
(thousand tonnes)
0 1000 2000 3000 4000 5000
1975
1980
1985
Sisal and cotton
Total area 1985 : 38 000 hectares
Sugar cane
Total area 1985 : 39 000 hectares
Fruit, Vegetable and Nut Production (thousand tonnes)
0 400 800 1200
1975
1980
1985
Bananas, coconuts, pineapples and cashewnuts
Market gardening (horticulture)
Coffee Production
(thousand tonnes)
0 50 100 150
1975
1980
1985
Coffee
Total area 1985 : 150 000 hectares
Tea Production
(thousand tonnes)
0 50 100 150
1975
1980
1985
Tea
Total area 1985 : 90 000 hectares
Note: Production figures from FAO Production Yearbooks
1 : 13 500 000
0 200 400 km

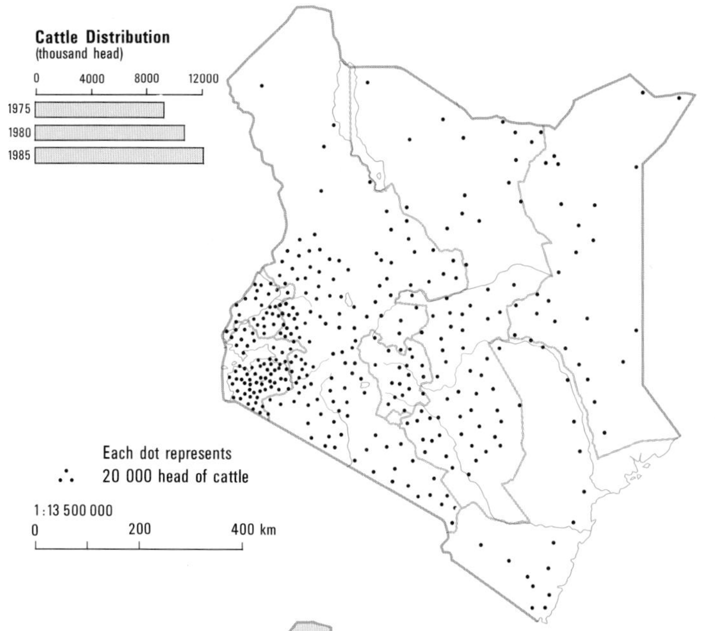

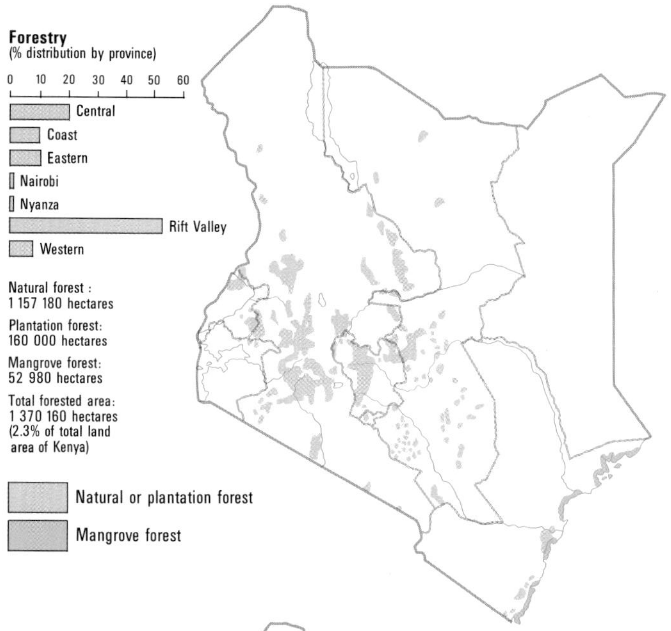

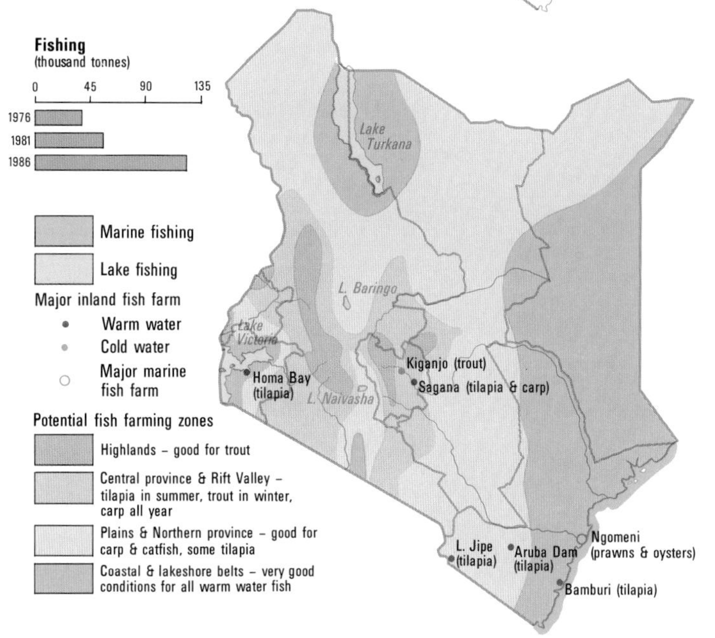

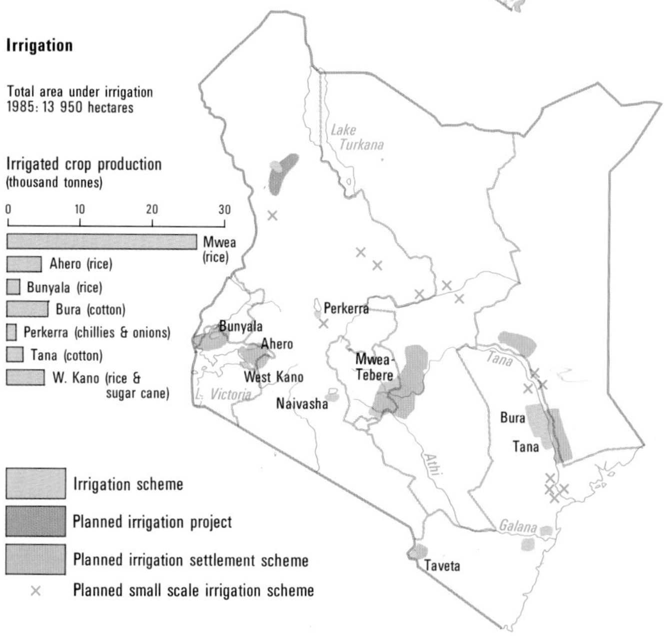

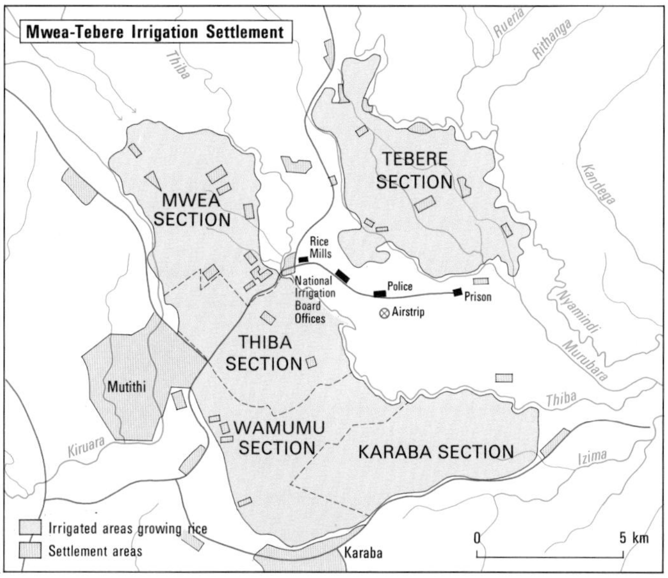

Mwea-Tebere Irrigation Scheme

Mwea-Tebera Irrigation	1965/66	1975/76	1985/6*
Area cropped (in hectares)	2 593	5 609	8 271
Number of plot holders	1 484	2 972	3 234
Rice paddy production (in tonnes)	14 454	32 060	26 407
Gross value of crops (in KShs)	6 373 240	37 020 000	84 240 000

* provisional figures

Under the Mwea-Tebere Irrigation Scheme, rice is cultivated on a large scale and the area is Kenya's largest rice supplier. The plot holders live in small villages located on or near the scheme; each tenant has 1·6 ha which is divided into 0·4 ha floodable basins. The plots are carefully maintained and cultivated by hand. The management provides the tenants with all the equipment needed for the crop cultivation – seeds, fertilisers, insecticides, tractors, transport and payment for the crop.

Rice cultivation at Mwea-Tebere Irrigation Scheme.

KENYA – MINERALS AND ENERGY

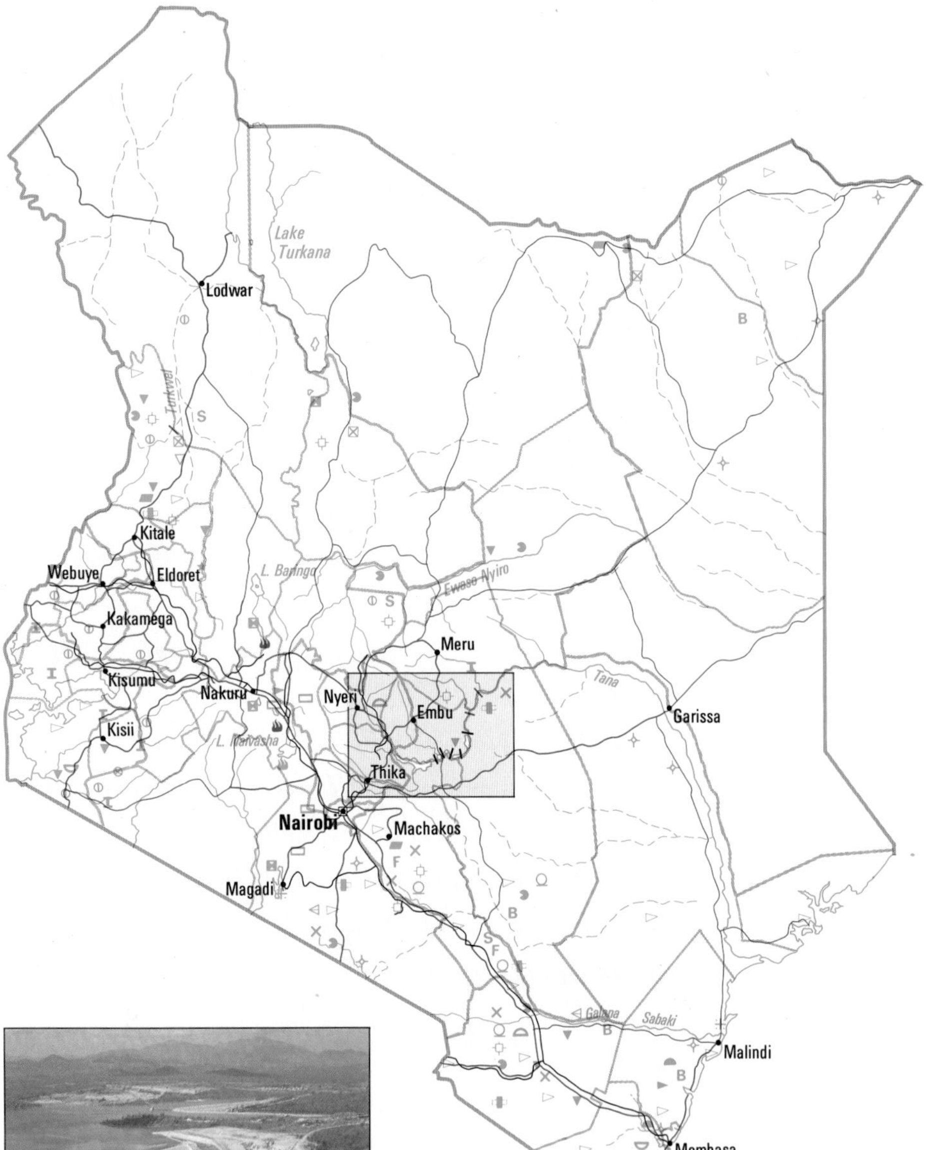

1 : 7 000 000

0 100 200 kilometres

Construction of the Kiambere dam and reservoir was completed in early 1988.

Principal Mineral Deposits

- Asbestos
- Barytes
- Chromite
- Copper
- Diatomite
- Felspar
- Fluorspar
- Garnets
- Gold and silver
- Graphite
- Gypsum
- Iron ores
- Kaolin
- Lead
- Limestone
- Manganese
- Mica
- Nickel ore
- Quartz
- Salt
- Sapphires
- Soapstone
- Soda ash
- Talc
- Wollastonite
- Zinc

Energy Resources

- Dam and hydro-electric power station
- Proposed dam and hydro-electric power station
- Geothermal power station
- Proposed geothermal power station

For power transmission lines and pipelines see page 25

Mineral Production 1980–1986

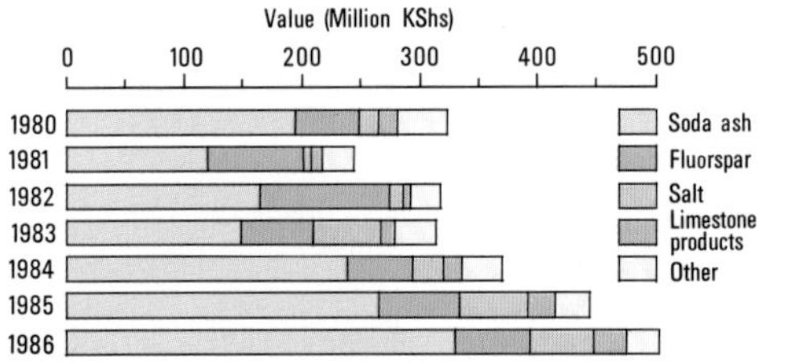

Generation of Electricity 1980–1986

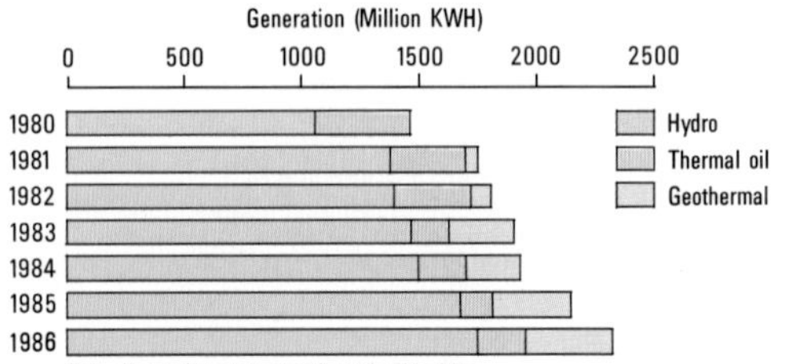

Note: Kenya imports approximately 5% of its electricity from Uganda

Kiambere Hydo-electric Scheme

At Kiambere the Tana River has been dammed to form a reservoir. Two dams have been constructed; a main dam which controls the flow of the river and an additional 'saddle dam' beneath which hydro-electric power is generated.
An intake structure controls the movement of water from the reservoir into the headrace tunnel which transfers water to the penstocks. These vertical penstocks maximise the speed at which the water reaches the turbines in the powerhouse where electricity is generated. The water is then returned to the Tana River, further downstream, via the tailrace tunnel.

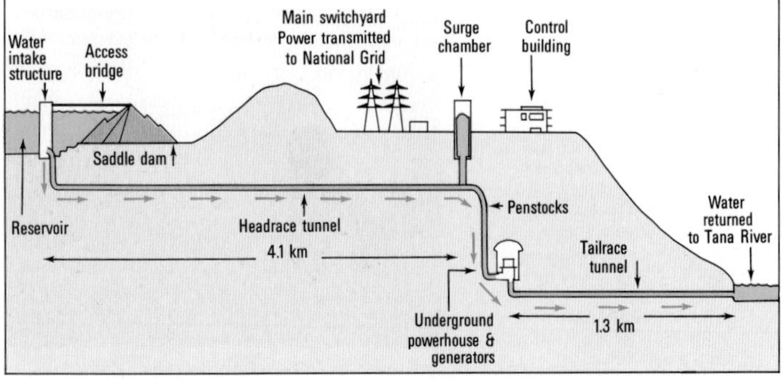

Tana River H.E.P.

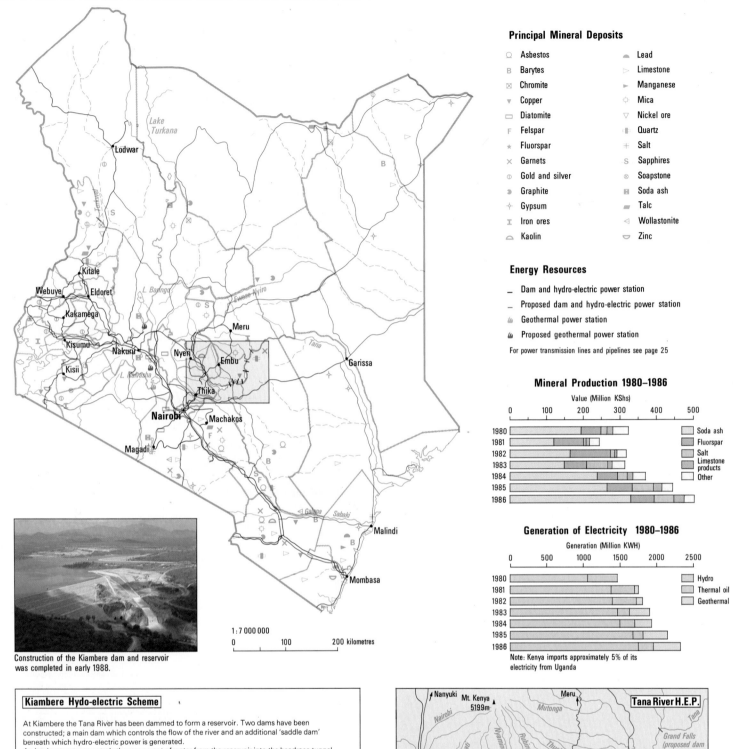

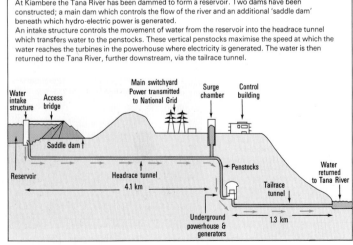

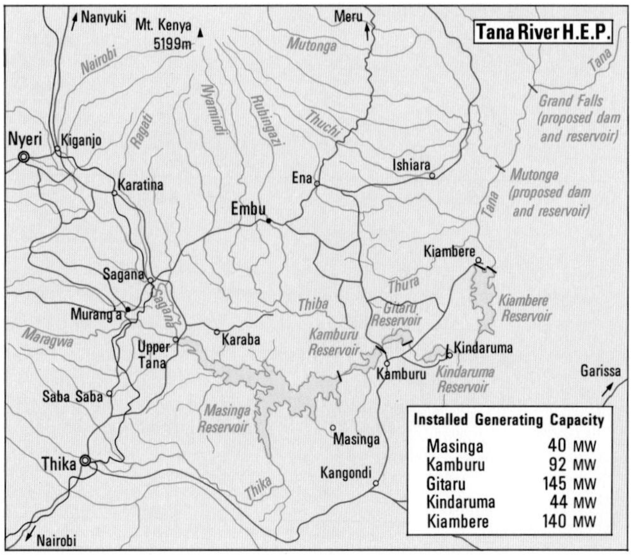

Installed Generating Capacity

Masinga	40 MW
Kamburu	92 MW
Gitaru	145 MW
Kindaruma	44 MW
Kiambere	140 MW

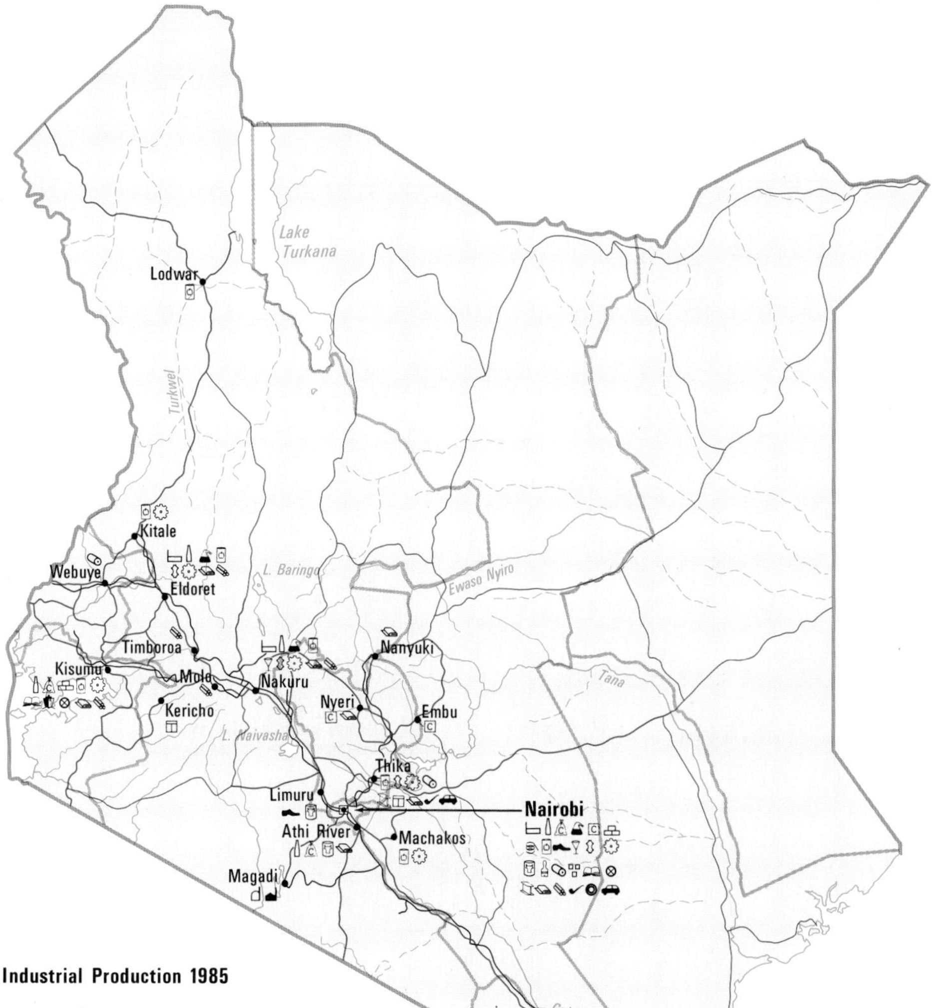

Location of Major Industries

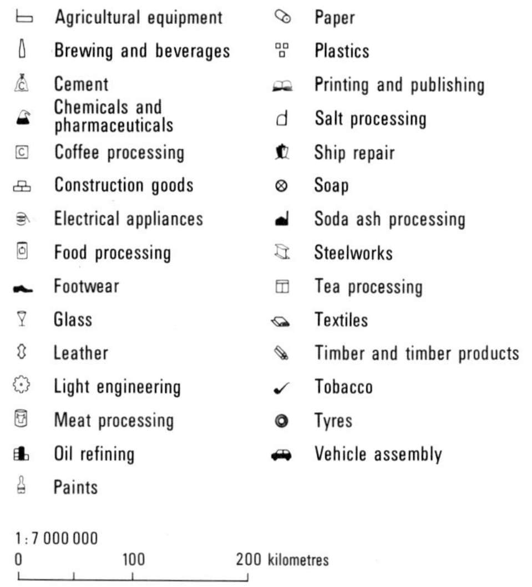

Employment by Industry 1985

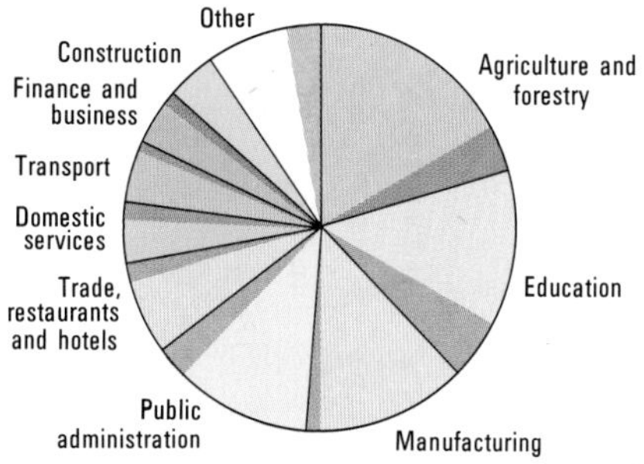

Shaded sectors represent female employees

Industrial Production 1985

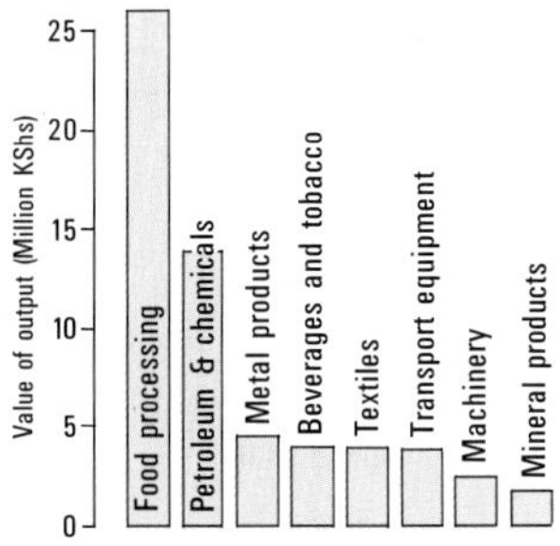

Thika - Industrial Location

Situated 40 kilometres north-east of Nairobi, Thika has recently developed into the second largest industrial centre in Kenya. Its location, in a rich farming area, initially made it an ideal centre for food processing and agriculture related industries. Today Thika has attracted more than 25 major industries which benefit from its proximity to the capital, its excellent road and rail communications to both Nairobi and Mombasa, availability of land for industrial growth, a ready supply of local labour, and ample water for industrial needs.

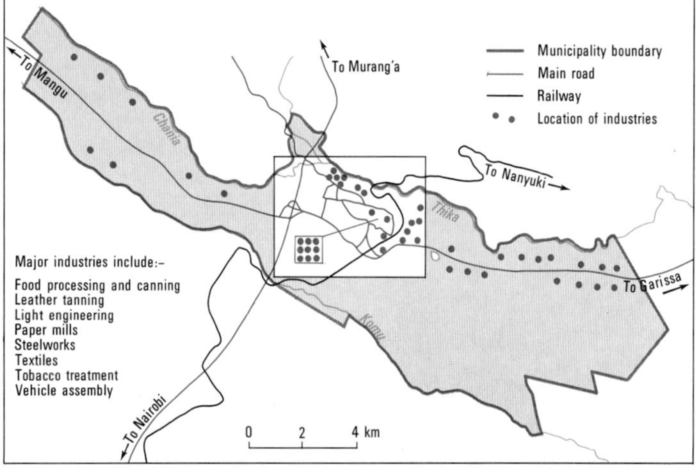

Thika - Urban Land Use

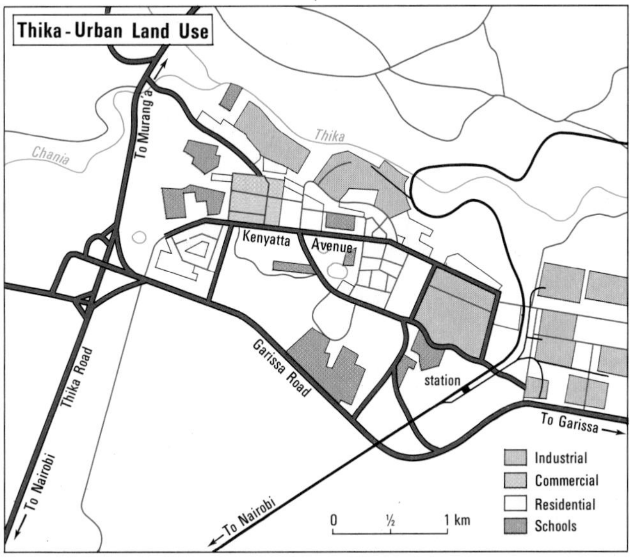

KENYA – TRADE

Principal Trading Partners

Direction of Trade 1986

- Country of destination for Kenya's exports
- Country of origin of Kenya's imports
- Country which has both import and export trade with Kenya

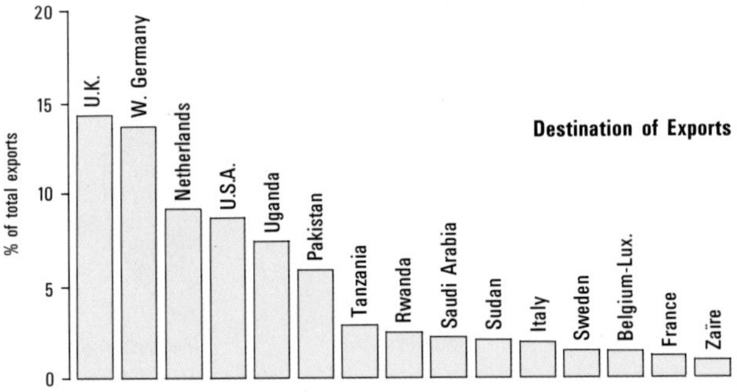

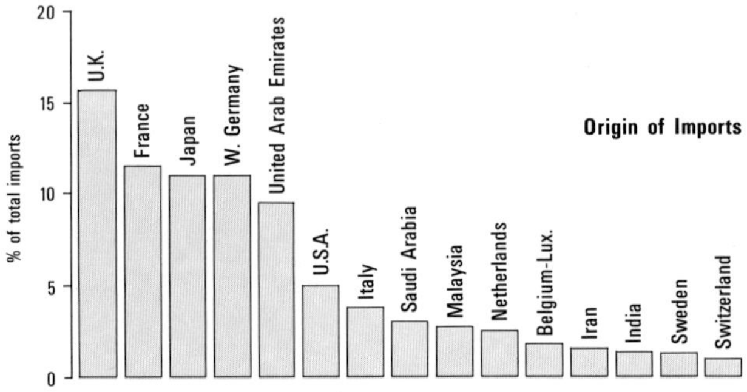

Principal Exports and Imports

Exports by Commodities 1986 Total Export Value: 19 159.4 million KShs

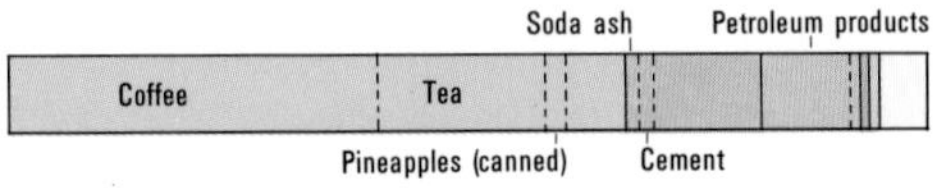

Imports by Commodities 1986 Total Import Value: 26 757.8 million KShs

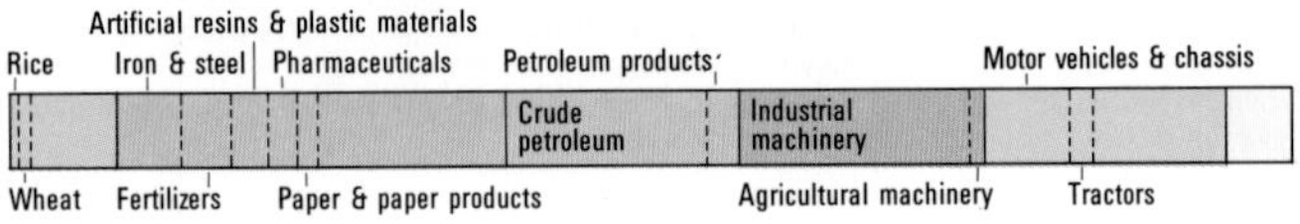

- Food and beverages
- Industrial supplies (non-food)
- Fuel and lubricants
- Machinery and other capital equipment
- Transport equipment
- Goods not elsewhere specified

Modern port facilities at Mombasa harbour.

PTA–Preferential Trade Area for East and Southern Africa

The PTA was established in December 1981 and began operating in July 1984. The treaty aims to encourage trade and economic co-operation between its members by the removal of trade barriers and the improvement of transport and communications within the area.

Trade with African Countries

- Preferential Trade Area members
- Other African countries

Exports 1982

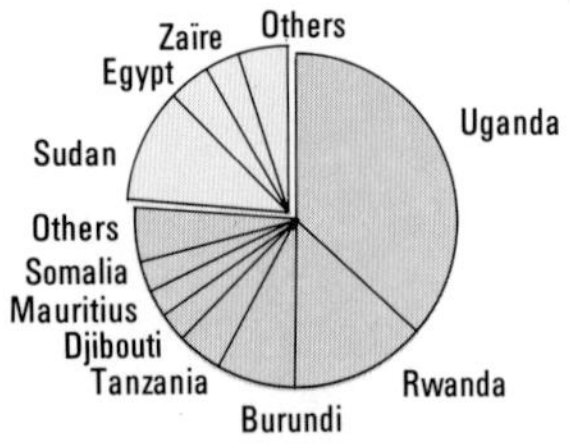

Total Export Value:
3 201 720 thousand KShs

Exports 1986

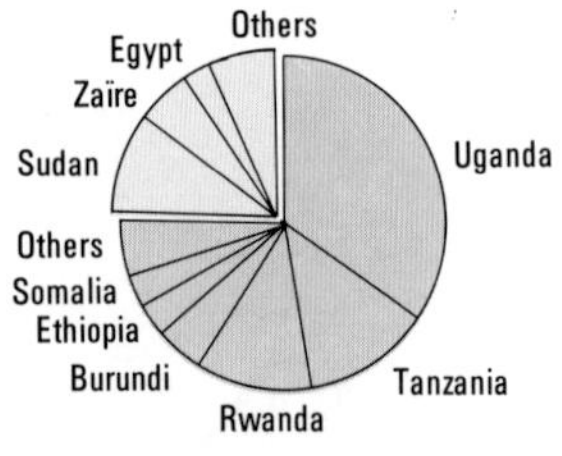

Total Export Value:
4 227 900 thousand KShs

Imports 1982

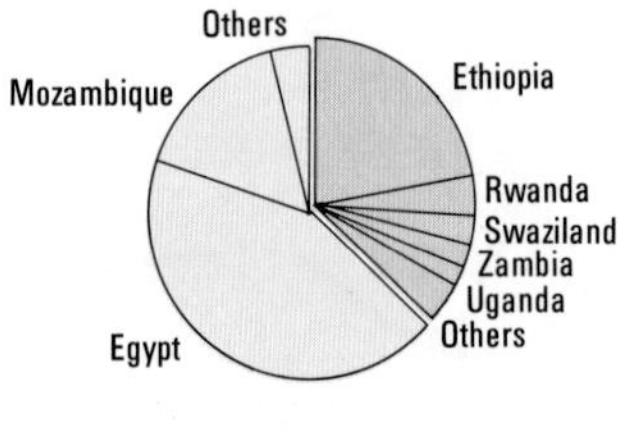

Total Import Value:
1 256 460 thousand KShs

Imports 1986

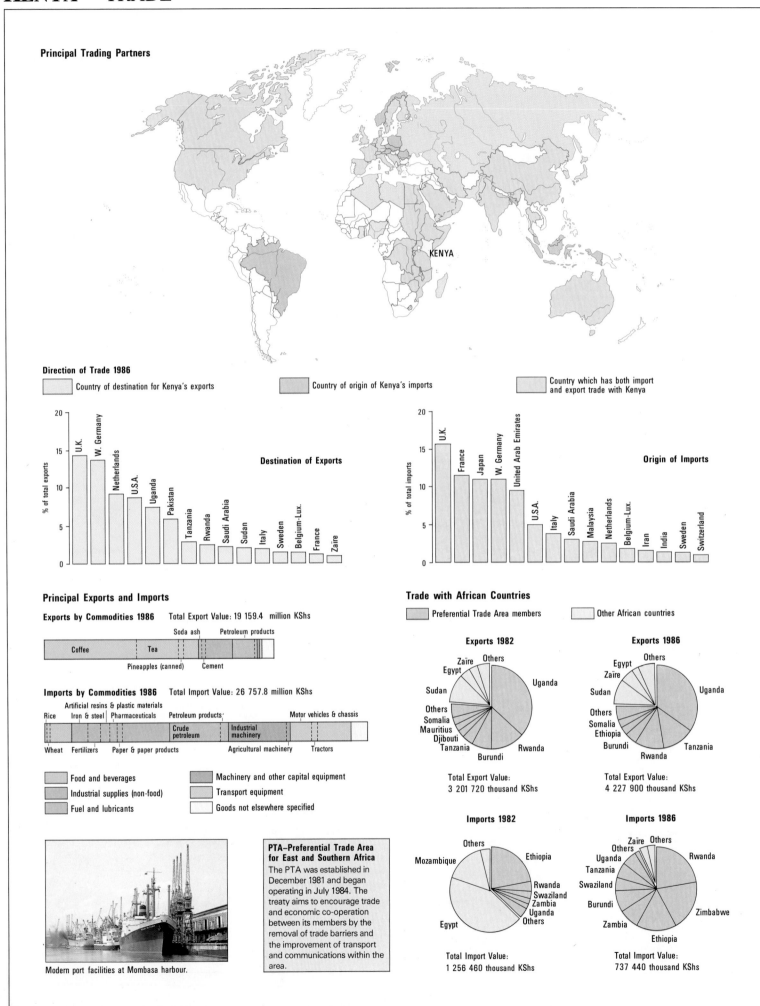

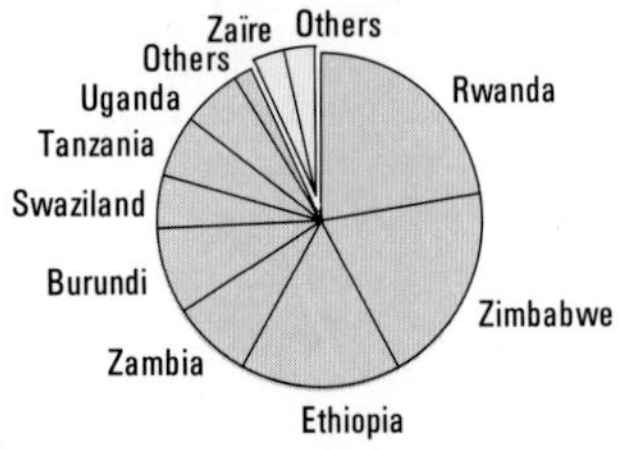

Total Import Value:
737 440 thousand KShs

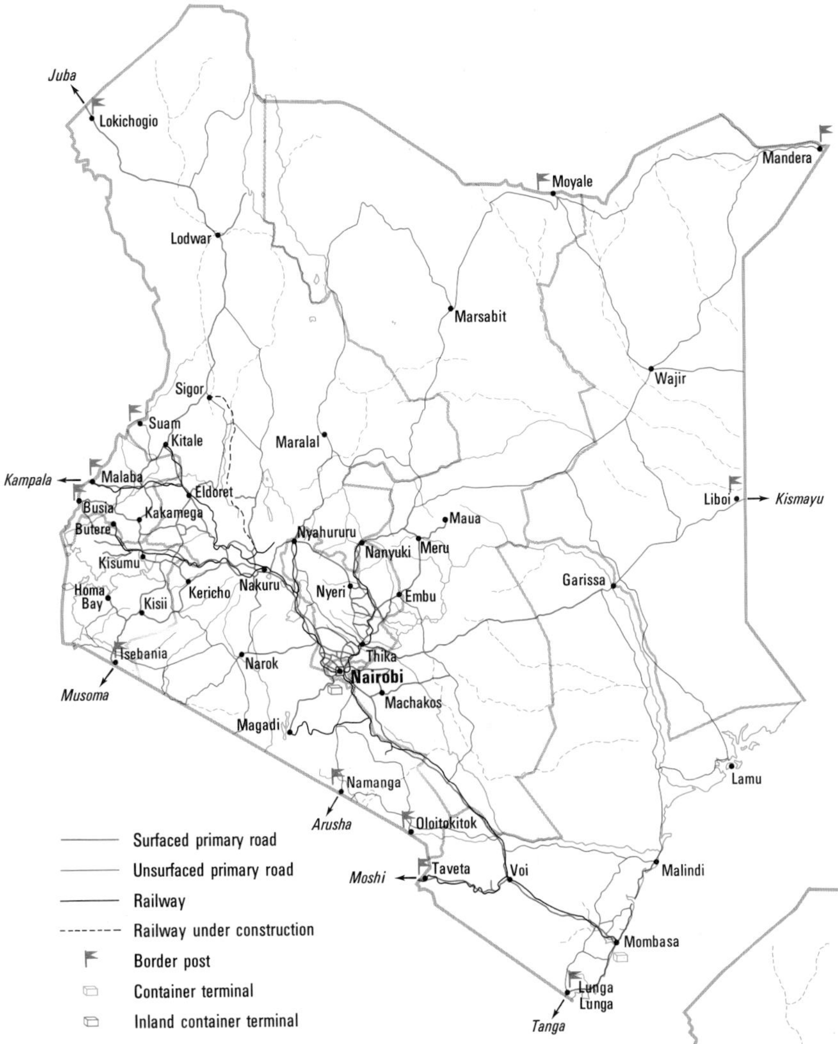

Earnings from Transport and Communications 1986

Total 15 580 million KShs

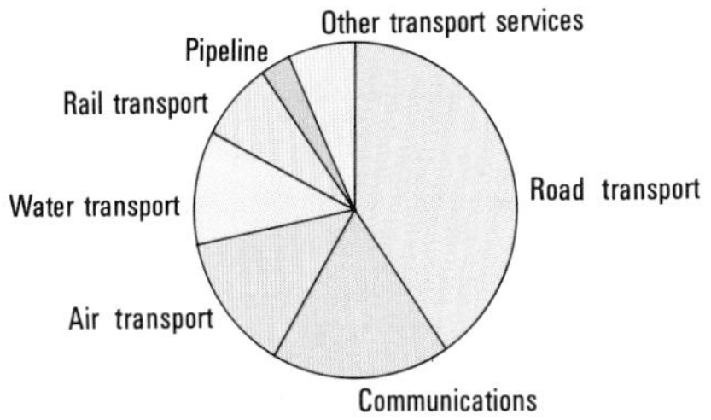

Freight Handled at Mombasa Harbour 1978–1986

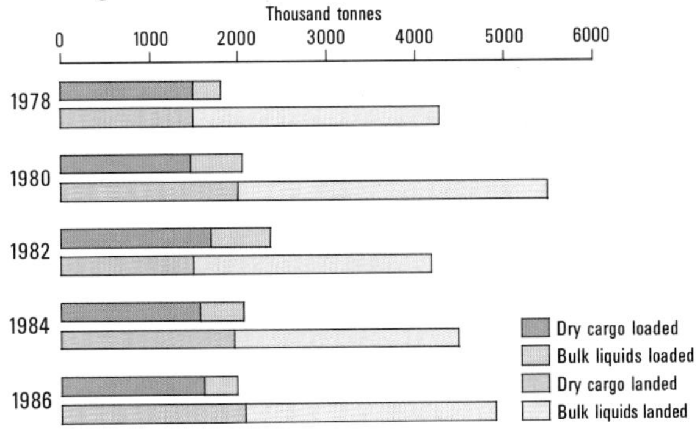

1:8 500 000

0 100 200 300 kilometres

Kenya Railways are used extensively for the transportation of both freight and passengers.

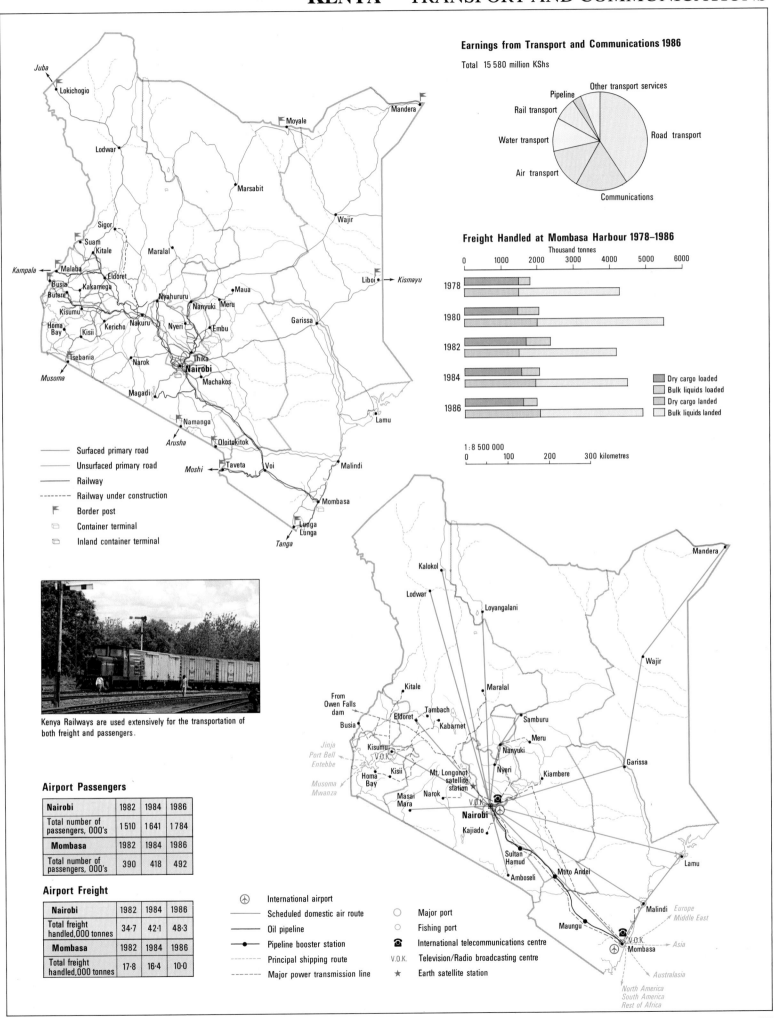

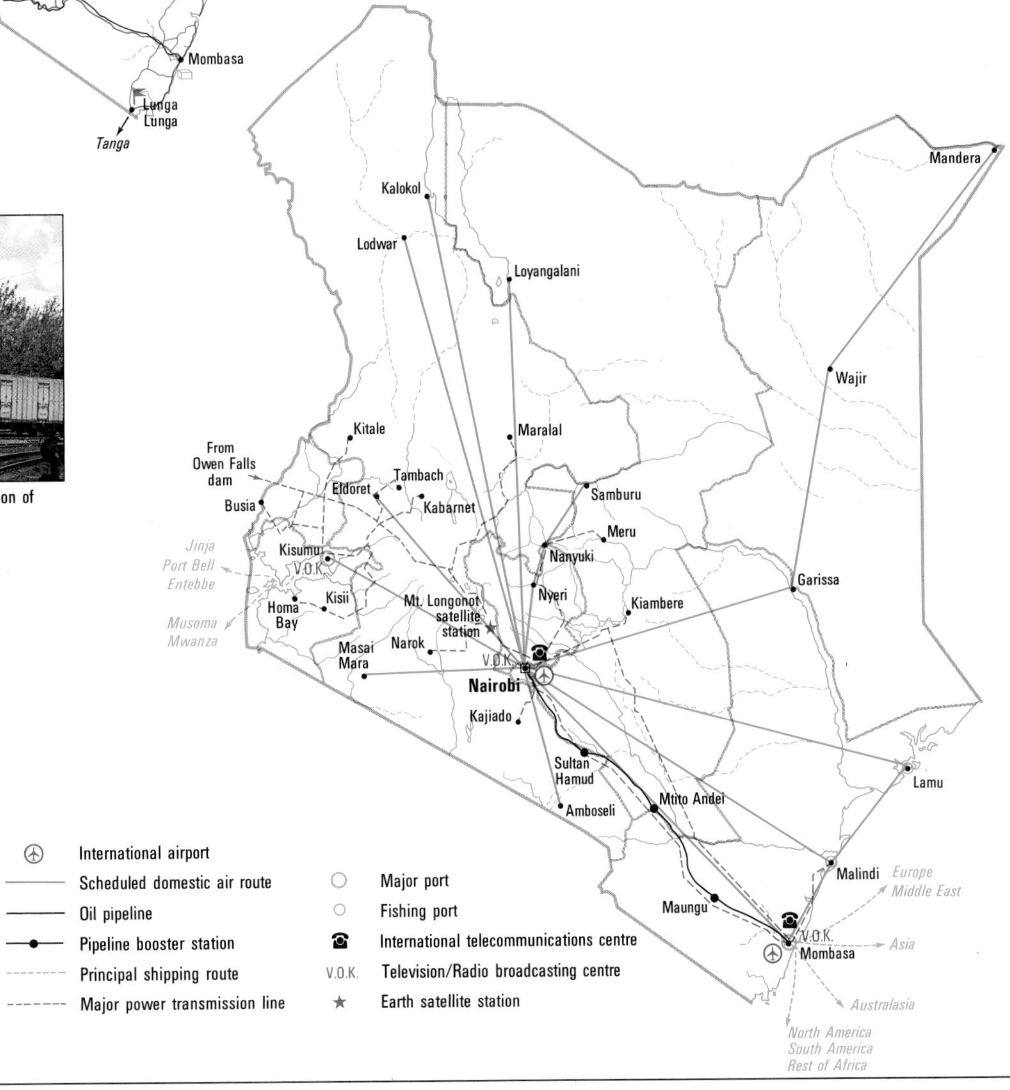

Airport Passengers

Nairobi	1982	1984	1986
Total number of passengers, 000's	1510	1641	1784
Mombasa	1982	1984	1986
Total number of passengers, 000's	390	418	492

Airport Freight

Nairobi	1982	1984	1986
Total freight handled,000 tonnes	34·7	42·1	48·3
Mombasa	1982	1984	1986
Total freight handled,000 tonnes	17·8	16·4	10·0

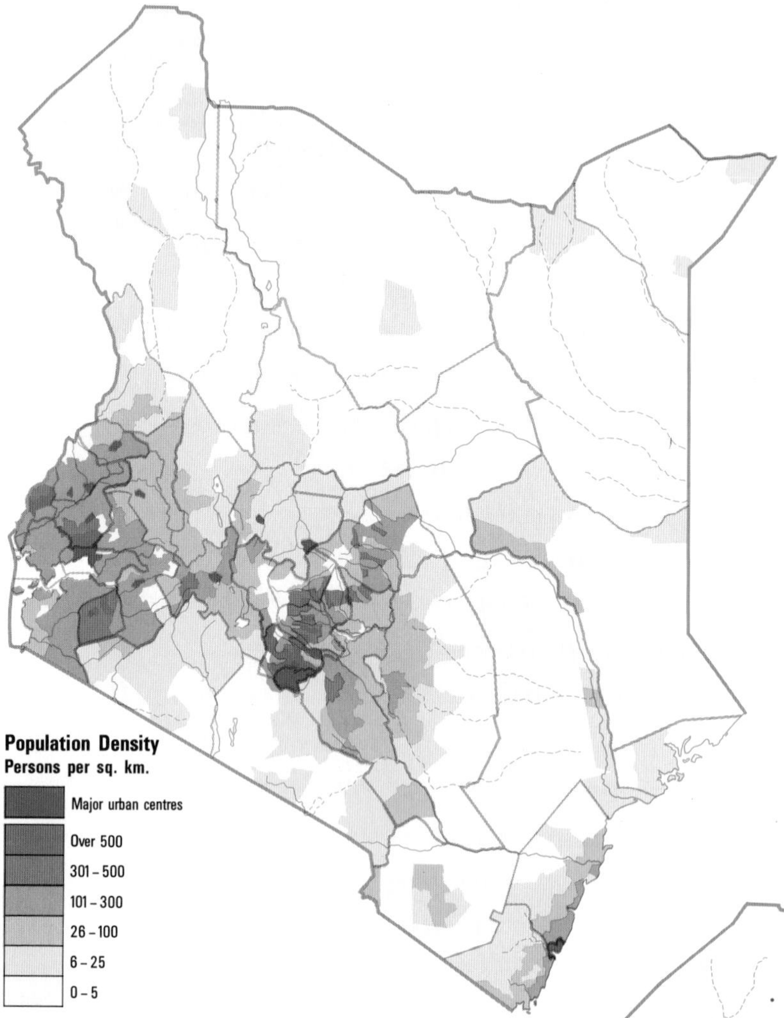

Population by Province

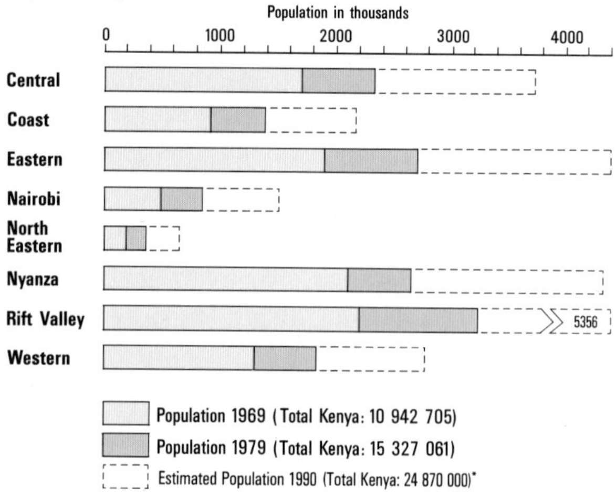

Area and Population Density by Province

Province	Area in sq. km.	Population density per sq. km. 1969	Population density per sq. km. 1979
Central	13 173	127	178
Coast	83 040	11	16
Eastern	155 759	12	17
Nairobi	648	746	1 210
North Eastern	126 902	2	2
Nyanza	12 525	168	211
Rift Valley	163 883	12	19
Western	8 196	161	223

1:8 500 000

0 100 200 300 kilometres

Population Growth

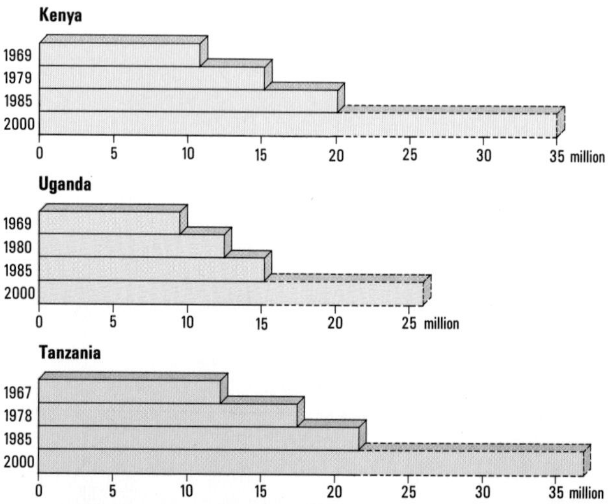

Note: Census dates vary between each country
1985 population figures are official estimates
2000 population figures are World Bank projections

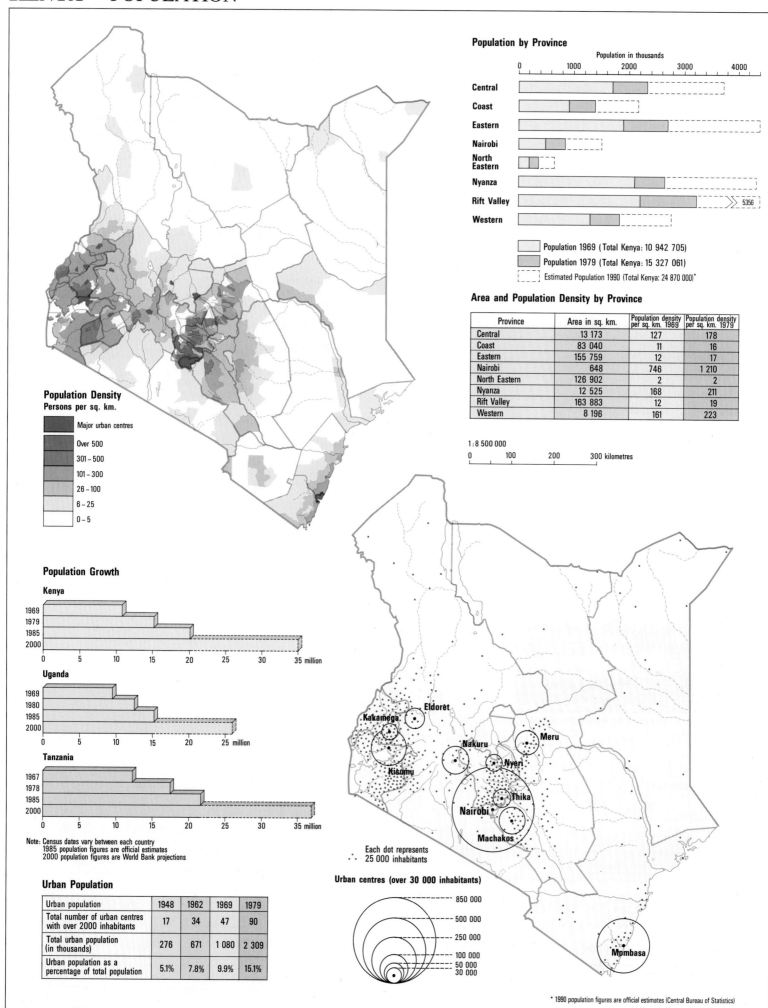

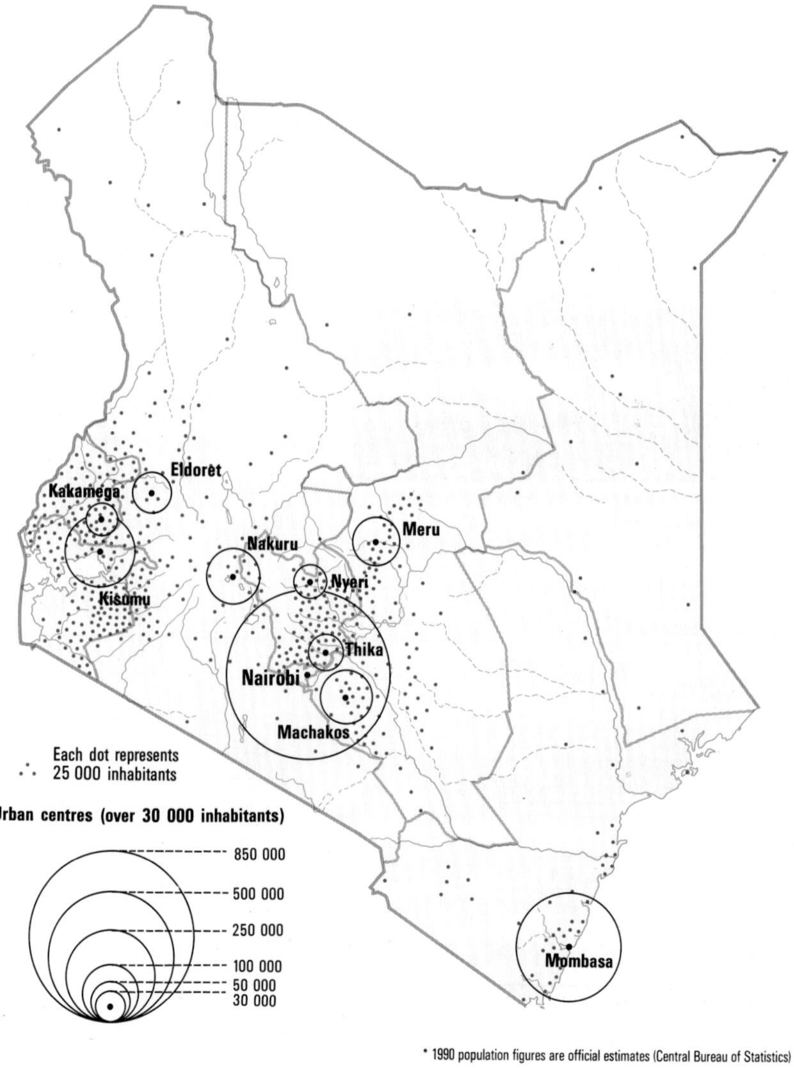

Urban Population

Urban population	1948	1962	1969	1979
Total number of urban centres with over 2000 inhabitants	17	34	47	90
Total urban population (in thousands)	276	671	1 080	2 309
Urban population as a percentage of total population	5.1%	7.8%	9.9%	15.1%

* 1990 population figures are official estimates (Central Bureau of Statistics)

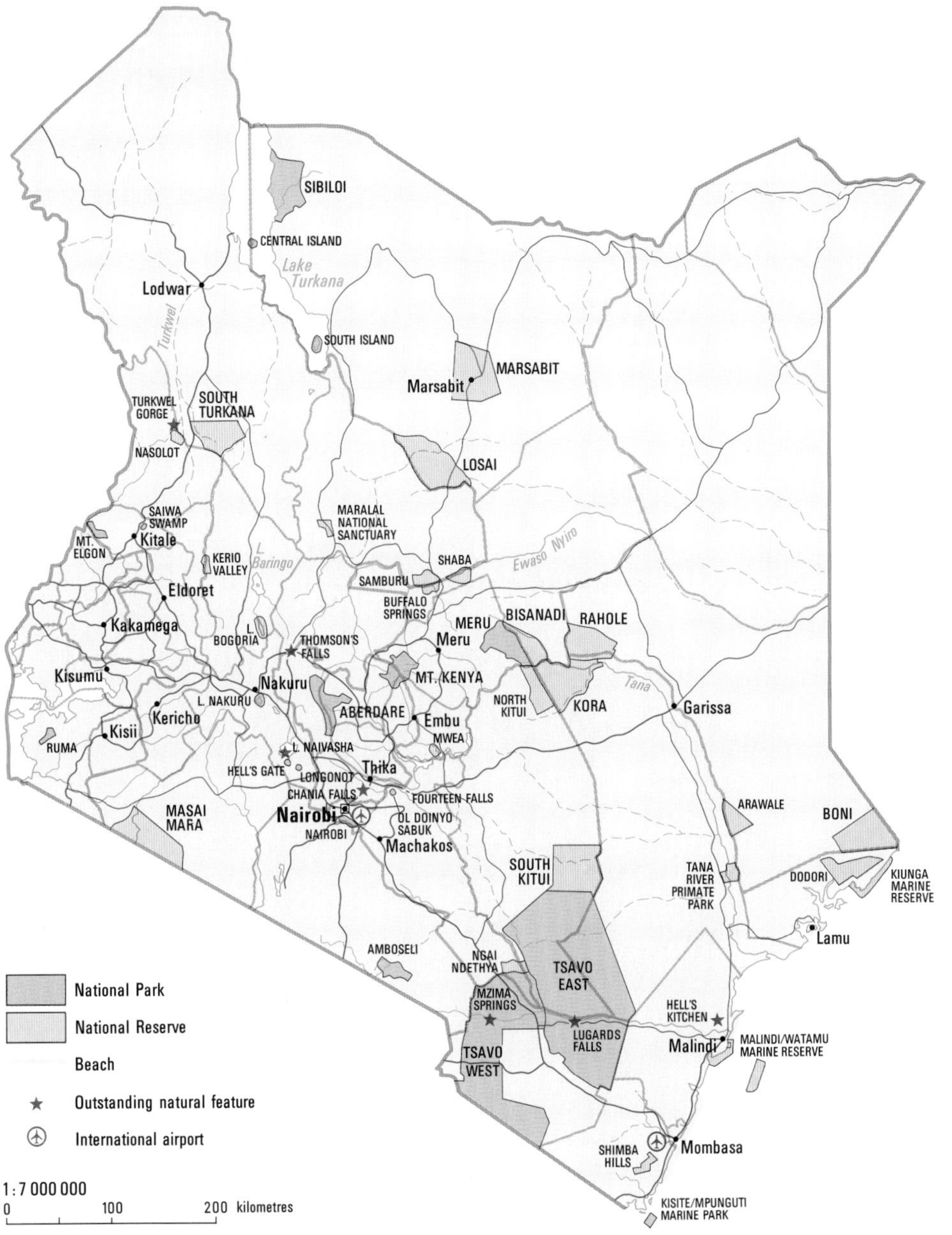

Visitors to National Parks and Reserves 1986

National Park/Reserve	Number of visitors
Amboseli	157 000
Lake Nakuru	127 900
Masai Mara	94 800
Nairobi	91 600
Tsavo (West)	82 900
Tsavo (East)	75 300
Animal Orphanage	73 000
Aberdare	42 500
Buffalo Springs	41 400
Malindi Marine	36 100
Lake Bogoria	25 600
Meru	20 400
Mount Kenya	16 600
Kisite/Mpunguti	12 200
Shimba Hills	10 900
Samburu	5 100
Mount Elgon	4 700
Marsabit	3 200
Ol-Doinyo Sabuk	1 700
Saiwa Swamp	1 200

Tourists by Country of Residence 1986

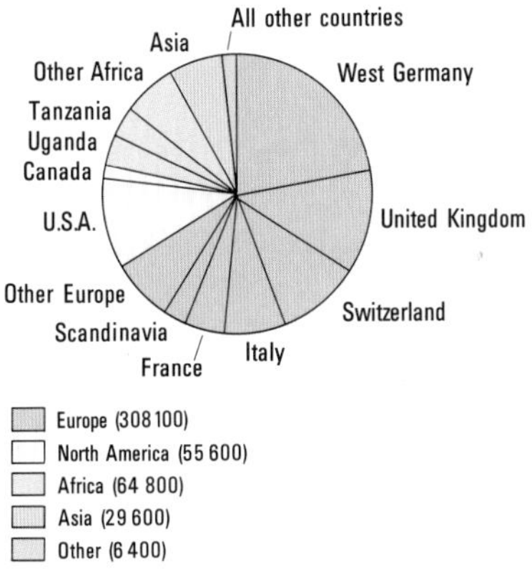

Cheetah in Masai Mara National Reserve.

Fort Jesus, Mombasa, built in 1593.

The tourist industry is of vital importance to the economy of Kenya and is second only to agriculture as a source of foreign exchange earnings. In 1986 receipts from tourism totalled KShs 4 940 million. Tourism has created employment for a large number of local people and has led to a general improvement of the country's infrastructure.

Visitors to Museums and Historic Sites 1986

Museum/Historic site	Number of visitors
National Museum	204 300
Snake Park (Nairobi)	188 100
Fort Jesus	168 800
Kisumu Museum	56 300
Kitale Museum	37 800
Gedi Ruins	32 800
Meru Museum	29 500
Lamu	28 000
Jumba la Mtwana	8 400
Olorgesailie	6 500
Kariandusi	4 100
Hyrax Hill	2 700

Museums and Historic Sites

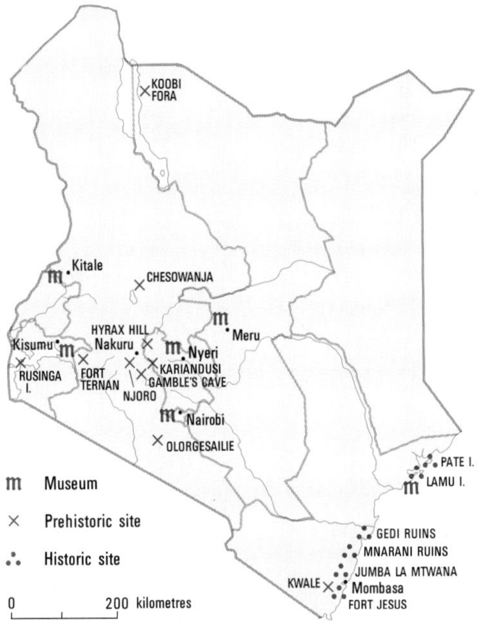

UGANDA – PHYSICAL AND POLITICAL

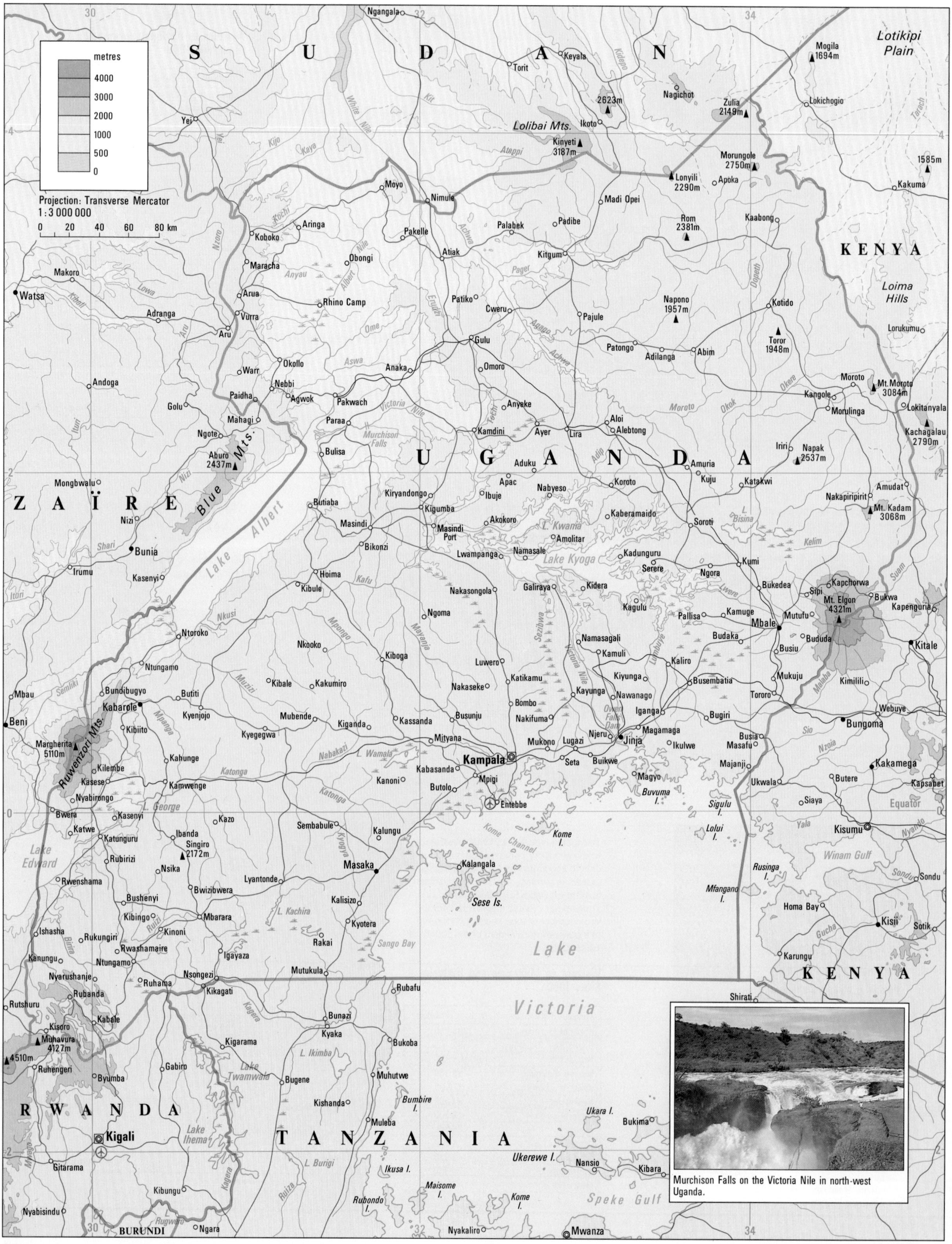

Murchison Falls on the Victoria Nile in north-west Uganda.

UGANDA – CLIMATE, NATURAL VEGETATION AND AGRICULTURE

Mean Annual Rainfall

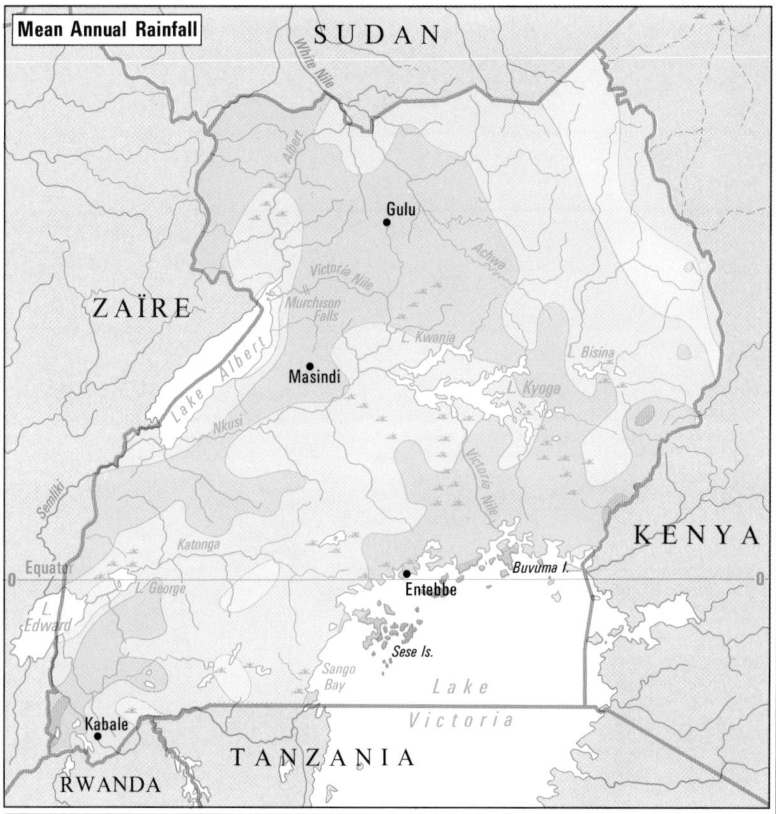

Rainfall in Millimetres

- over 2000
- 1600–2000
- 1200–1600
- 800–1200
- 600–800
- 400–600

1 : 6 500 000

0 100 200 km

Climate Statistics

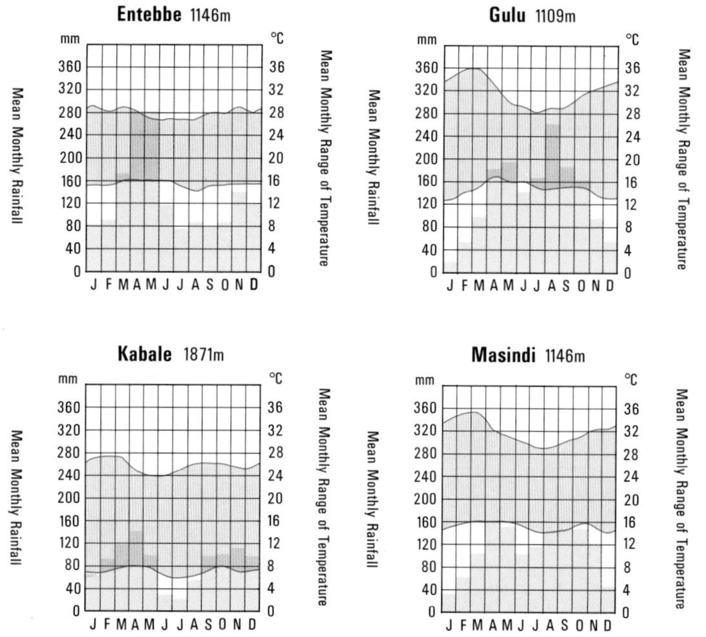

Climate station	Number of days when rainfall over 0.2 mm	Maximum fall of rain in 24hours (millimetres)	Sunshine hours per day	Relative humidity (%) 0800hrs	Relative humidity (%) 1400hrs
Entebbe	170	283	6.6	86	66
Gulu	161	105	7.9	79	51
Kabale	154	76	5.0	95	61
Masindi	132	93	5.6	81	56

Natural Vegetation

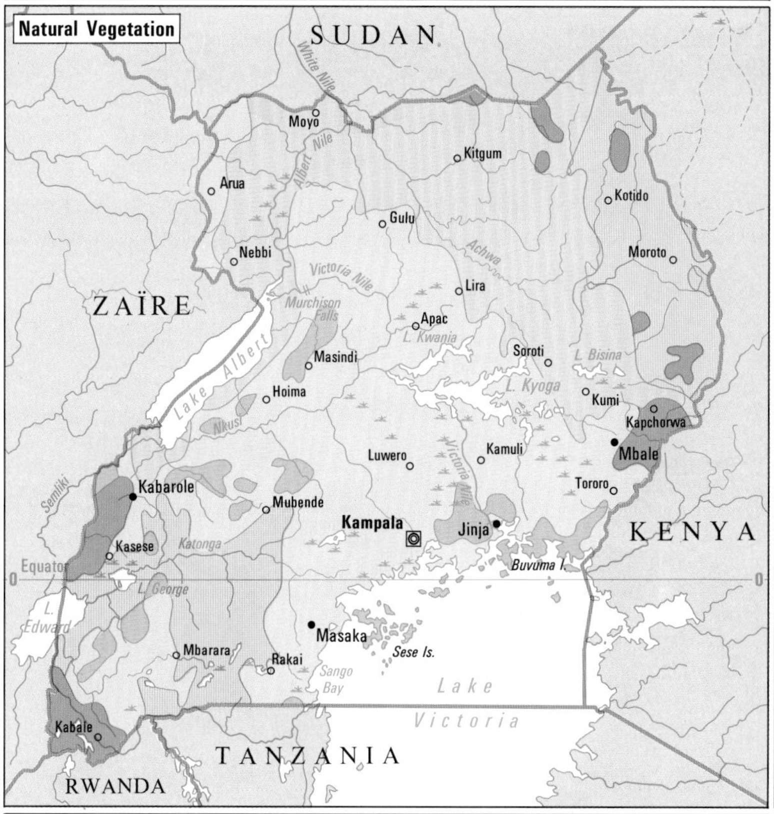

- Rain forest
- Mosaic of lowland rain forest and secondary grassland
- Montane forest and scrub
- Undifferentiated woodland
- Mosaic of East African evergreen bushland and secondary wooded grassland
- Deciduous bushland and thicket

Agriculture

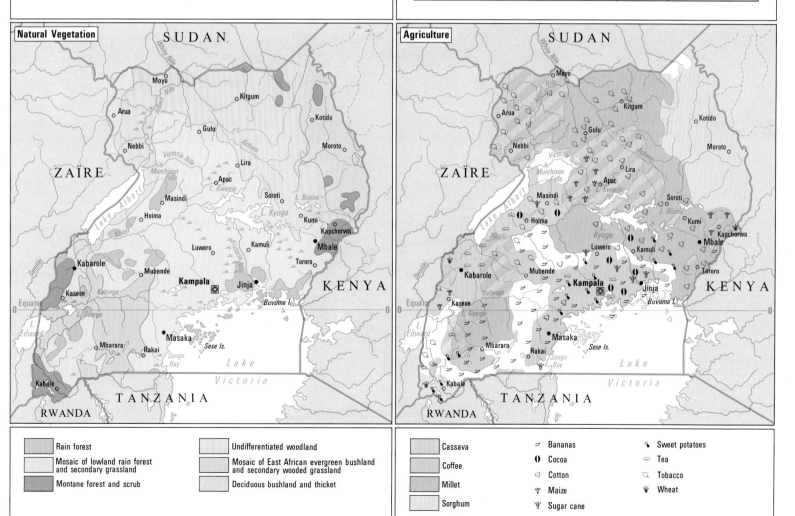

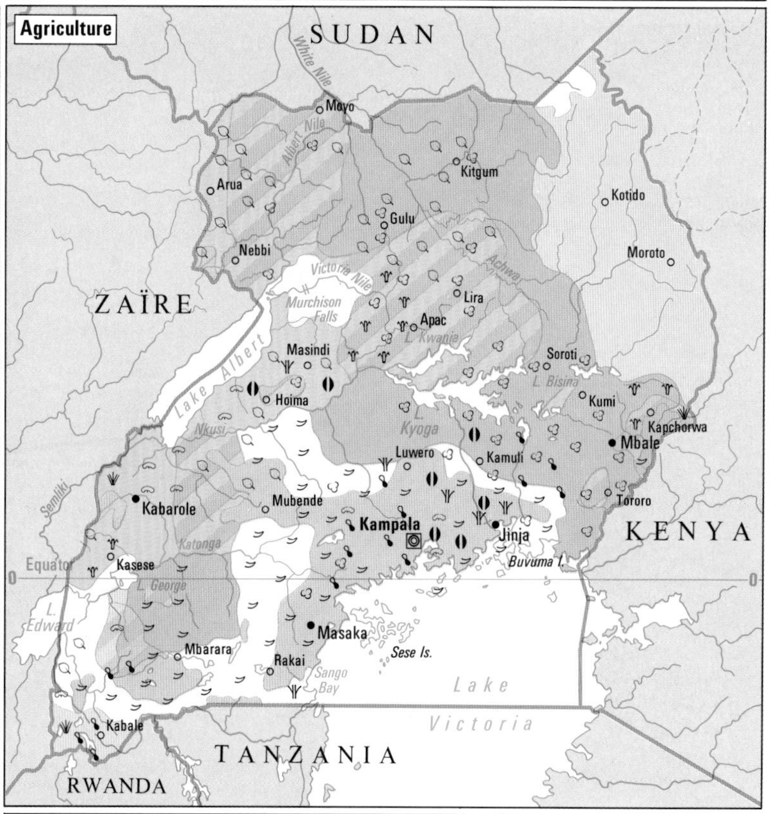

- Cassava
- Coffee
- Millet
- Sorghum
- Bananas
- Cocoa
- Cotton
- Maize
- Sugar cane
- Sweet potatoes
- Tea
- Tobacco
- Wheat

UGANDA – ADMINISTRATIVE AND ECONOMIC

International boundary
District boundary
■ Capital city
● District headquarters
1 : 6 500 000
0 100 200 km

Population Density by District (1980)
Persons per sq. km.

- over 1000
- 200–1000
- 150–200
- 100–150
- 50–100
- 0–50

Population of Towns

- over 40 000
- 20 000–40 000
- 10 000–20 000
- 2 000–10 000

Main road
Other road
Railway
Waterway
International airport
Airfield
Ferry

Minerals

- Beryllium
- Copper
- Diamonds
- Gold
- Iron ore
- Limestone
- Phosphates
- Salt
- Sand
- Tin

Industry

- Cement
- Engineering
- Fishing
- Food processing
- HEP station
- Metal processing
- Pulp and paper
- Textiles
- Tobacco
- Vehicle assembly

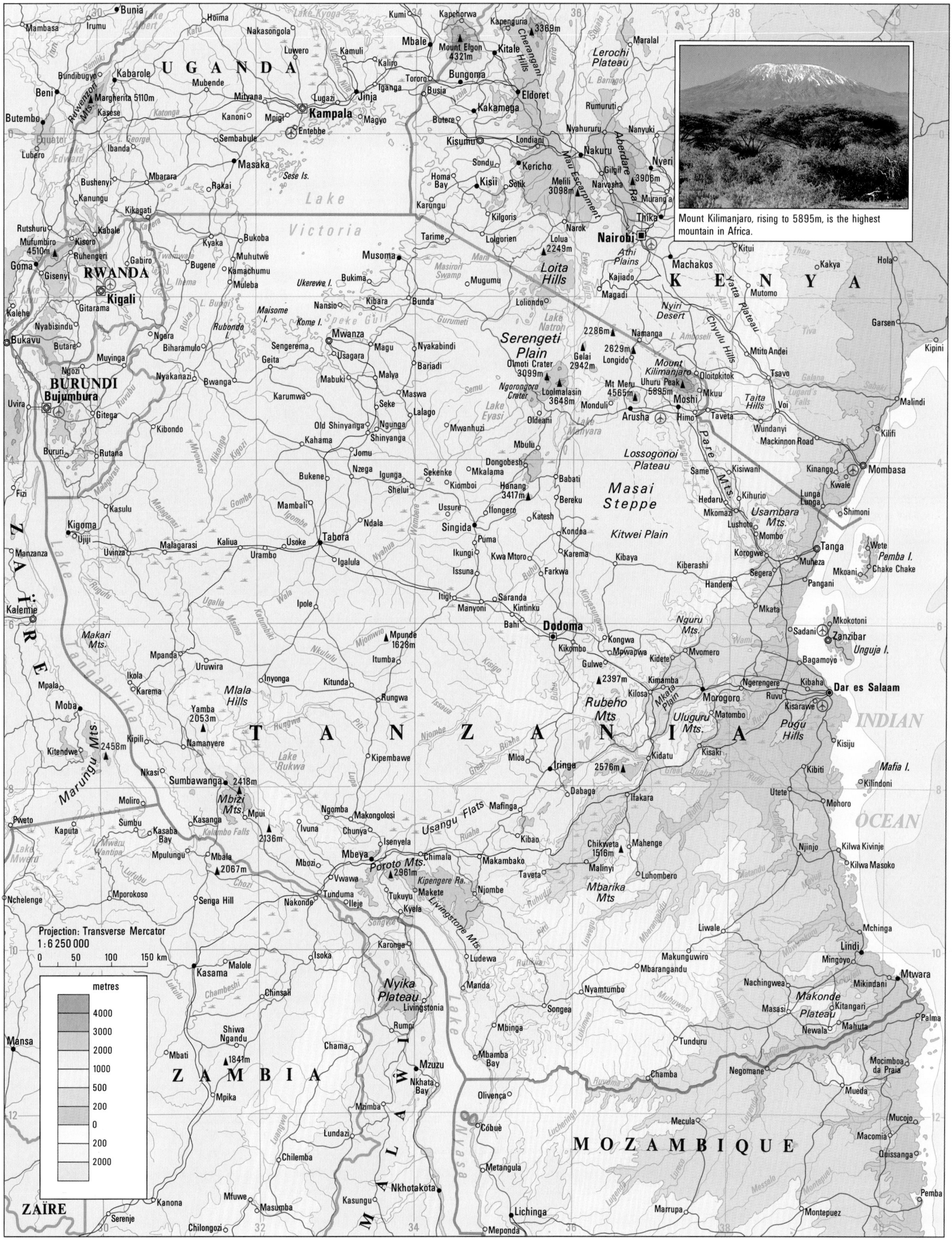

Mount Kilimanjaro, rising to 5895m, is the highest mountain in Africa.

UGANDA

KENYA

RWANDA

BURUNDI

ZAÏRE

TANZANIA

ZAMBIA

MALAWI

MOZAMBIQUE

ZAÏRE

Mambasa
Bunia
Irumu
Hoima
Nakasongola
Kumi
Kapchorwa
Kapenguria ▲3369m
Maralal
Lerochi Plateau
Butembo
Beni
Bundibugyo
Kabarole
Mubende
Luwero
Kamuli
Mbale
Mount Elgon 4321m
Kitale
Cherangani Hills
Margherita 5110m
Kasese
Mpigi
Mityana
Lugazi
Jinja
Iganga
Tororo
Busia
Bungoma
Butere
Kakamega
Eldoret
Rumuruti
Nyahururu
L. Baringo
Kampala
Entebbe
Sembabule
Magyo
Kisumu
Londiani
Nanyuki
Aberdare Ra.
Masaka
Mbarara
Rakai
Sese Is.
Kericho
Sondu
Homa Bay
Kisii
Sotik
Melili 3098m
Naivasha
Gilgil
Nakuru
Nyeri ▲3906m
Murang'a
Thika
Kanungu
Kikagati
Lake Victoria
Ukerewe I.
Karungu
Kilgoris
Narok
Lolua ▲2249m
Nairobi
Athi Plains
Machakos
Kitui
Kakya
Hola
Rutshuru
Kisoro
Kabale
Bukoba
Tarime
Loliondo
Loita Hills
Kajiado
Magadi
Mutomo
Mufumbiro 4510m
Ruhengeri
Gabiro
Kyaka
Muhutwe
Musoma
Mugumu
Nyiri Desert
Namanga
Garsen
Goma
Giseny
Gitarama
Bugene
Kamachumu
Muleba
Maisome I.
Kome I.
Nansio
Kibara
Bunda
Serengeti Plain
L. Natron
Longido ▲2629m
Mount Kilimanjaro
Chyulu Hills
Mto Andei
Tsavo
Kipini
Kigali
Nyabisindu
Ngera
Biharamulo
Rubondo I.
Nyakabindi
Gelai 2942m
2286m ▲
Oloitokitok
Mkuu
Galana
Bukavu
Butare
Muyinga
Mwanza
Sengerema
Usagara
Magu
Bariadi
Olmoti Crater 3099m
Loolmalasin 3648m
Ngorongoro Crater
Mt Meru 4565m
Uhuru Peak 5895m
Moshi
Taita Hills
Voi
Malindi
BURUNDI
Bujumbura
Ngozi
Nyakanazi
Bwanga
Geita
Mabuki
Maswa
Mwanhuzi
Lake Eyasi
Monduli
Arusha
Himo
Taveta
Wundanyi
Mackinnon Road
Kilifi
Uvira
Gitega
Kibondo
Kahama
Ngunga
Shinyanga
Old Shinyanga
Mbulu
Oldeani
Lake Manyara
Bururi
Rutana
Jomu
Nzega
Dongobesh
Mkalama
Babati
Lossogonoi Plateau
Same
Kisiwani
Kihurio
Usambara Mts.
Fizi
Igunga
Sekenke
Hanang 3417m
Bereku
Masai Steppe
Hedaru
Lushoto
Mombo
Kasulu
Kigoma
Ujiji
Kiomboi
Shelui
Ilongero
Ussure
Katesh
Kondoa
Kiberashi
Korogwe
Mkomazi
Manzanza
Mambali
Ndala
Singida
Puma
Kwa Mtoro
Karema
Kibaya
Handeni
Segera
Muheza
Tanga
Kalemie
Malagarasi
Kaliua
Usoke
Tabora
Igalula
Ikungu
Issuna
Farkwa
Kibirashi
Mkata
Pemba I.
Wete
Urambo
Itigi
Saranda
Kintinku
Manyoni
Bahi
Dodoma
Kikombo
Kongwa
Mpwapwa
Kidete
Nguru Mts.
Mvomero
Sadani
Mkoani
Chake Chake
Ikola
Karema
Mpanda
Uruwira
Mpunde 1628m
Itumba
Kisigo
Gulwe
Kimamba ▲2397m
Kilosa
Morogoro
Ngerengere
Ruvu
Kibaha
Zanzibar
Unguja I.
Mpala
Moba
Kipili
Namanyere
Inyonga
Kitunda
Rungwa
Uluguru Mts.
Matombo
Kisarawe
Dar es Salaam
Makari Mts.
2458m ▲
Yamba 2053m
Mlala Hills
Rubeho Mts.
Mkata Plain
Kisaki
Pugu Hills
Kisiju
INDIAN
Kitendwe
Nkasi
Sumbawanga ▲2418m
Iringa ▲2576m
Kidatu
Mafia I.
Kilindoni
Mbizi Mts. 2136m
Mpui
Kasanga
Ngomba
Makongolosi
Mafinga
Dabaga
Ifakara
Kibao
Utete
Mehoro
Mpulungu
Mbala ▲2067m
Mbozi
Vwawa
Chunya
Isenyela
Ruaha
Taveta
Njinjo
Kilwa Kivinje
Mporokoso
Senga Hill
Tunduma
Ileje
Mbeya
Poroto Mts. ▲2961m
Chimala
Makambako
Malinyi
Chikweta 1516m
Mahenge
Luhombero
Kilwa Masoko
Nchelenge
Nakonde
Karonga
Tukuyu
Kyela
Makete
Njombe
Mbarika Mts.
Kasama
Malole
Isoka
Ludewa
Manda
Mchinga
Lindi
Mingoyo
Kanona
Chinsali
Nyika Plateau
Livingstonia
Rumphi
Nyamtumbo
Songea
Makunguwiro
Mbarangandu
Nachingwea
Mikindani
Mtwara
Mbati
Shiwa Ngandu ▲1841m
Chama
Mbinga
Makonde Plateau
Masasi
Kitangari
Palma
Mansa
Mbamba Bay
Tunduru
Newala
Mahuta
Mpika
Mzuzu
Nkhata Bay
Olivença
Mbeya
Songea
Negomane
Mocimboa da Praia
Mpulungu
Mwambe
Chamba
Mueda
Mucojo
Mfuwe
Masumba
Kasungu
Nkhotakota
Lichinga
Meponda
Ruvuma
Mecula
Macomia
Serenje
Chilongozi
Kanona
Chilemba
Mzimba
Lundazi
Cóbuè
Metangula
Marrupa
Qissanga
Macomia
Mocojo
Chilwa

Projection: Transverse Mercator
1 : 6 250 000

0 50 100 150 km

metres
4000
3000
2000
1000
500
200
0
200
2000

TANZANIA – CLIMATE, NATURAL VEGETATION AND AGRICULTURE

Mean Annual Rainfall

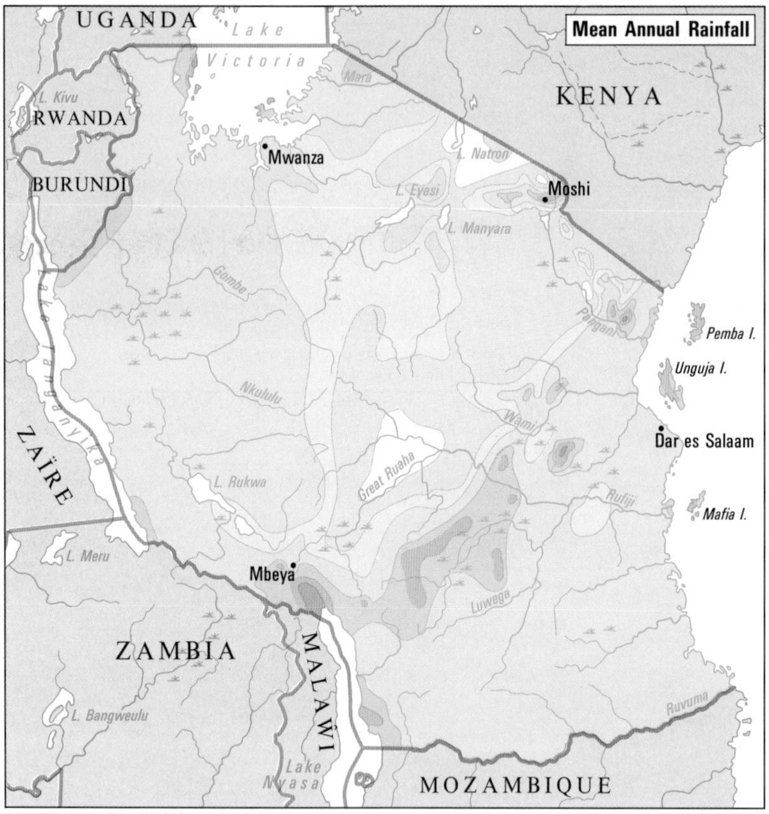

Rainfall in Millimetres

- Over 2000
- 1600–2000
- 1200–1600
- 800–1200
- 600–800
- 400–600
- 200–400

1:12 500 000

0 200 400km

Climate Statistics

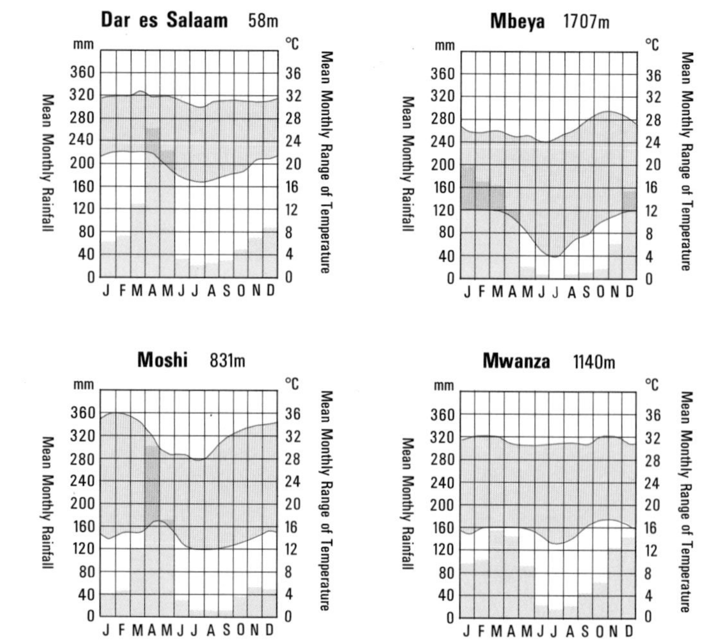

Climate station	Number of days when rainfall over 0.2mm	Maximum fall of rain in 24hours (millimetres)	Sunshine hours per day	Relative humidity(%) 0800hrs	Relative humidity(%) 1400hrs
Dar es Salaam	113	137	7·8	83	65
Mbeya	117	59	6·9	74	58
Moshi	77	178	7·1	78	48
Mwanza	111	151	8·4	73	59

Natural Vegetation

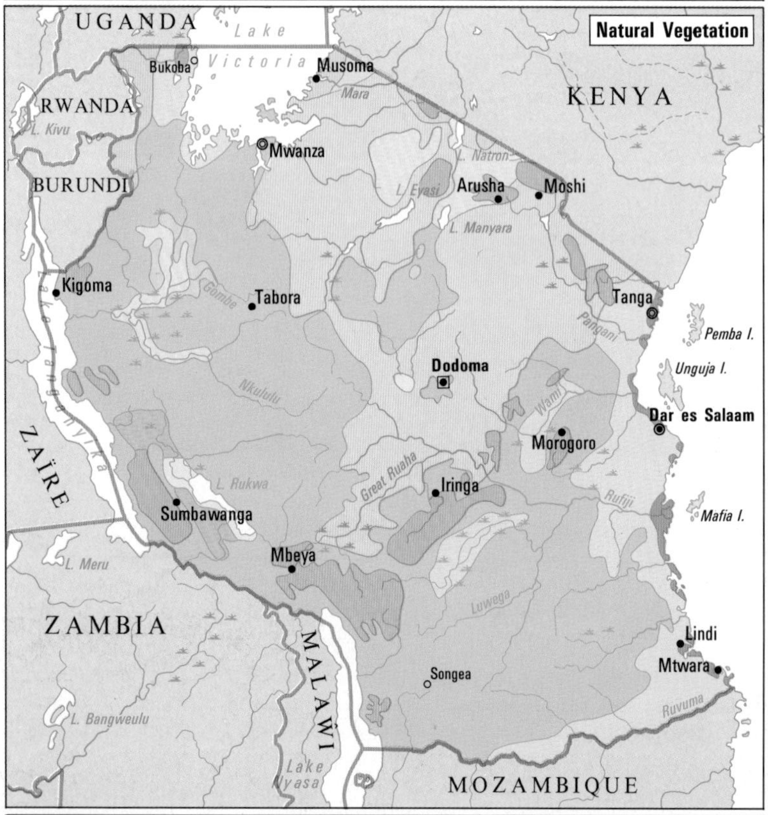

- Rain forest
- Rain forest with grassland
- Mangrove forest
- Montane forest and scrub
- Miombo woodland
- Undifferentiated woodland
- Wooded grassland
- Grassland
- Evergreen bushland and secondary wooded grassland
- Deciduous bushland and thicket
- Semi–desert grassland
- Salt–flat vegetation

Agriculture

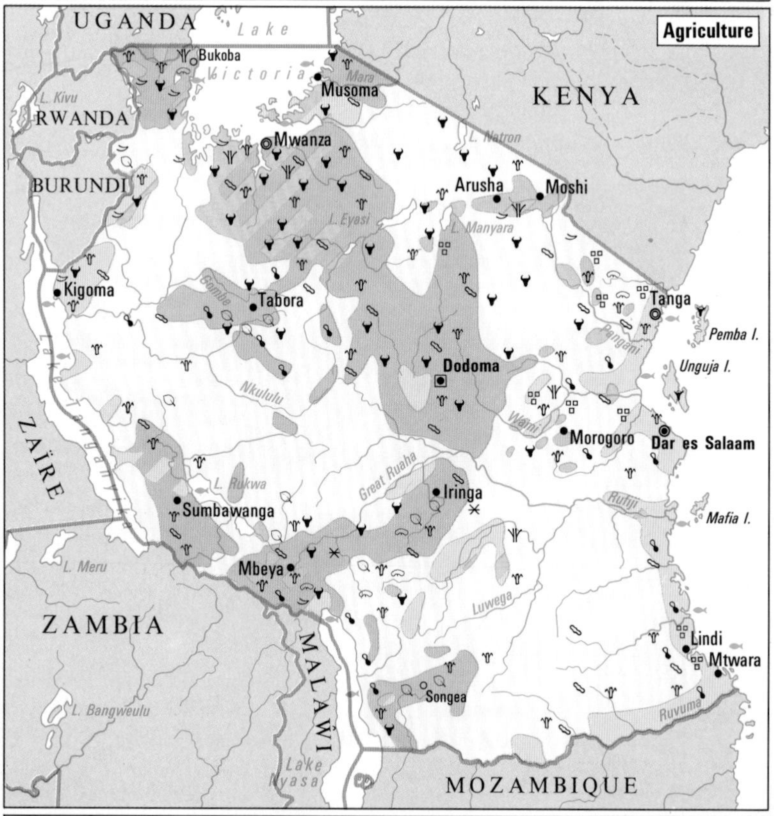

- Coffee
- Cotton
- Millet
- Rice
- Bananas
- Cassava
- Cloves
- Groundnuts
- Maize
- Pyrethrum
- Sisal
- Sugar cane
- Tea
- Tobacco
- Cattle
- Fishing

TANZANIA – ADMINISTRATIVE AND ECONOMIC

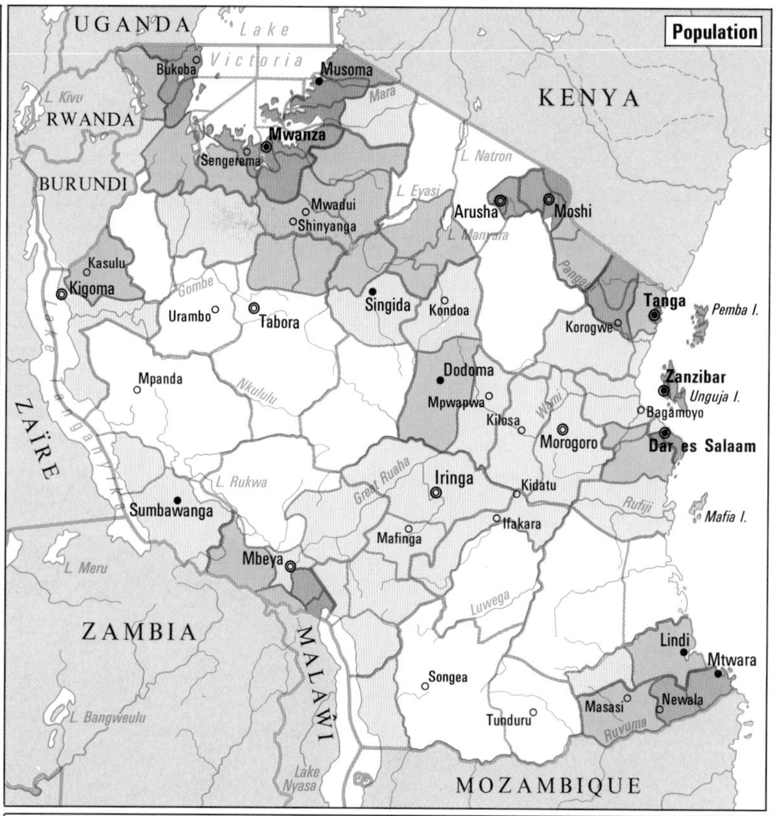

International boundary
Regional boundary

■ Capital city
● Regional headquarters

1:12 500 000
0 200 400 km

Population Density by District
Persons per sq. km.

- over 500
- 100–500
- 50–100
- 25–50
- 10–25
- 0–10

Population of Towns

- over 100 000
- 50 000–100 000
- 25 000–50 000
- 10 000–25 000

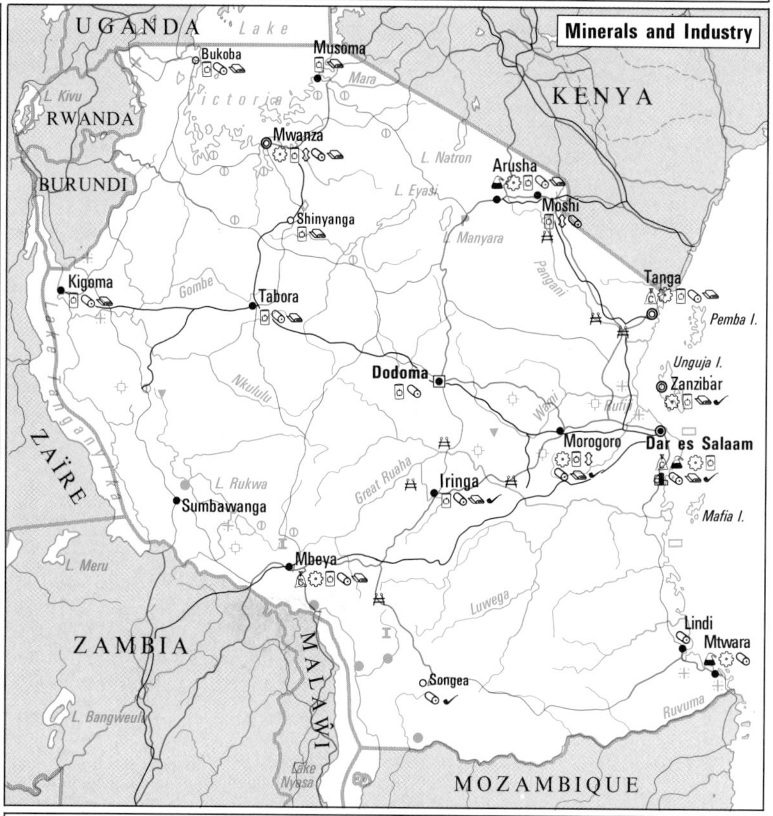

- Main road
- Other road
- Railway
- Waterway
- International airport
- Airfield
- F Ferry

Minerals

- Coal
- Copper
- Diamonds
- Gold
- Iron ore
- Mica
- Natural gas
- Phosphates
- Salt
- Tin

Industry

- Cement
- Chemicals
- Engineering
- Food processing
- HEP station
- Leather
- Oil refining
- Paper and pulp
- Textiles
- Tobacco

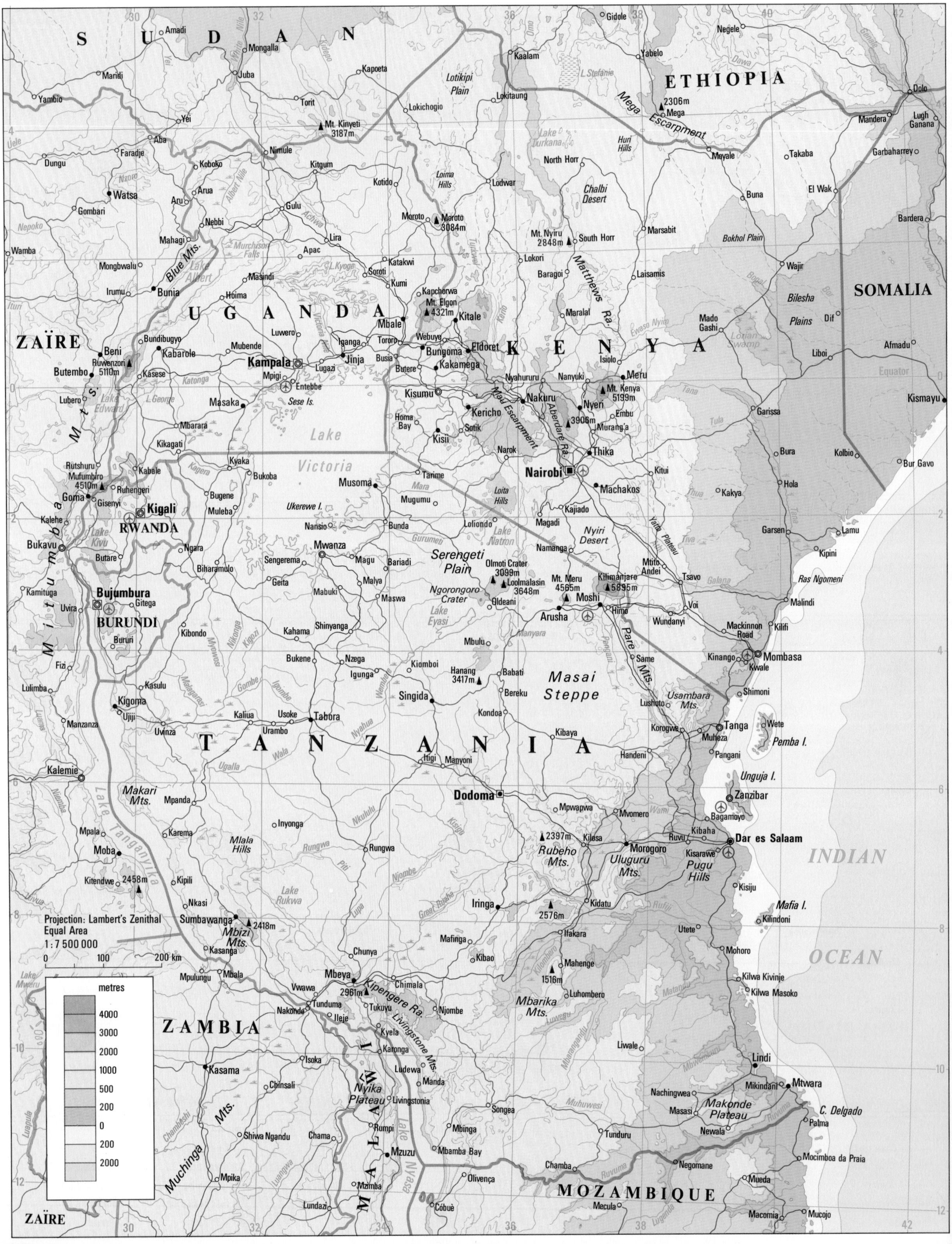

SUDAN
ETHIOPIA
SOMALIA
KENYA
UGANDA
ZAÏRE
RWANDA
BURUNDI
TANZANIA
ZAMBIA
MALAWI
MOZAMBIQUE
INDIAN OCEAN
Amadi
Mongalla
Maridi
Juba
Kapoeta
Yambio
Torit
Yei
Mt. Kinyeti 3187m
Lokichogio
Lotikipi Plain
Lokitaung
Kaalam
L. Stefanie
Gidole
Negele
Yabelo
Mega Escarpment
2306m Mega
Moyale
Dolo
Mandera
Lugh Ganana
Takaba
Garbaharrey
Aba
Faradje
Dungu
Nimule
Koboko
Kitgum
Lake Turkana
Huri Hills
North Horr
Loima Hills
Kotido
Lodwar
Chalbi Desert
Buna
El Wak
Watsa
Arua
Aru
Gombari
Gulu
Nebbi
Moroto
Moroto 3084m
Marsabit
Bardera
Bokhol Plain
Wamba
Mahagi
Murchison Falls
Lira
Apac
Mt. Nyiru 2848m
South Horr
Mongbwalu
Blue Mts.
Lake Albert
Katakwi
Soroti
Kumi
Lokori
Baragoi
Matthews Ra.
Laisamis
Wajir
Masindi
Irumu
Bunia
Hoima
Kapchorwa
Mt. Elgon 4321m
Kitale
Maralal
Bilesha Plains
Dif
Mado Gashi
Lorian Swamp
Mbale
Luwero
Bundibugyo
Beni
Mubende
Kabarole
Iganga
Tororo
Jinja
Webuye
Eldoret
Bungoma
Isiolo
Liboi
Afmadu
Ruwenzori 5110m
Butembo
Kasese
Kampala
Mpigi
Lugazi
Busia
Butere
Kakamega
Nyahururu
Nanyuki
Meru
Mt. Kenya 5199m
Equator
Kismayu
Entebbe
Kisumu
Lubero
L. George
Lake Edward
Masaka
Sese Is.
Nakuru
Nyeri
Embu
Kericho
Mau Escarpment
Aberdare Ra.
3906m
Murang'a
Garissa
Homa Bay
Sotik
Mbarara
Kisii
Lake Victoria
Narok
Thika
Kikagati
Bura
Kolbio
Bur Gavo
Rutshuru
Mufumbiro 4510m
Kabale
Kyaka
Bukoba
Tarime
Nairobi
Kitui
Hola
Ruhengeri
Goma
Gisenyi
Kigali
Bugene
Muleba
Musoma
Mugumu
Machakos
Kakya
Loita Hills
Ukerewe I.
Kajiado
Kalehe
Lake Kivu
Nansio
Bunda
Loliondo
Lake Natron
Magadi
Nyiri Desert
Yatta Plateau
Garsen
Lamu
Bukavu
Ngara
Butare
Mwanza
Serengeti Plain
Namanga
Kipini
Sengerema
Magu
Bariadi
Olmoti Crater 3099m
Biharamulo
Geita
Malya
Loolmalasin 3648m
Mt. Meru 4565m
Kilimanjaro 5895m
Mtito Andei
Tsavo
Ras Ngomeni
Kamituga
Bujumbura
Gitega
Mabuki
Maswa
Ngorongoro Crater
Oldeani
Moshi
Arusha
Himo
Voi
Malindi
Uvira
Lake Eyasi
Kahama
Shinyanga
Wundanyi
Kilifi
Mackinnon Road
Kibondo
Buruni
Mbulu
Mitumba Mts.
Pare Mts.
Same
Fizi
Bukene
Nzega
Kiomboi
Igunga
Hanang 3417m
Babati
Masai Steppe
Kinango
Mombasa
Kwale
Lulimba
Kasulu
Kigoma
Ujiji
Singida
Bereku
Shimoni
Usambara Mts.
Lushoto
Kaliua
Usoke
Tabora
Kondoa
Manzanza
Uvinza
Urambo
Kibaya
Korogwe
Tanga
Wete
Muheza
Pemba I.
Itigi
Manyoni
Pangani
Handeni
Kalemie
Unguja I.
Makari Mts.
Mpanda
Dodoma
Zanzibar
Mpwapwa
Mvomero
Bagamoyo
Lake Tanganyika
Inyonga
Mpala
Karema
2397m
Kilosa
Morogoro
Ruvu
Kibaha
Dar es Salaam
Moba
Mlala Hills
Rungwa
Rubeho Mts.
Uluguru Mts.
Kisarawe
Pugu Hills
Kitendwe 2458m
Kipili
Lake Rukwa
Kisiju
Nkasi
Iringa
2576m
Kidatu
Mafia I.
Kilindoni
Projection: Lambert's Zenithal Equal Area
1:7 500 000
0 100 200 km
Sumbawanga
2418m
Mbizi Mts.
Kasanga
Ifakara
Utete
Mafinga
Mohoro
Kibao
Chunya
Mahenge
1516m
Mbala
Mpulungu
Mbeya
2961m
Chimala
Kipengere Ra.
Vwawa
Tunduma
Nakonde
Tukuyu
Ileje
Njombe
Livingstone Mts.
Luhombero
Mbarika Mts.
Kilwa Kivinje
Kilwa Masoko
metres
4000
3000
2000
1000
500
200
0
200
2000
Kyela
Karonga
Liwale
Lindi
Kasama
Isoka
Ludewa
Chinsali
Manda
Nyika Plateau
Livingstonia
Mikindani
Mtwara
Nachingwea
Masasi
Makonde Plateau
Newala
C. Delgado
Palma
Songea
Muchinga Mts.
Shiwa Ngandu
Chama
Rumphi
Mbinga
Tunduru
Mzuzu
Mbamba Bay
Lake Nyasa
Chamba
Negomane
Mocimboa da Praia
Mpika
Mzimba
Olivença
Mueda
Lundazi
Cóbuè
Mecula
Macomia
Mucojo

Tectonic Plates

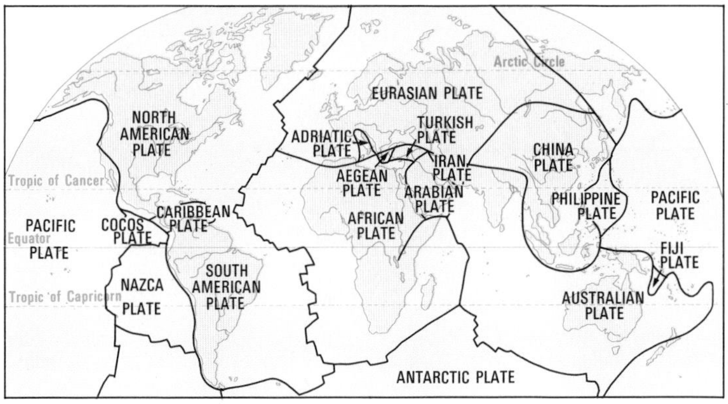

Tectonic plate boundary

Normal Faulting-Tension Theory

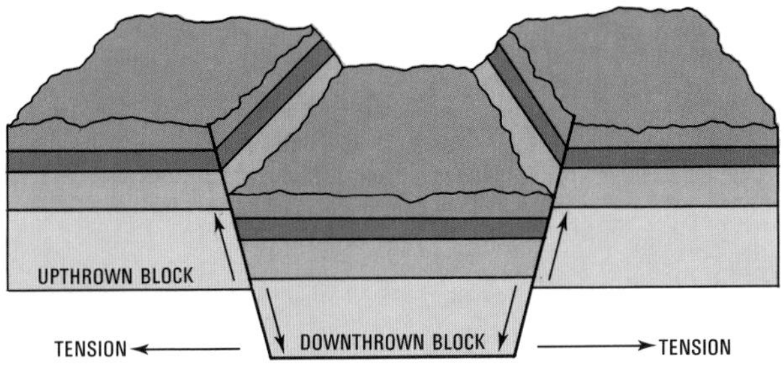

Reverse Faulting-Compression Theory

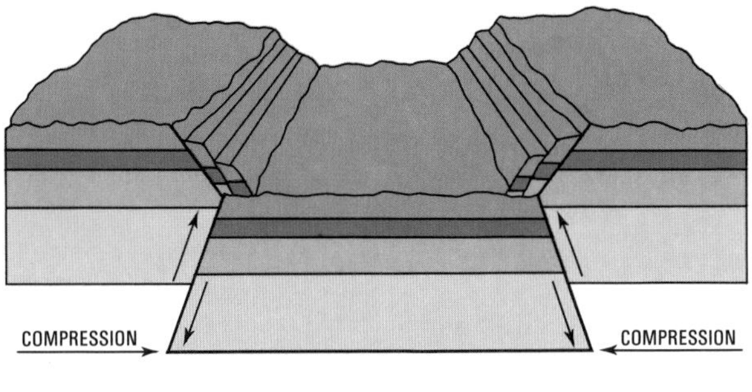

- Pleistocene volcanics
- Miocene volcanics
- Tertiary basalts
- Mesozoic limestone
- Mesozoic sandstone
- Pre-cambrian basement

The East African Rift Valley

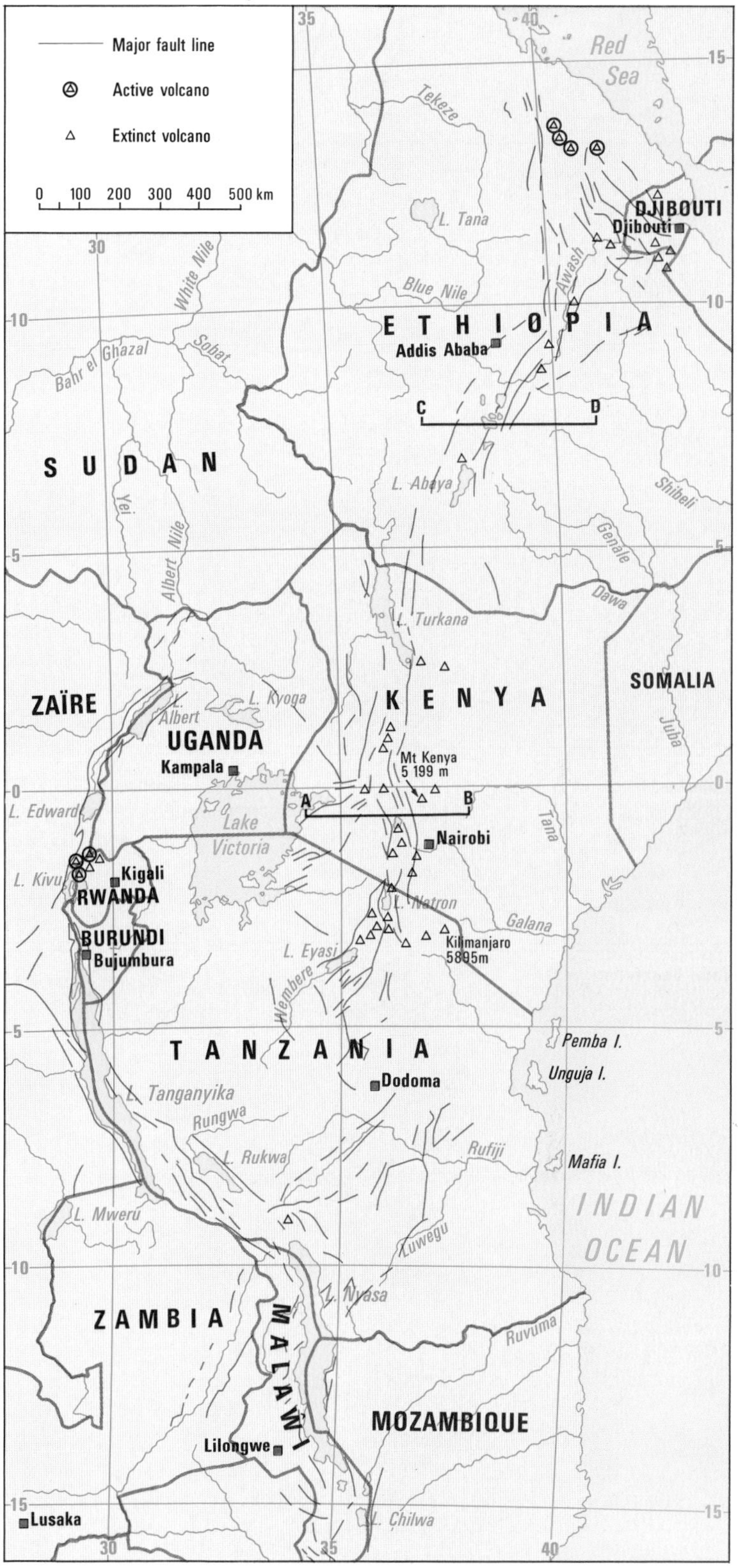

Latitudinal Section Across Kenyan Rift Valley at 0° 30′S

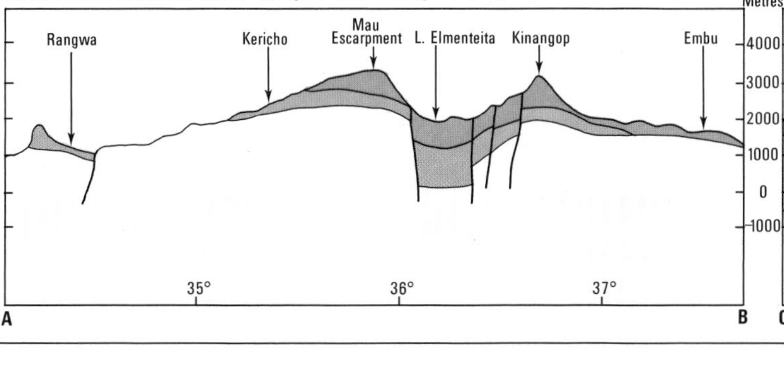

Latitudinal Section Across Ethiopian Rift Valley at 7° 30′N

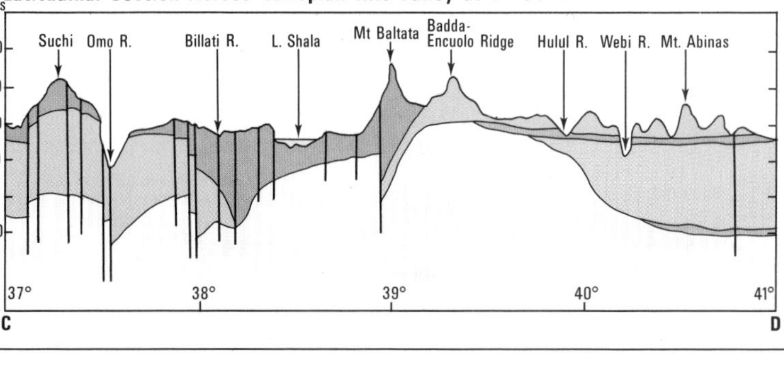

EAST AFRICA – CLIMATE, NATURAL VEGETATION AND DISEASES

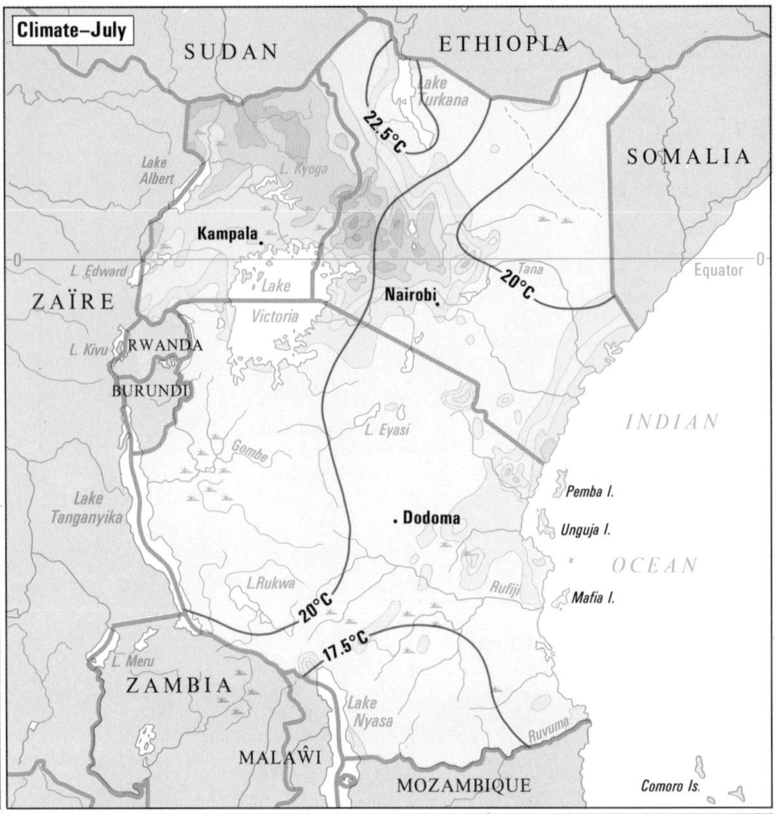

Climate–January

Mean Monthly Rainfall in Millimetres

- 300–400
- 200–300
- 150–200
- 100–150
- 50–100
- 25–50
- 10–25
- 0–10

Mean Daily Temperature

—25°C— Isotherm

Climate–July

Mean Monthly Rainfall in Millimetres

- 200–300
- 150–200
- 100–150
- 50–100
- 25–50
- 10–25
- 0–10

Mean Daily Temperature

—20°C— Isotherm

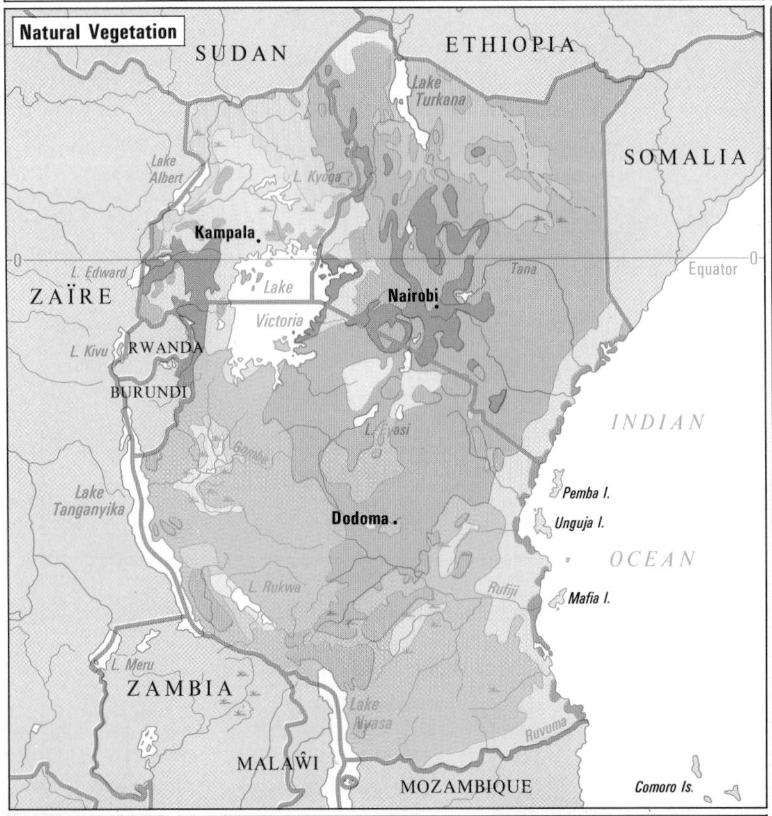

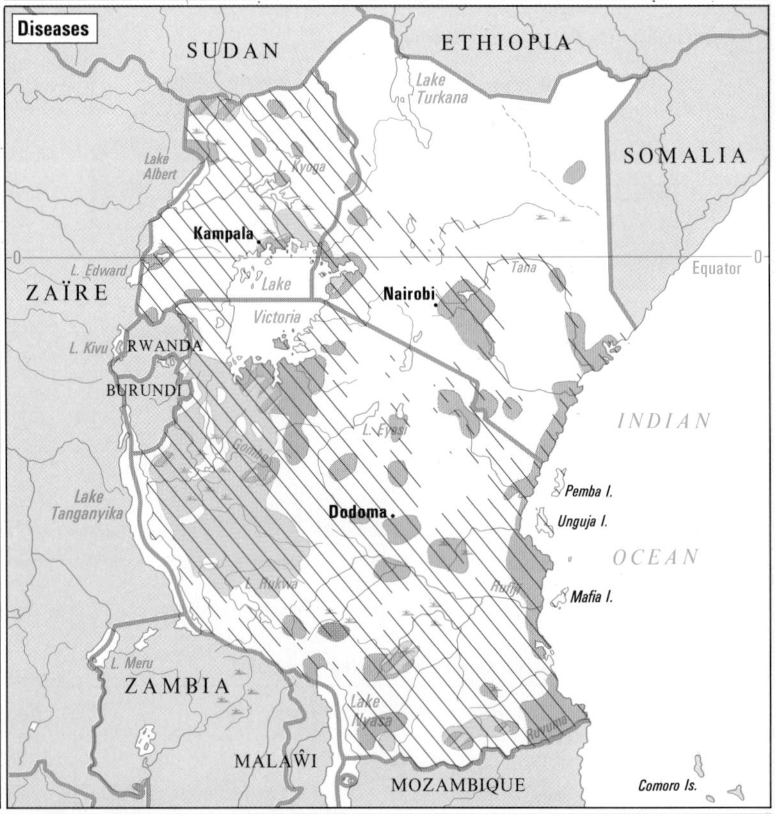

Natural Vegetation

- Rain forest
- Rain forest with grassland
- Mangrove forest
- Montane forest and scrub
- Woodland
- Miombo woodland
- Wooded grassland
- Grassland
- Evergreen bushland and secondary wooded grassland
- Deciduous bushland and thicket
- Semi-desert grassland
- Salt-flat vegetation

Diseases

- Areas where malarial infection may occur
- Areas where schistosomiasis (bilharzia) may occur
- Areas where trypanosomiasis (sleeping sickness) may occur

1:19 000 000

0 400 800 km

Commercial farming

Subsistence farming - food crops and livestock

Non-intensive livestock rearing - with scattered food crops

Desert - some nomadic herding

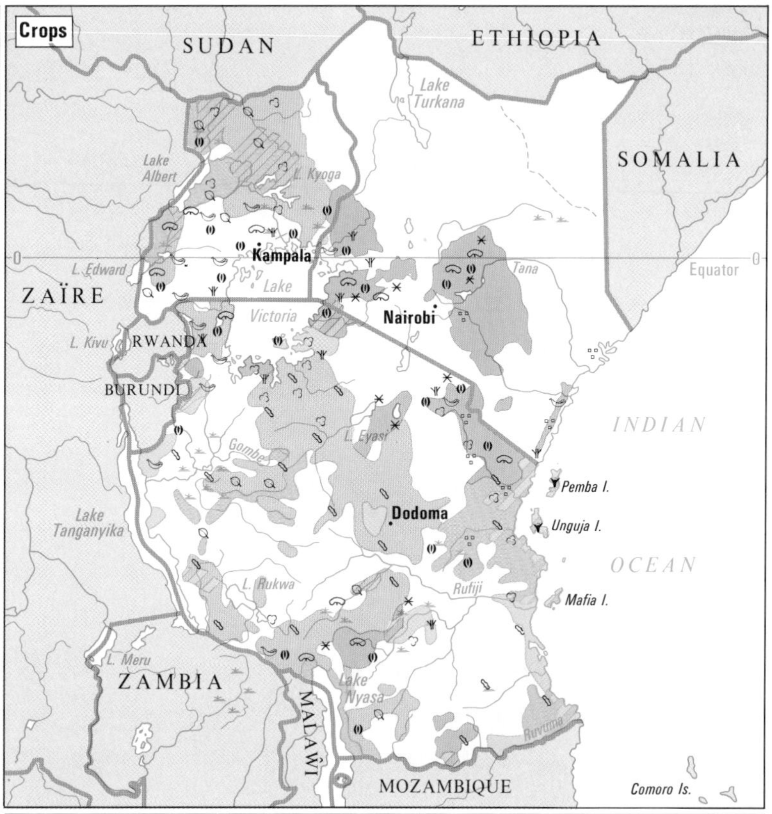

Cassava

Maize

Millet

Rice

Bananas

Cloves

Coffee

Cotton

Groundnuts

Pyrethrum

Sisal

Sugar cane

Tea

Tobacco

Crop Production

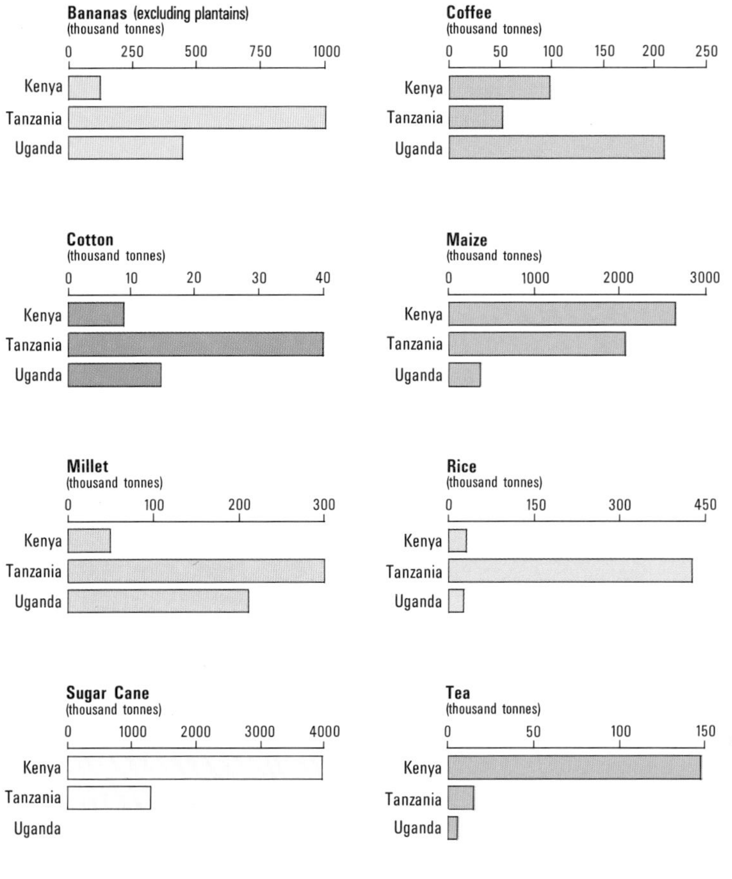

Note: 1985 Production figures from FAO Production Yearbook

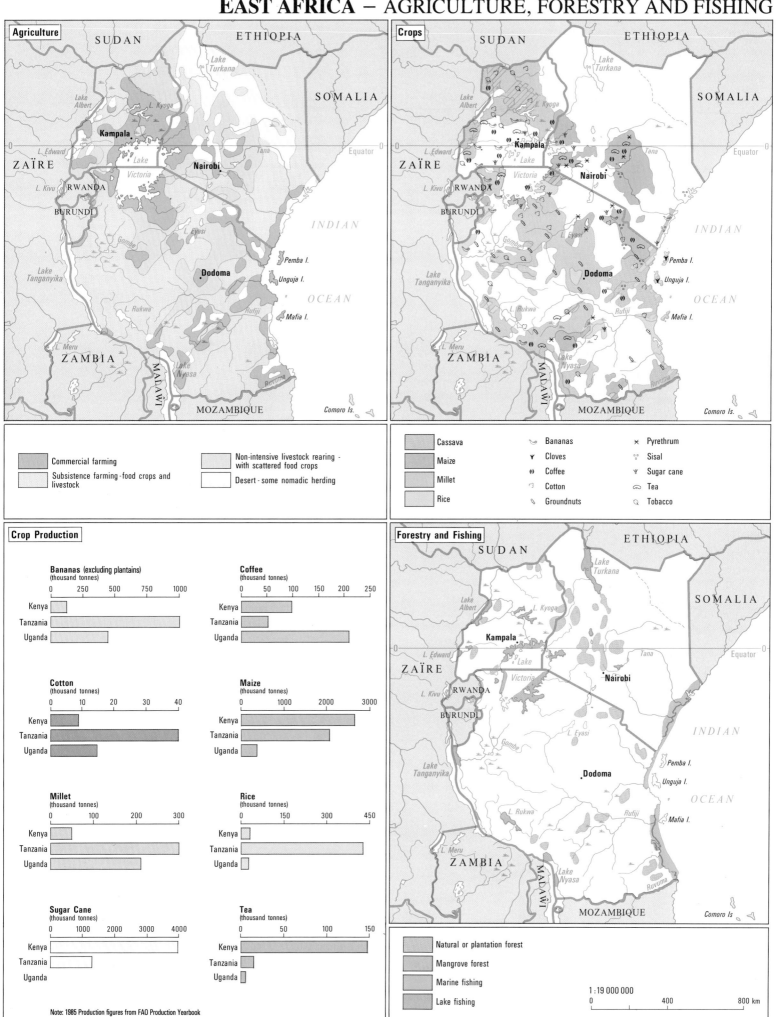

Natural or plantation forest

Mangrove forest

Marine fishing

Lake fishing

1 : 19 000 000

0 400 800 km

EAST AFRICA – ECONOMIC

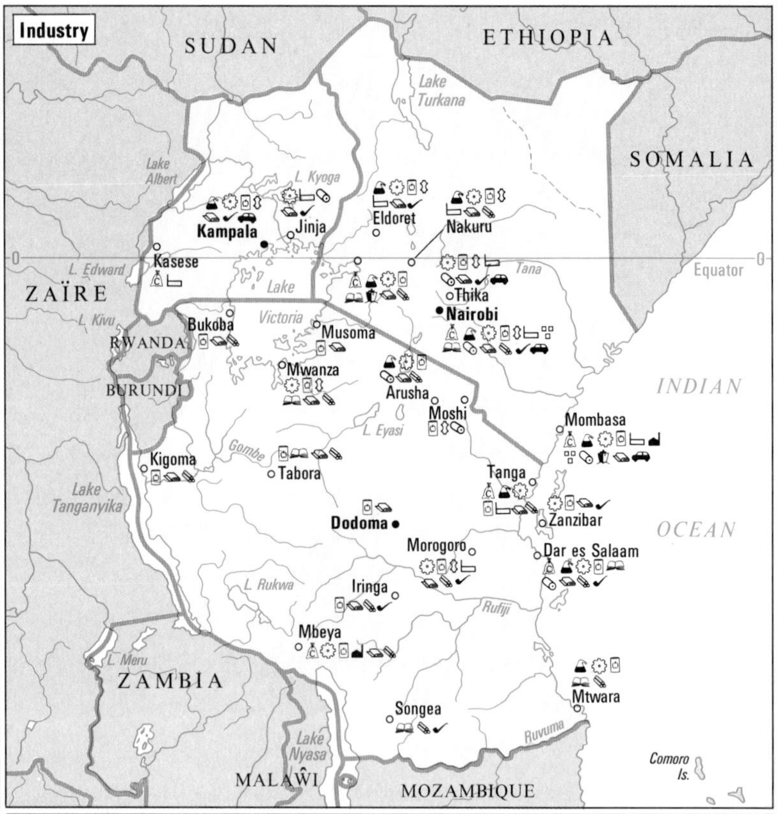

Copper	Gypsum	Salt
Diamonds	Iron ore	Soda ash
Fluorspar	Limestone	Tin
Gemstones	Mica	
Gold	Phosphates	

Cement	Metallurgy	Ship repairing
Chemicals	Mineral processing	Textiles
Engineering	Plastics	Timber
Food processing	Printing and publishing	Tobacco
Leather	Pulp and paper	Vehicle assembly

Coal	Oil refinery
Natural gas	Oil pipeline
Geothermal power station	
Hydro-electric power station	1:19 000 000
Thermal power station	0 400 800 km

Percentage Gross Domestic Product by Economic Activity

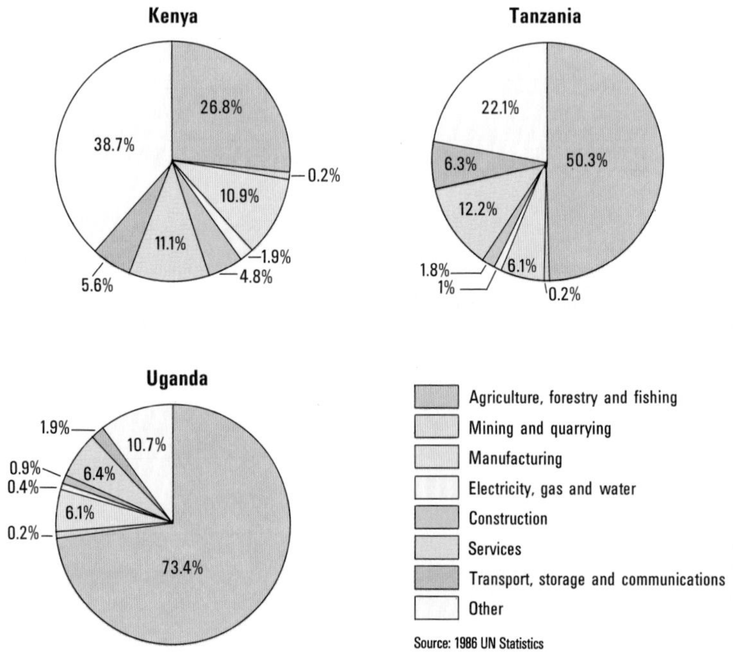

Source: 1986 UN Statistics

Generation of Electricity (Million KWH)

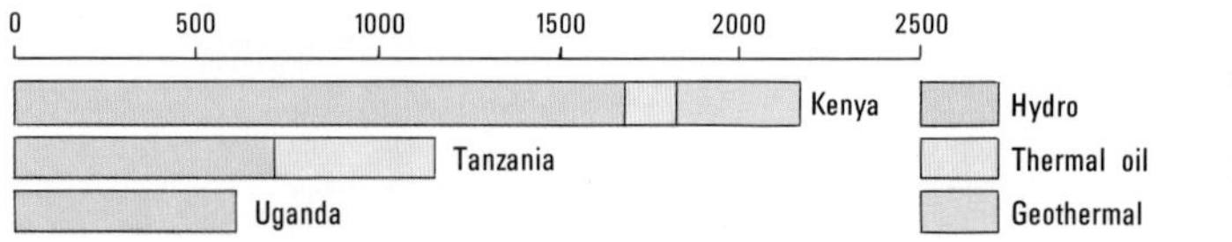

Source: latest available official statistics for each country

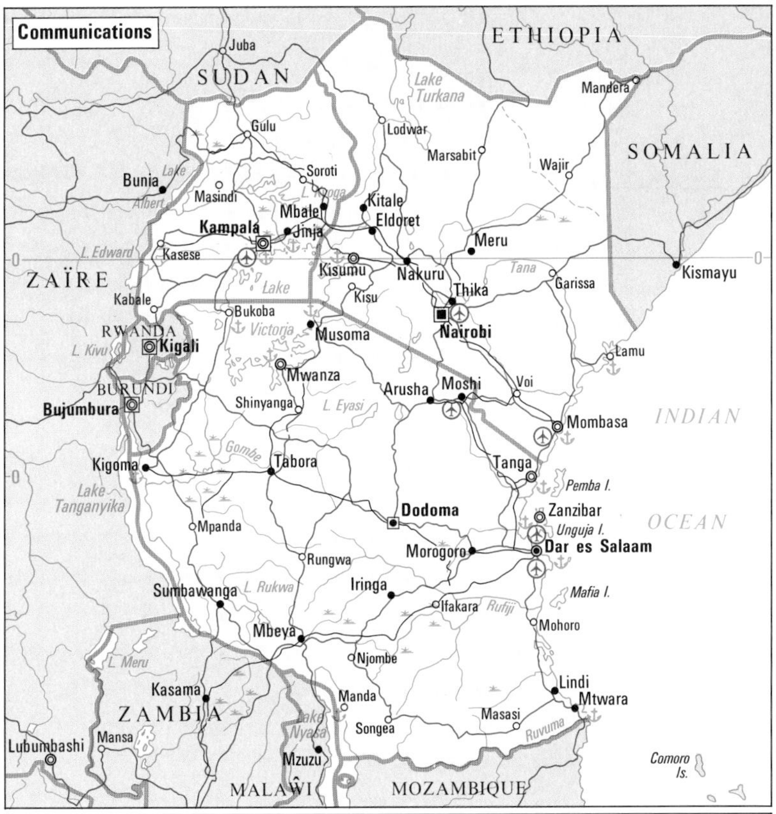

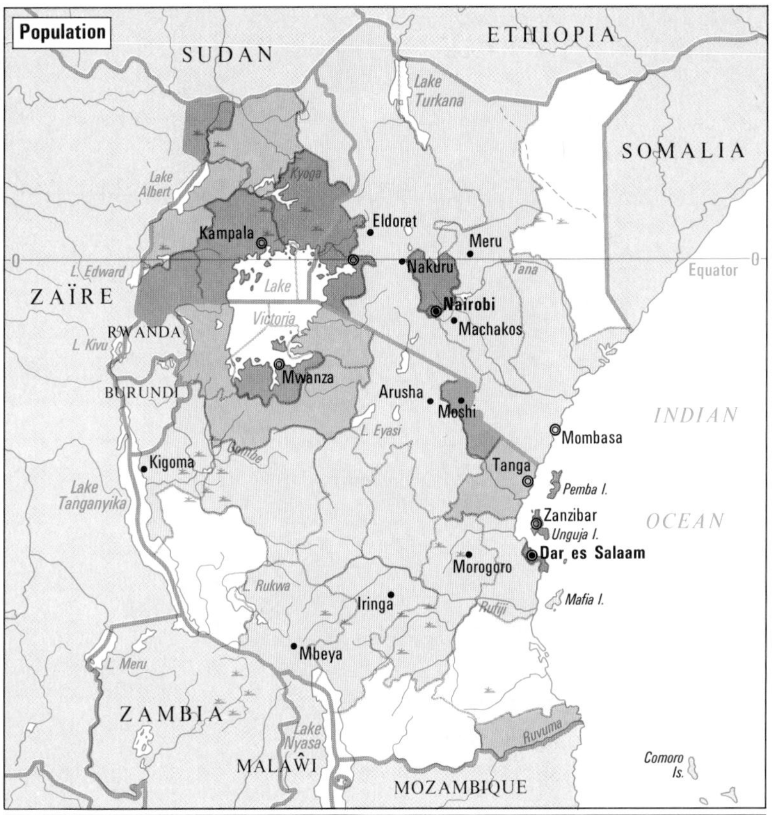

Road

Railway

International airport

Port

1 : 19 000 000

0 400 800 km

Population Density

Persons per sq. km.

- over 500
- 100–500
- 50–100
- 25–50
- 10–25
- 0–10

Population of Towns

- over 500 000
- 100 000–500 000
- 50 000–100 000

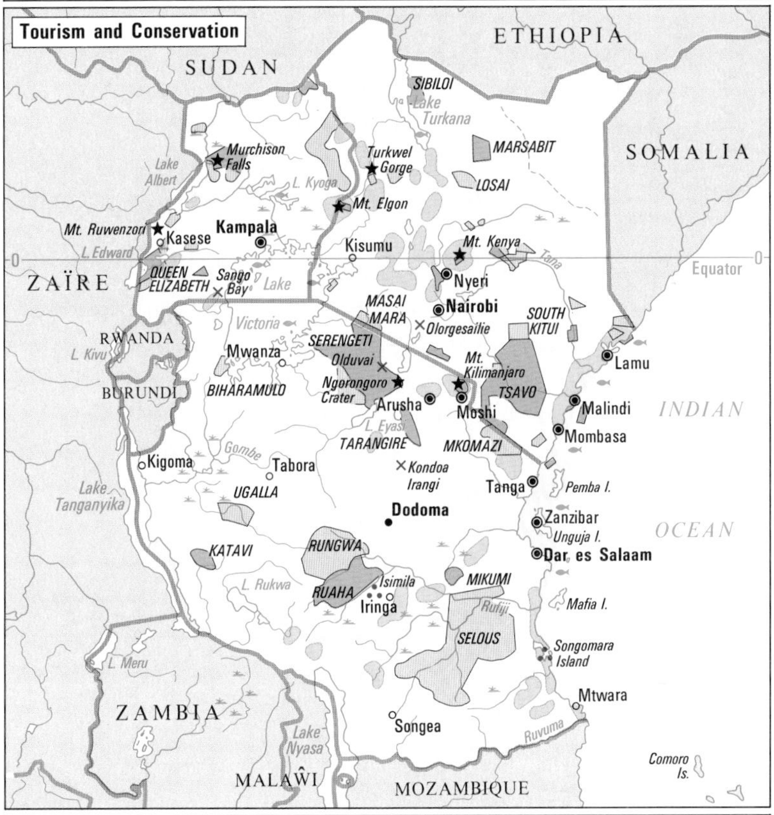

- Forest
- National Park
- National Reserve/ Game Reserve
- Beach
- Tourist centre
- Place of interest
- Prehistoric site
- Historic site
- Recreational fishing

Area and Population

Country	Kenya	Tanzania	Uganda
Land area (sq. km.)	564 162	883 663	197 044
Total population	15 327 061	23 174 336	12 636 179
Population density (persons per sq. km.)	27.2	26.2	64.1
Annual population growth rate	3.8%	3.2%	2.8%

Urban Population

Country	Kenya	Tanzania	Uganda
Total number of urban centres with over 2000 inhabitants	90	107	55
Total urban population (thousands)	2 309	2 413	930
Urban population as a percentage of total population	15.1%	13.8%	7.4%

Note: latest population census data used for each country.

Male and female lion on the savanna grasslands of East Africa

Wildebeest migration from Serengeti National Park to Masai Mara National Reserve.

SOUTHERN AFRICA – PHYSICAL AND POLITICAL

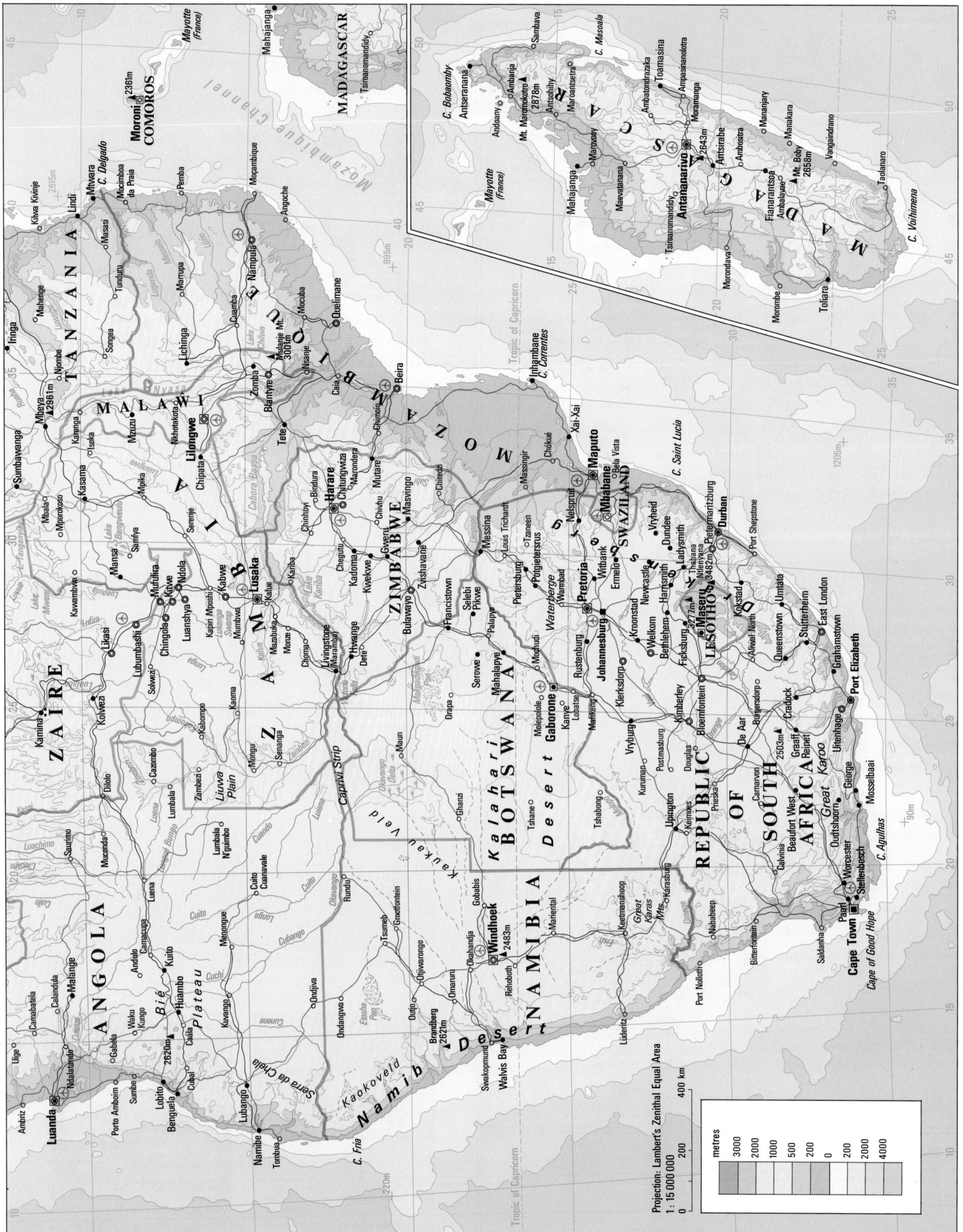

SOUTHERN AFRICA – CLIMATE, NATURAL VEGETATION AND AGRICULTURE

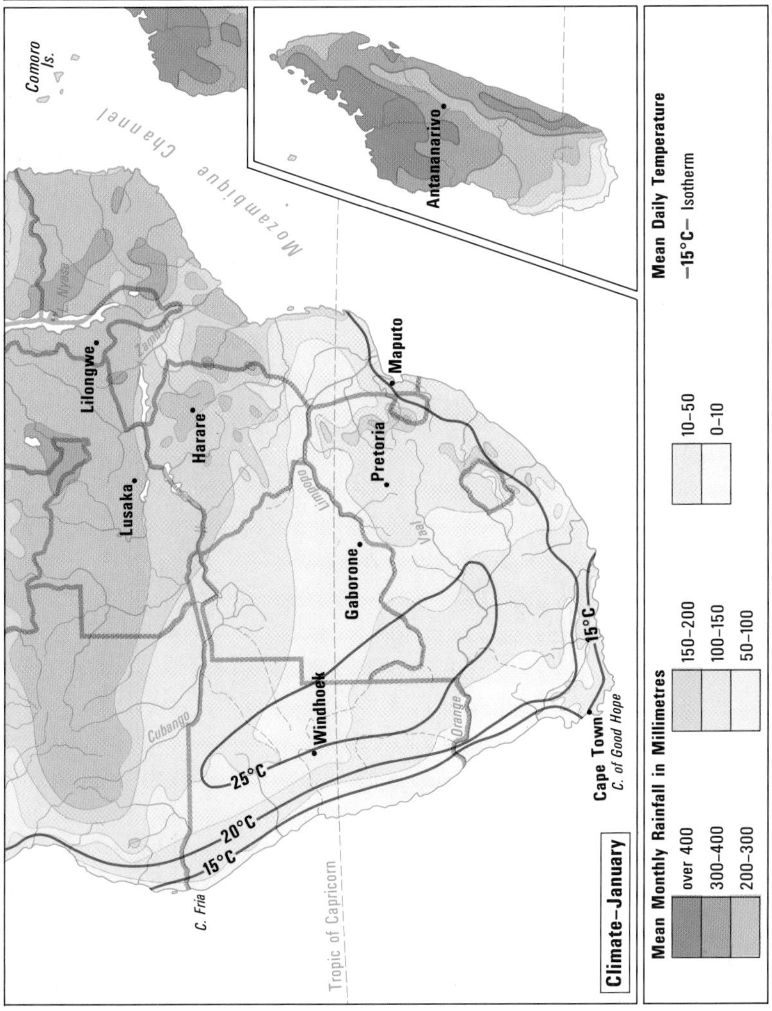

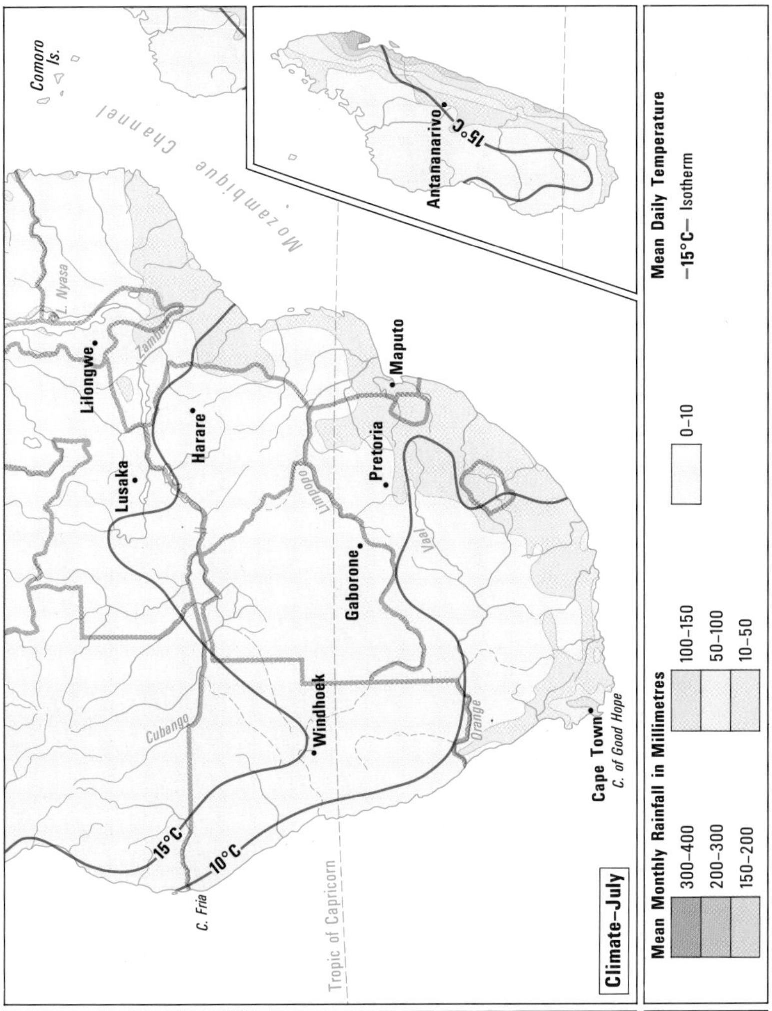

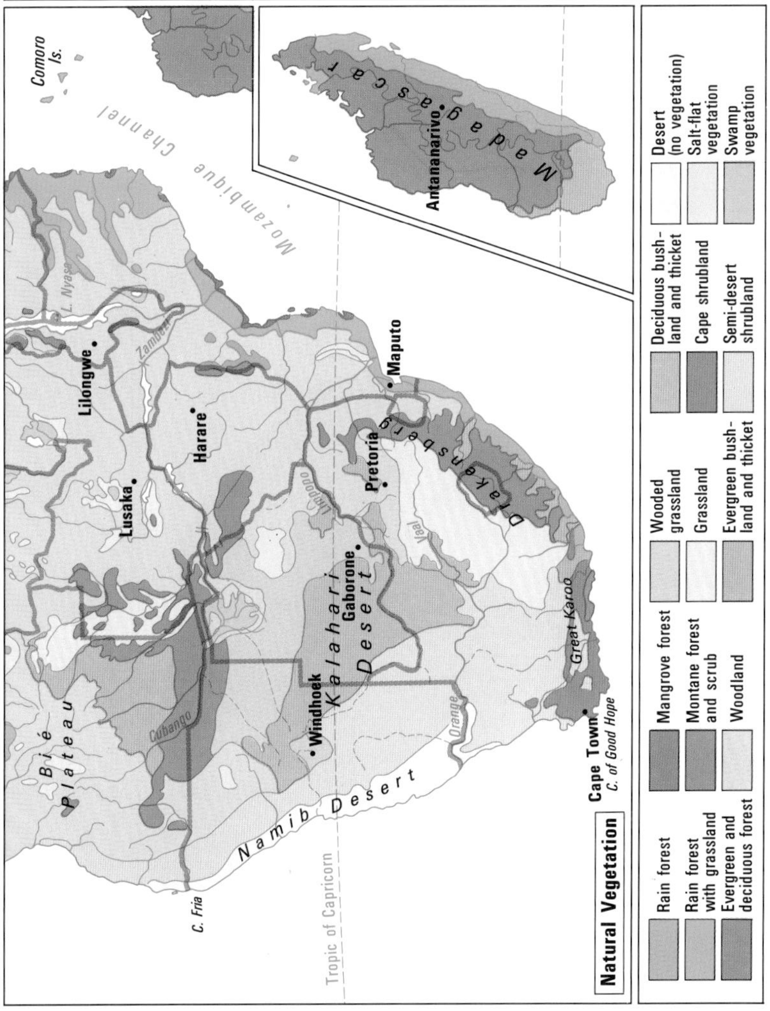

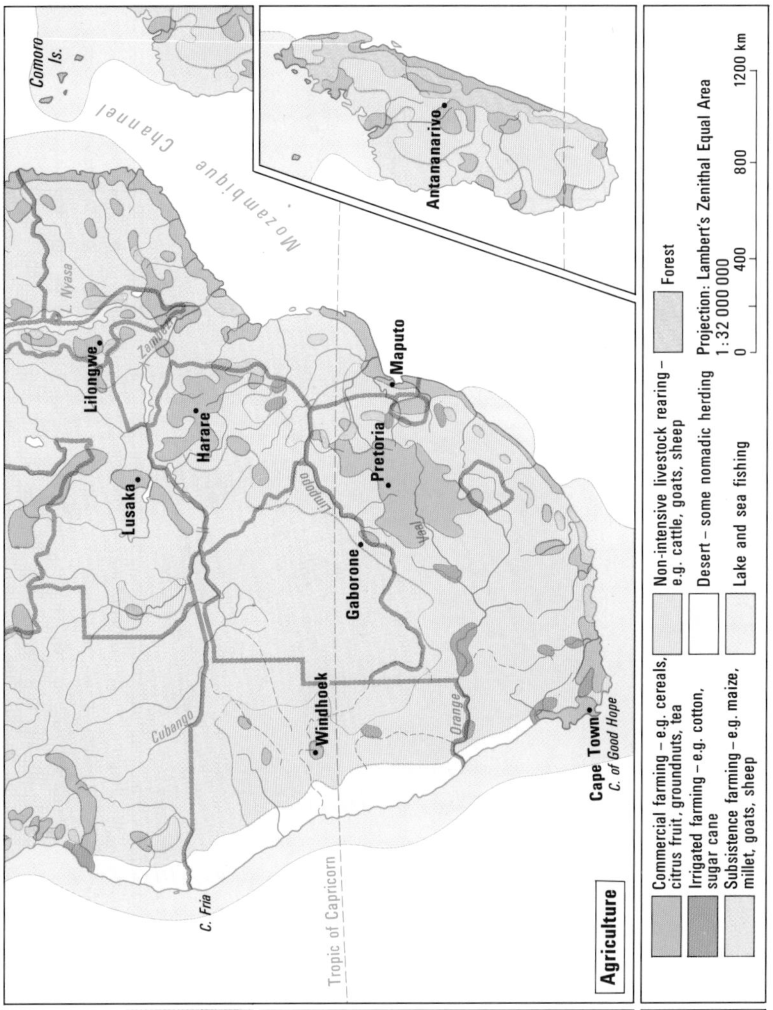

SOUTHERN AFRICA – ECONOMIC

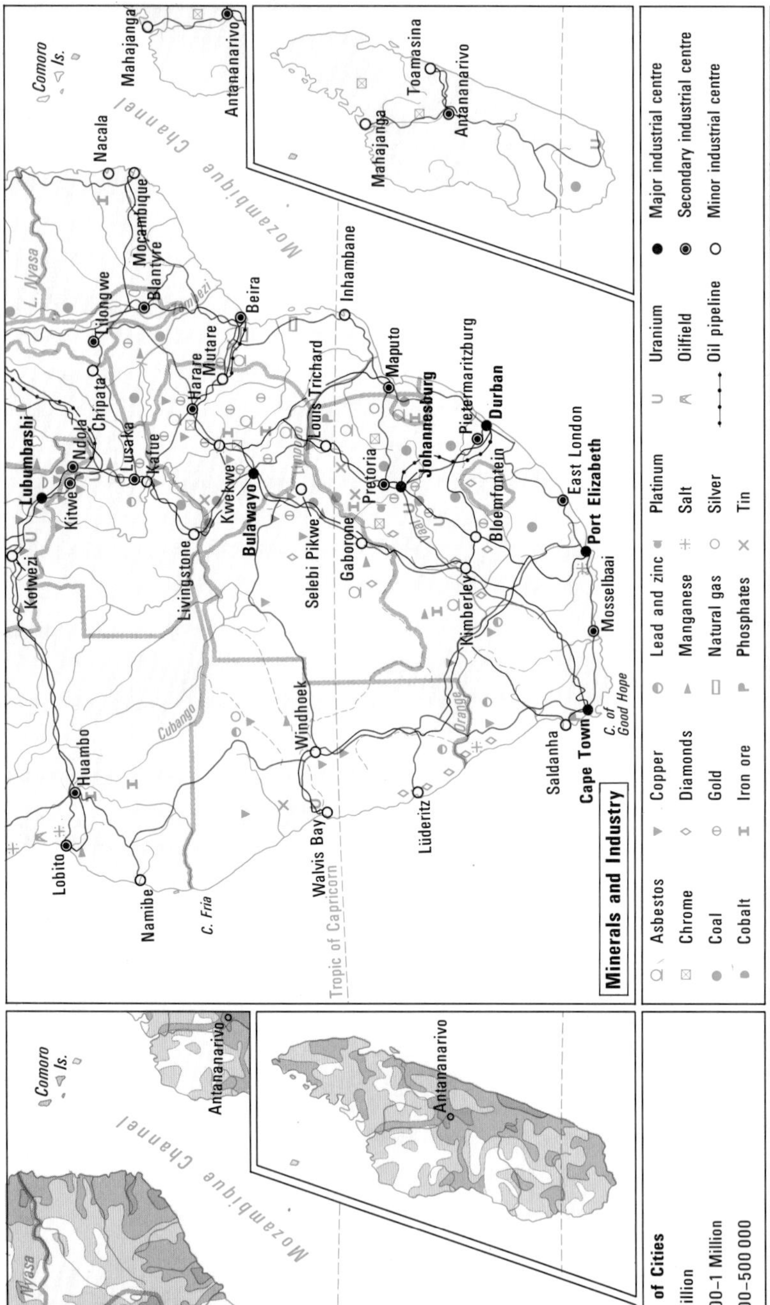

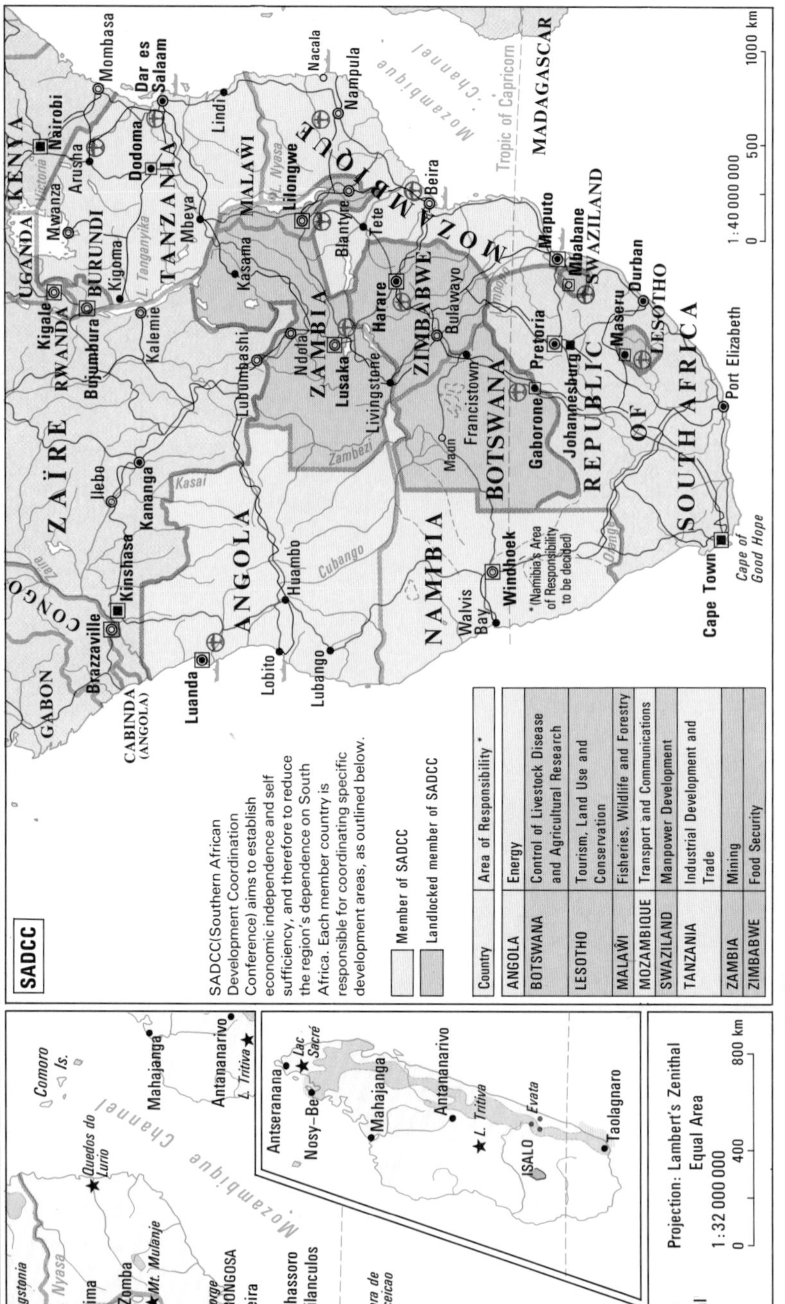

SADCC (Southern African Development Coordination Conference) aims to establish economic independence and self sufficiency, and therefore to reduce the region's dependence on South Africa. Each member country is responsible for coordinating specific development areas, as outlined below.

Country	Area of Responsibility*
ANGOLA	Energy
BOTSWANA	Control of Livestock Disease and Agricultural Research
LESOTHO	Tourism, Land Use and Conservation
MALAWI	Fisheries, Wildlife and Forestry
MOZAMBIQUE	Transport and Communications
SWAZILAND	Manpower Development
TANZANIA	Industrial Development and Trade
ZAMBIA	Mining
ZIMBABWE	Food Security

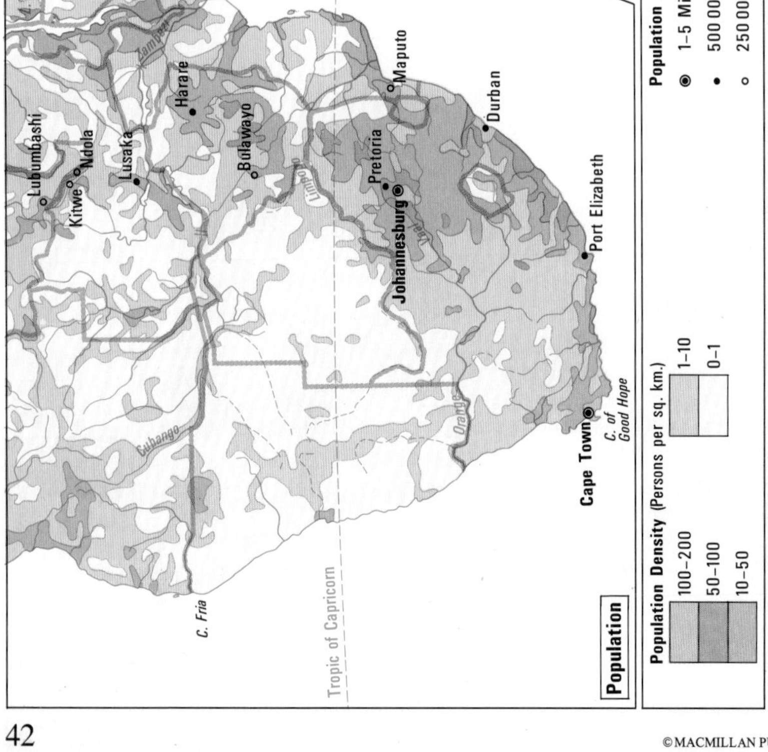

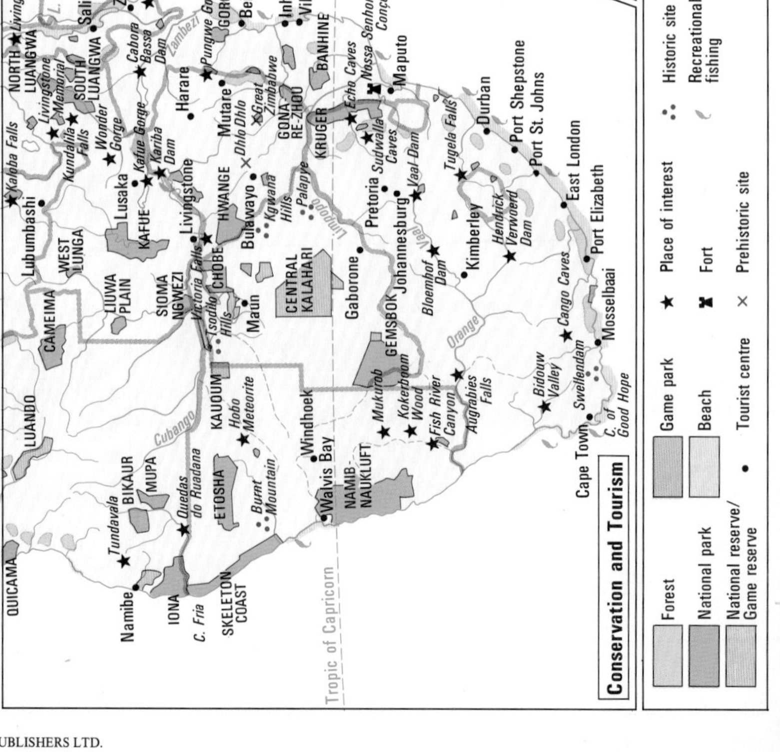

NIGER
CHAD
SUDAN
NIGERIA
CAMEROON
CENTRAL AFRICAN REPUBLIC
EQUATORIAL GUINEA
SÃO TOMÉ AND PRÍNCIPE
GABON
CONGO
ZAÏRE
UGANDA
RWANDA
BURUNDI
TANZANIA
ANGOLA
ZAMBIA
CABINDA (ANGOLA)
ATLANTIC OCEAN
Aïr (Azbine)
Ténéré
Azaouagh
Mt. Bagzane 1920m
Agadez
Bilma
Largeau
Bodélé
Mt. Momou 1311m
Fada
Massif Ennedi
Wadi Howar
Teiga Plateau
Karima
Baiyuda Desert
Shendi
Khartoum North
Omdurman
Khartoum
Tahoua
Tanout
Manga
Nguigmi
Lake Chad
Mao
Biltine
Massif du Maraoné
Dārfūr
Kutum
Mt. Gurgei 2397m
Wad Medani
Ed Dueim
Sennar
Birni-Nkonni
Maradi
Zinder
Goure
Diffa
Abéché
Oum Hadjer
Ati
El Geneina
El Fasher
Sodiri
El Obeid
Kosti
Sokoto
Katsina
Nguru
Gusau
Kano
Maiduguri
N'Djaména
Mongo
Massif du Guéra
Mt. Marra 3070m
Zalingei
Nyala
En Nahud
Abu Zabad
Dilling
Nuba Mts. 1324m
Birnin Tudu
Potiskum
Massenya
Melfi
Am Timan
El Muglad
Kadugli
Bin Yauri
Zaria
Kaduna
Bauchi
Biu
Maroua
Mandara Mts.
Bongor
Minna
Jos
Jos 1698m
Mubi
Birao
Malakal
Bida
Abuja
Plateau
Langtang
Yola
Garoua
Pala
Laï
Sarh
Kyabé
Mt. Toussoro 1330m
Ilorin
Lokoja
Vogel Peak 2042m
Moundou
Doba
Ndélé
Aweil
Raga
Bahr el Ghazal
Sudd
Makurdi
Adamawa Highlands
Gotel Mts.
Ngaoundére
Banyo
Wau
Owo
Enugu
Ogoja
Kaga Bandoro
Rumbek
Bor
Benin City
Onitsha
Bamenda
Bouar
Bossangoa
Bria
Warri
Foumban
Bafoussam
Sibut
Bambari
Mingala
Mboki
Obo
Calabar
Port Harcourt
Kumba
Nkongsamba
Mt. Cameroon 4095m
Buea
Limbe
Douala
Edéa
Yaoundé
Bertoua
Berbérati
Bangui
Bangassou
Juba
Yambio
Torit
Bight of Bonny
Malabo
Bioko
Mbalmayo
Bosobolo
Libenge
Gemena
Businga
Bondo
Dungu
Aba
Faradje
Nimule
Mt. Kinyeti 3187m
Kribi
Ebolowa
Buta
Isiro
Watsa
Arua
Gulu
Aketi
Lisala
Bumba
Wamba
Apac
Bata
Oyem
Príncipe
Ouesso
Impfondo
Bongandanga
Bunia
Lake Albert
Hoima
Blue Mts.
Yangambi
Bafwasende
Libreville
São Tomé
Congo Basin
Kisangani
Kabarole
Kampala
Beni
Butembo
Ruwenzori 5110m
Jinja
Entebbe
Kasese
Masaka
Mbarara
Equator
Mbandaka
Boende
Ubundu
Lake Edward
Lake Victoria
Booué
Owando
Lambaréné
C. Lopez
Port Gentil
Annobón
Mt. Iboundji 1575m
Masuku
Ikela
Bukoba
4510m
Kigali
Mouila
Gamboma
Inongo
Lomela
Kalima
Bukavu
Mwanza
Tchibanga
Djambala
Bolobo
Kindu
Mossendjo
Mushie
Kutu
Bandundu
Lodja
Bujumbura
Shinyanga
Kahama
Brazzaville
Madingou
Loubomo
Kinshasa
Mangai
Ilebo
Kasongo
Kenge
Bulungu
Mweka
Lusambo
Pointe-Noire
Tshela
Mbanza Ngungu
Kikwit
Luebo
Demba
Sentele
Kigoma
Tabora
Cabinda
Boma
Matadi
Kananga
Kabinda
Popokabaka
Soyo
Mbanza Congo
Tshikapa
Mbuji-Mayi
Kabalo
Kalemie
Mpanda
Gandajika
Mwene Ditu
Moba
4720m
Uíge
Lóvua
Manono
Kabongo
2458m
Mitumba Mts.
Ambriz
Camabatela
Sumbawanga
Projection: Lambert's Zenithal Equal Area
1 : 15 000 000
0
200
400 km
Luanda
Kamina
Calandula
Mbala
Mbeya
Ndalatando
Saurimo
Mporokoso
Malange
Kawambwa
metres
4000
3000
2000
1000
500
200
0
200
2000
4000
Kasama
Isoka
Porto Amboim
Gabela
Muconda
Dilolo
Kolwezi
Likasi
Sumbe
Waku Kungo
Andulo
Camacupa
Mansa
Lake Bangweulu
Samfya
Lubumbashi
Lobito
Benguela
2620m
Bié
Kuito
Luena
Cazombo
Solwezi
Mufulira
Chingola
Kitwe
Mpika
Mzuzu
Huambo
Cubal
Caála
Plateau
Lumbala
Luanshya
Ndola
Serenje
Lilongwe
Chipata
Zambezi
Kabompo
Kapiri Mposhi
Lukanga Swamp
Kabwe
Mumbwa
Kuvango
Menongue
Lumbala N'guimbo
Liuwa Plain
Kaoma
Lubango
Namibe
Cuito Cuanavale
Mongu
Lusaka
Kafue
Cabora Bassa
Serra da Chela
Tombua
Senanga
Mazabuka
Monze
Kariba
Tete
Choma
Lake Kariba
Chinhoyi
Bindura
Ondjiva
Livingstone (Maramba)
Caprivi Strip
Ondangwa
Harare

EQUATORIAL AFRICA – CLIMATE, NATURAL VEGETATION AND AGRICULTURE

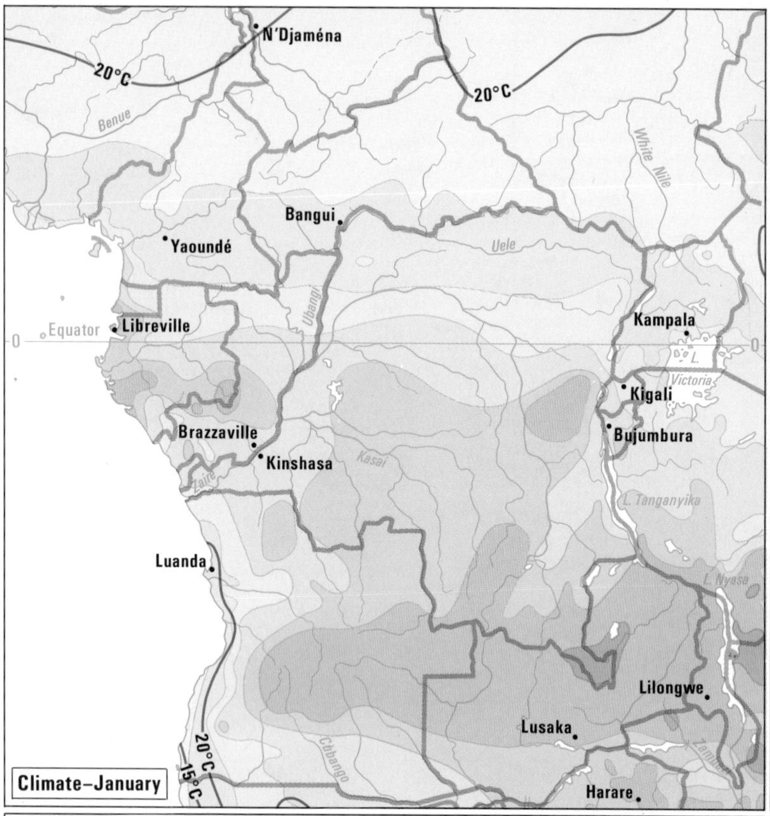

Climate–January

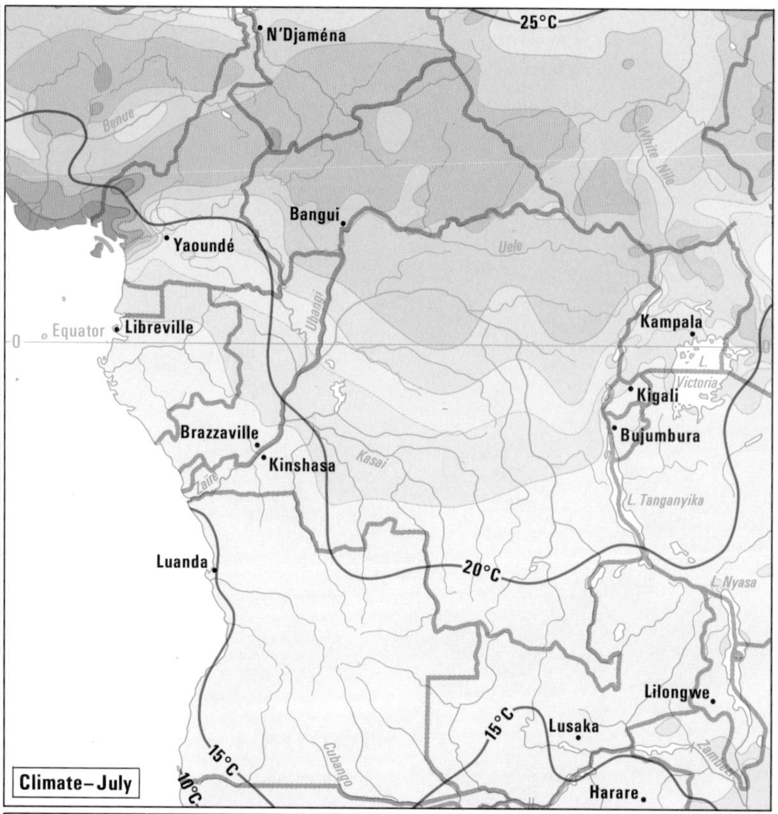

Climate–July

Mean Monthly Rainfall in Millimetres (January)

- 300–400
- 200–300
- 150–200
- 100–150
- 50–100
- 10–50
- 0–10

Mean Daily Temperature

–20°C– Isotherm

Mean Monthly Rainfall in Millimetres (July)

- over 400
- 300–400
- 200–300
- 150–200
- 100–150
- 50–100
- 10–50
- 0–10

Mean Daily Temperature

–15°C– Isotherm

Projection: Lambert's Zenithal Equal Area
1 : 32 000 000
0 400 800 km

Natural Vegetation

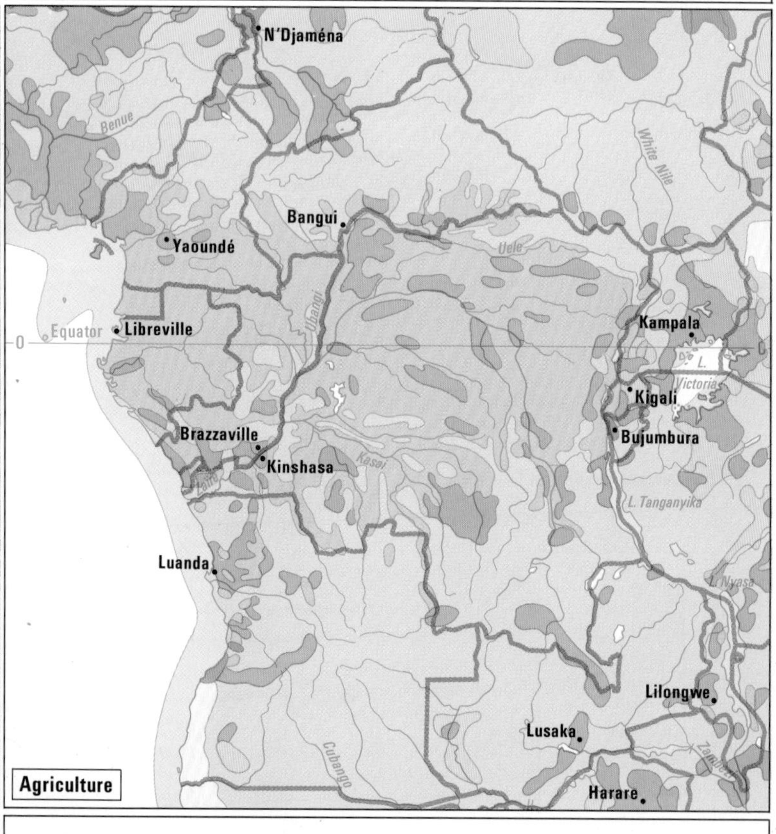

Agriculture

Natural Vegetation key

- Rain forest
- Rain forest with grassland
- Evergreen and deciduous forest
- Mangrove forest
- Montane forest and scrub
- Woodland
- Wooded grassland
- Grassland
- Evergreen bush-land and thicket
- Deciduous bush-land and thicket
- Semi-desert shrubland
- Semi-desert grassland
- Desert (no vegetation)
- Salt-flat vegetation
- Swamp vegetation

Agriculture key

- Commercial farming – e.g. cocoa, coffee, cotton, oil palm
- Irrigated farming – e.g. sugar cane
- Subsistence farming – e.g. bananas, cassava, maize, cattle, goats
- Non-intensive livestock rearing – e.g. cattle, goats, sheep
- Desert – some nomadic herding
- Lake and sea fishing
- Forest

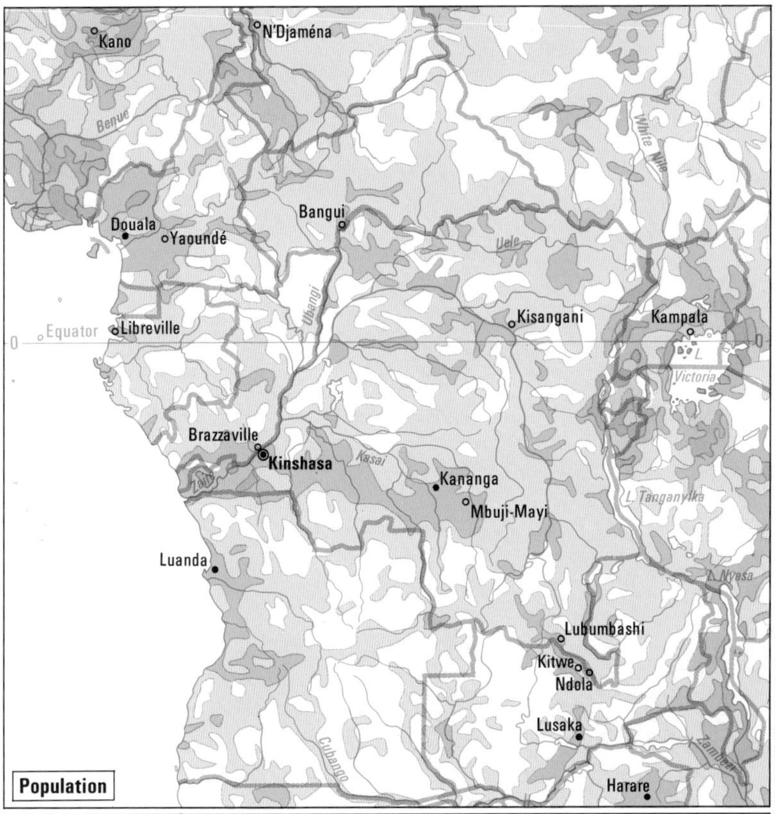

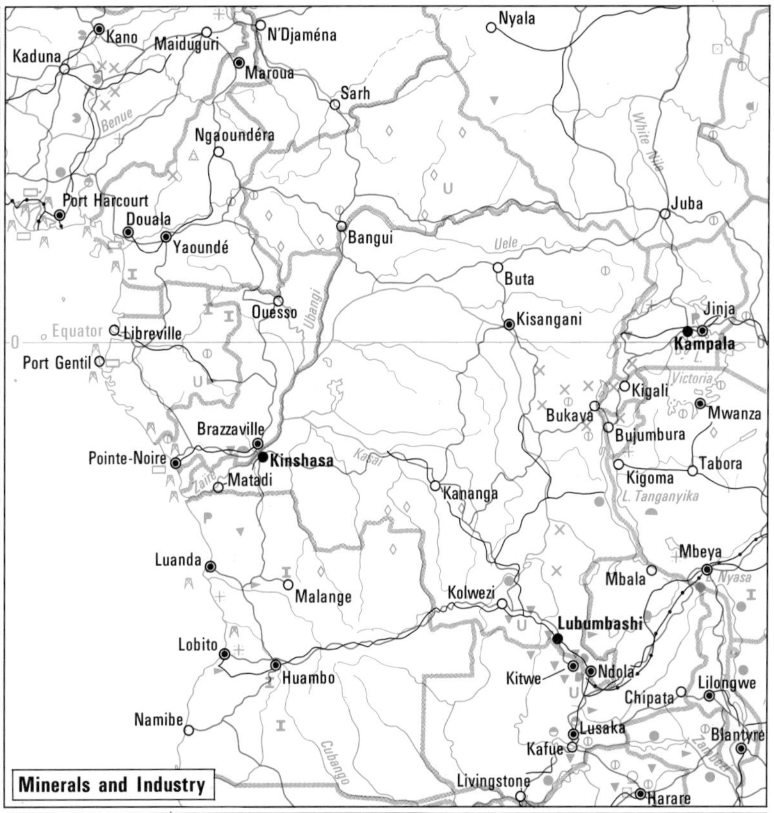

Population

Kano · N'Djaména · Douala · Yaoundé · Bangui · Kisangani · Kampala · Libreville · Equator · Brazzaville · **Kinshasa** · Kananga · Mbuji-Mayi · Luanda · Lubumbashi · Kitwe · Ndola · Lusaka · Harare

Benue · White Nile · Uele · Ubangi · Kasai · L. Victoria · L. Tanganyika · L. Nyasa · Cubango · Zambezi

Population Density
(persons per sq. km.)

Over 200	10–50
100–200	1–10
50–100	0–1

Population of Cities

- ◉ 1–5 million
- ● 500 000–1 million
- ○ 250 000–500 000

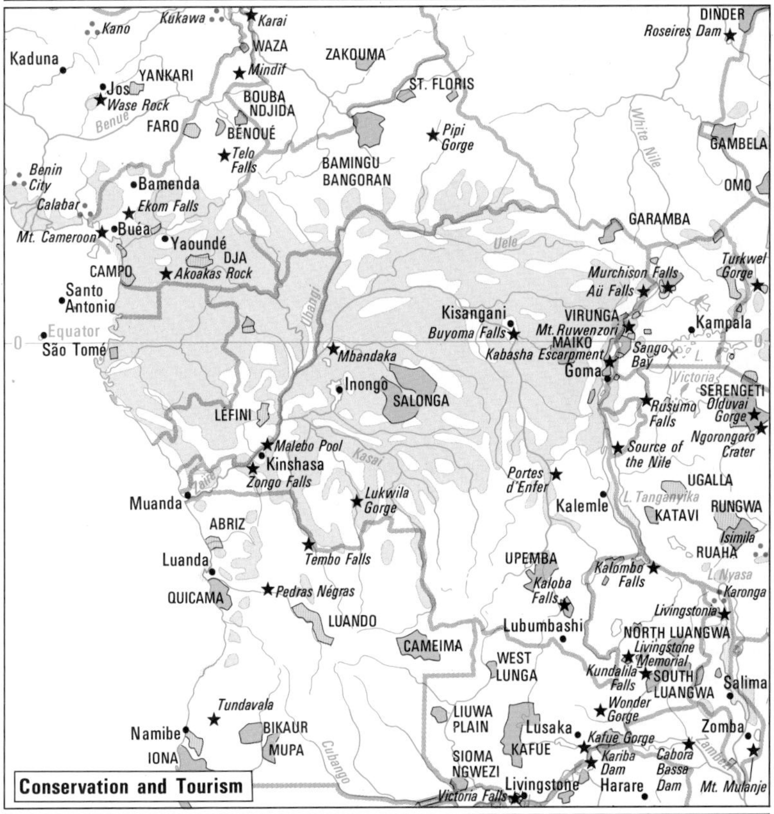

Minerals and Industry

Kano · Maiduguri · Kaduna · N'Djaména · Maroua · Nyala · Sarh · Ngaoundéra · Port Harcourt · Douala · Yaoundé · Bangui · Buta · Juba · Ouesso · Kisangani · Jinja · Libreville · Equator · **Kampala** · Port Gentil · Kigali · Mwanza · Bukava · Bujumbura · Brazzaville · **Kinshasa** · Tabora · Pointe-Noire · Matadi · Kananga · Kigoma · L. Tanganyika · Mbeya · Luanda · Malange · Nyasa · Kolwezi · Mbala · **Lubumbashi** · Lobito · Kitwe · Ndola · Chipata · Lilongwe · Huambo · Kafue · Lusaka · Blantyre · Namibe · Livingstone · Harare

◠ Asbestos	▼ Copper	▭ Natural gas	⋀ Oilfield
△ Bauxite	◇ Diamonds	ℙ Phosphates	┼─ Oil pipeline
▢ Chrome	◔ Gold	◀ Platinum	● Major industrial centre
◖ Coal	I Iron ore	+ Salt	◎ Secondary industrial centre
◓ Cobalt	⊟ Lead and zinc	✕ Tin	○ Minor industrial centre
◔ Columbite	▶ Manganese	U Uranium	

Conservation and Tourism

Kano · Kukawa · Karai · DINDER · Kaduna · YANKARI · WAZA · ZAKOUMA · Roseires Dam · Jos · Wase Rock · Mindif · ST. FLORIS · BOUBA NDJIDA · FARO · BÉNOUÉ · BAMINGU BANGORAN · Pipi Gorge · GAMBELA · Benin City · Bamenda · Telo Falls · OMO · Calabar · Ekom Falls · GARAMBA · Mt. Cameroon · Buéa · Yaoundé · DJA · Murchison Falls · Turkwel Gorge · CAMPO · Akoakas Rock · Aü Falls · Santo Antonio · Kisangani · VIRUNGA · Kampala · São Tomé · Equator · Buyoma Falls · MAIKO · Mt. Ruwenzori · Kabasha Escarpment · Sango Bay · Mbandaka · Goma · SERENGETI · Olduvai Gorge · Inongô · SALONGA · Rusumo Falls · Ngorongoro Crater · LEFINI · Source of the Nile · Malebo Pool · Kinshasa · Portes d'Enfer · UGALLA · Zongo Falls · Lukwila Gorge · Kalemle · L. Tanganyika · KATAVI · RUNGWA · Muanda · ABRIZ · UPEMBA · Isimila · RUAHA · Luanda · Tembo Falls · Kalombo Falls · Kaloba Falls · L. Nyasa · Karonga · QUICAMA · Pedras Negras · LUANDO · Livingstonia · CAMEIMA · NORTH LUANGWA · Livingstone Memorial · Salima · Namibe · Tundavala · BIKAUR · Lubumbashi · Kundalila Falls · SOUTH LUANGWA · Zomba · IONA · MUPA · WEST LUNGA · Wonder Gorge · LIUWA PLAIN · Lusaka · Kafue Gorge · Kariba Dam · Cabora Bassa Dam · SIOMA NGWEZI · KAFUE · Livingstone · Harare · Mt. Mulanje · Victoria Falls

Forest	Beach	Recreational fishing
National park	● Tourist centre	
National reserve/ Game reserve	★ Place of interest	
Game park	✕ Prehistoric site	
	⋮ Historic site	

Projection: Lambert's Zenithal Equal Area
1 : 32 000 000
0   400   800 km

The Copper Belt

Kasenga · Kanzene · L. N'zilo · Tenke · Kienge · Kolwezi · Kambove · Likasi · L. Changanele · Minga · Mansa · Kengere · Lushia · ZAÏRE · ZAMBIA · Lubumbashi · Chembe · Kipushi · Chisasa · Solwezi · Kasumbalesa · Tshinsenda · Mokambo · Ntambu · Chililabombwe · Kawana · Mufulira · Sakania · Chingola · Kalulushi · Kitwe · Ndola · Mapunga · Luanshya · Miengwe · ZAMBIA · Ngabwe · Kashitu · Kapiri Mposhi · Mulungushi Dam · Lungu · Chibwe · Mita Hills Dam · Kabwe · Lubungu · Mumbwa · Keembe · Chisamba · **Lusaka** · Matala · Kafue · Mazabuka

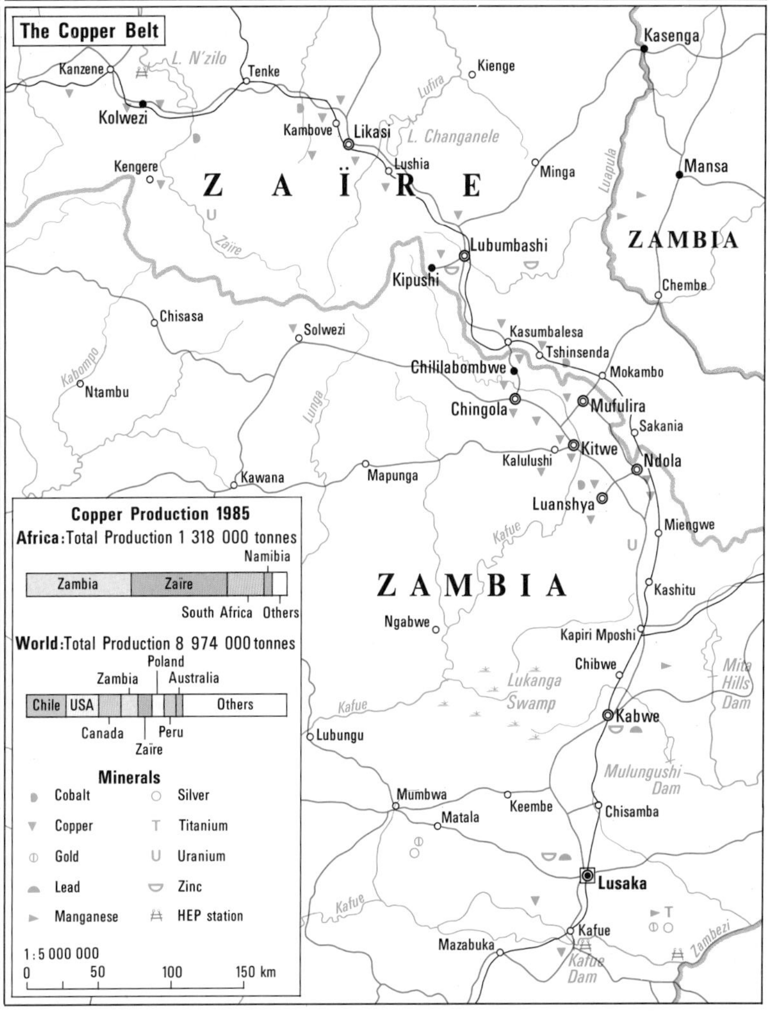

Copper Production 1985
Africa: Total Production 1 318 000 tonnes

Zambia · Zaire · Namibia · South Africa · Others

World: Total Production 8 974 000 tonnes

Chile · USA · Zambia · Poland · Australia · Others · Canada · Zaire · Peru

Minerals

◣ Cobalt	○ Silver	
▽ Copper	T Titanium	
◑ Gold	U Uranium	
⬡ Lead	▽ Zinc	
▶ Manganese	⌂ HEP station	

1 : 5 000 000
0   50   100   150 km

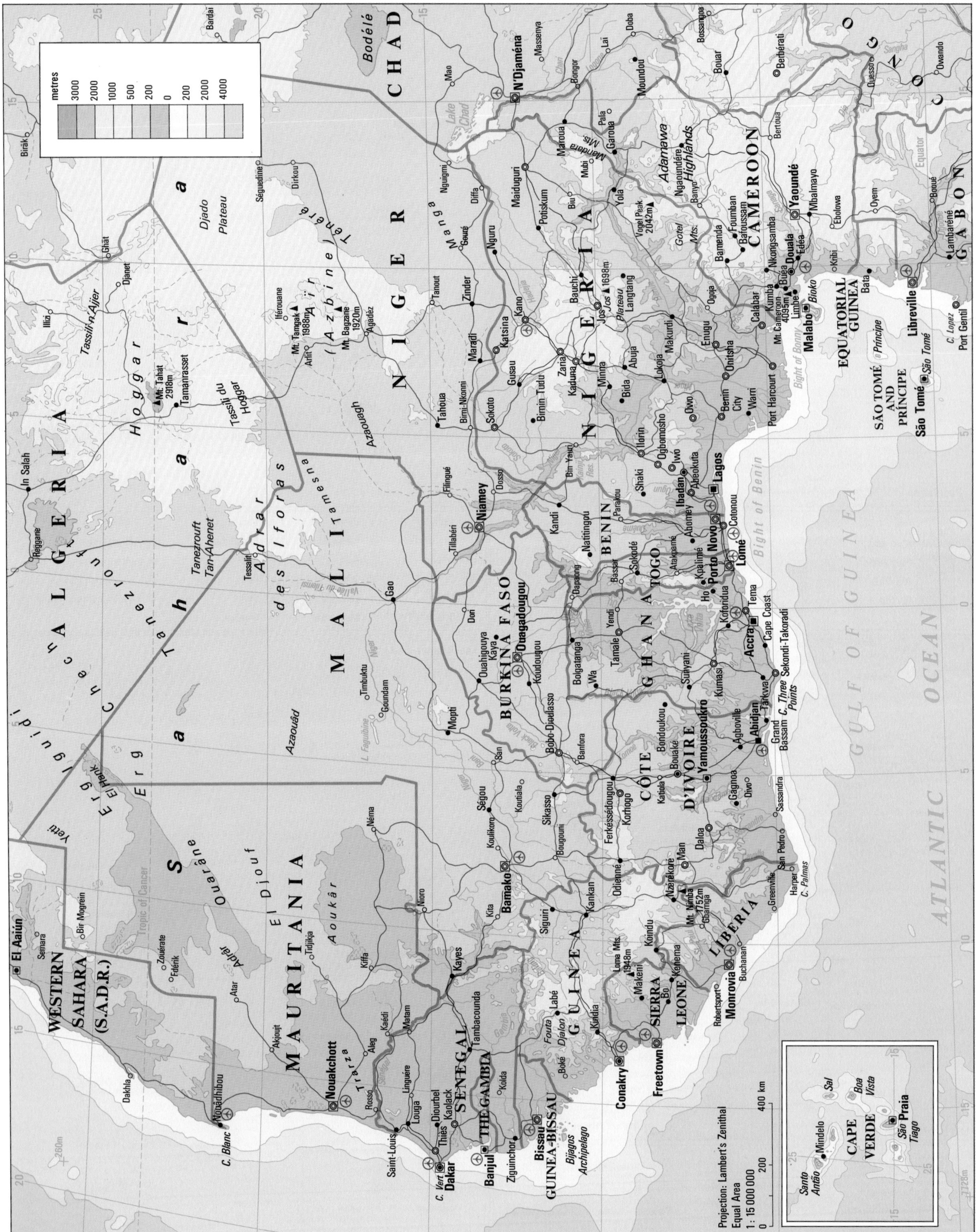
metres
3000
2000
1000
500
200
0
200
2000
4000
ALGERIA
MALI
NIGER
CHAD
MAURITANIA
WESTERN SAHARA (S.A.D.R.)
SENEGAL
THE GAMBIA
GUINEA-BISSAU
GUINEA
SIERRA LEONE
LIBERIA
CÔTE D'IVOIRE
BURKINA FASO
GHANA
TOGO
BENIN
NIGERIA
CAMEROON
EQUATORIAL GUINEA
SÃO TOMÉ AND PRÍNCIPE
GABON
CONGO
CAPE VERDE
Sahara
Hoggar
Tanezrouft
Adrar des Iforas
Aïr (Azbine)
Ténéré
Djado Plateau
Erg Chech
Erg Iguidi
El Djouf
Aoukâr
Adamawa Highlands
Jos Plateau
Fouta Djalon
Bijagós Archipelago
Lake Chad
Bight of Benin
Bight of Bonny
GULF OF GUINEA
ATLANTIC OCEAN
Tropic of Cancer
Equator
Nouakchott
Dakar
Banjul
Bissau
Conakry
Freetown
Monrovia
Yamoussoukro
Abidjan
Accra
Lomé
Porto Novo
Cotonou
Lagos
Abuja
Niamey
Ouagadougou
Bamako
N'Djaména
Yaoundé
Douala
Malabo
Libreville
São Tomé
Praia
Mindelo
El Aaiún
Timbuktu
Gao
Kano
Ibadan
Kumasi
Mt. Cameroon 4095m
Mt. Tahat 2918m
Mt. Nimba 1752m
Projection: Lambert's Zenithal Equal Area
1: 15 000 000
0
200
400 km

WEST AFRICA – CLIMATE, NATURAL VEGETATION AND AGRICULTURE

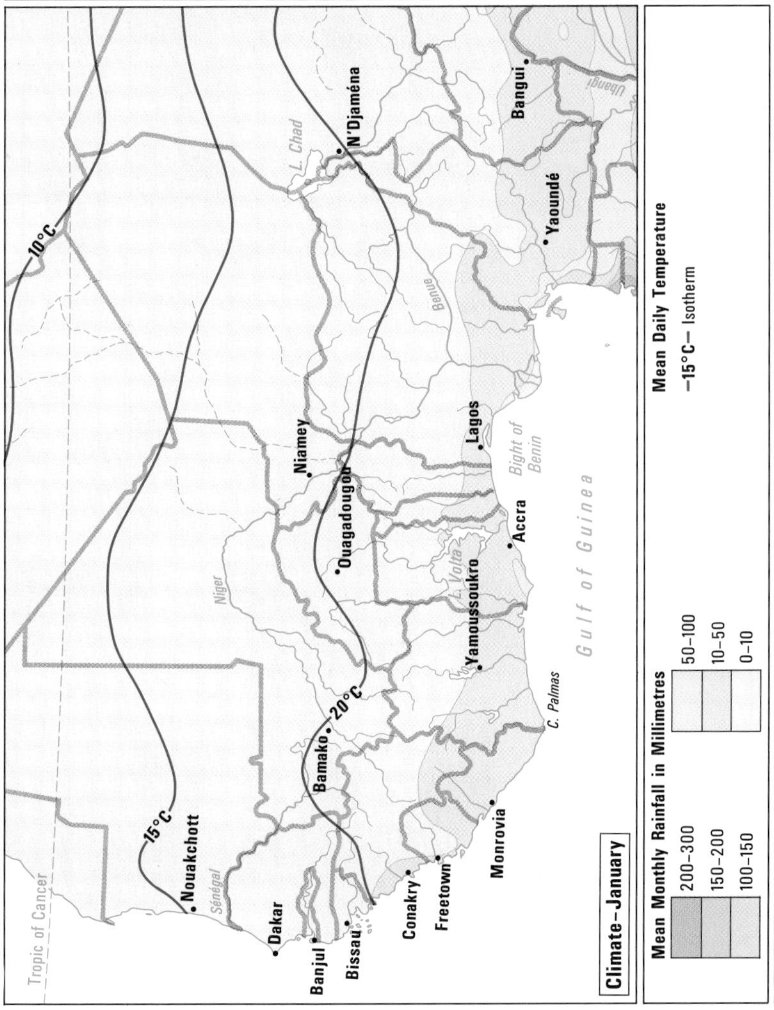

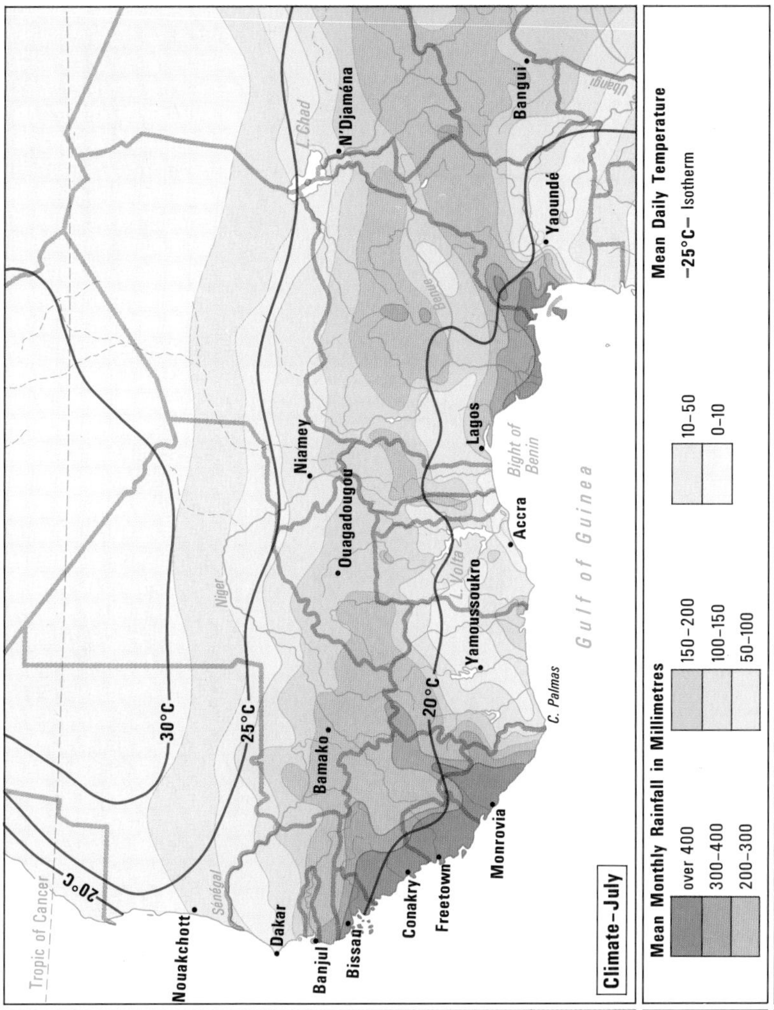

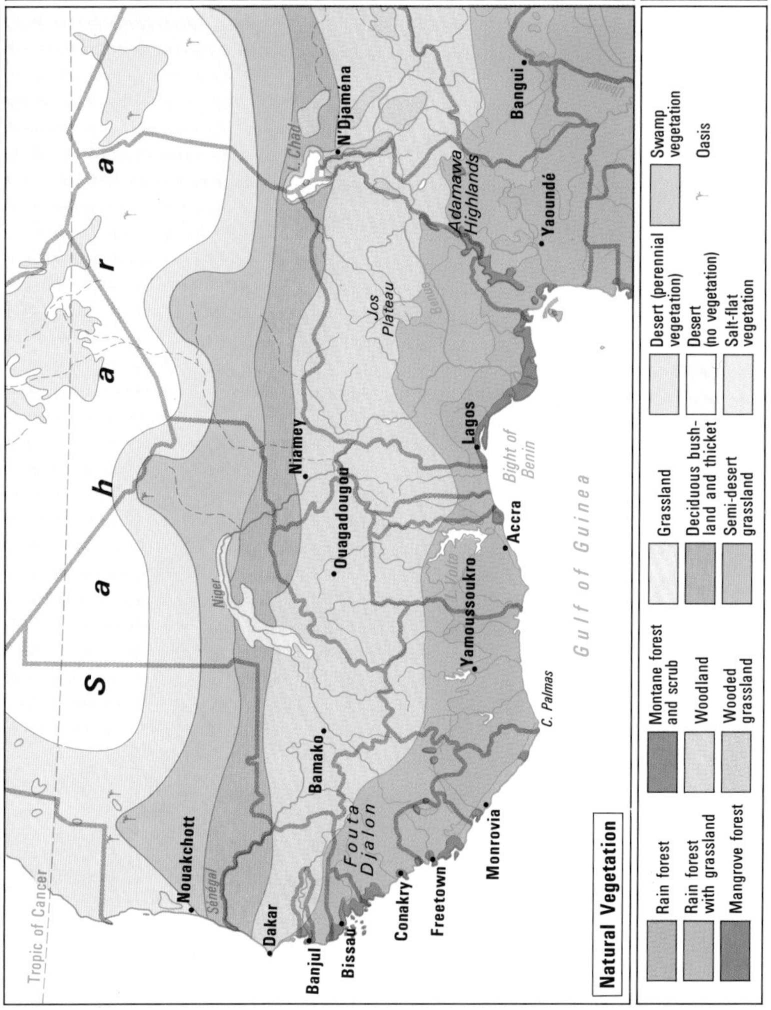

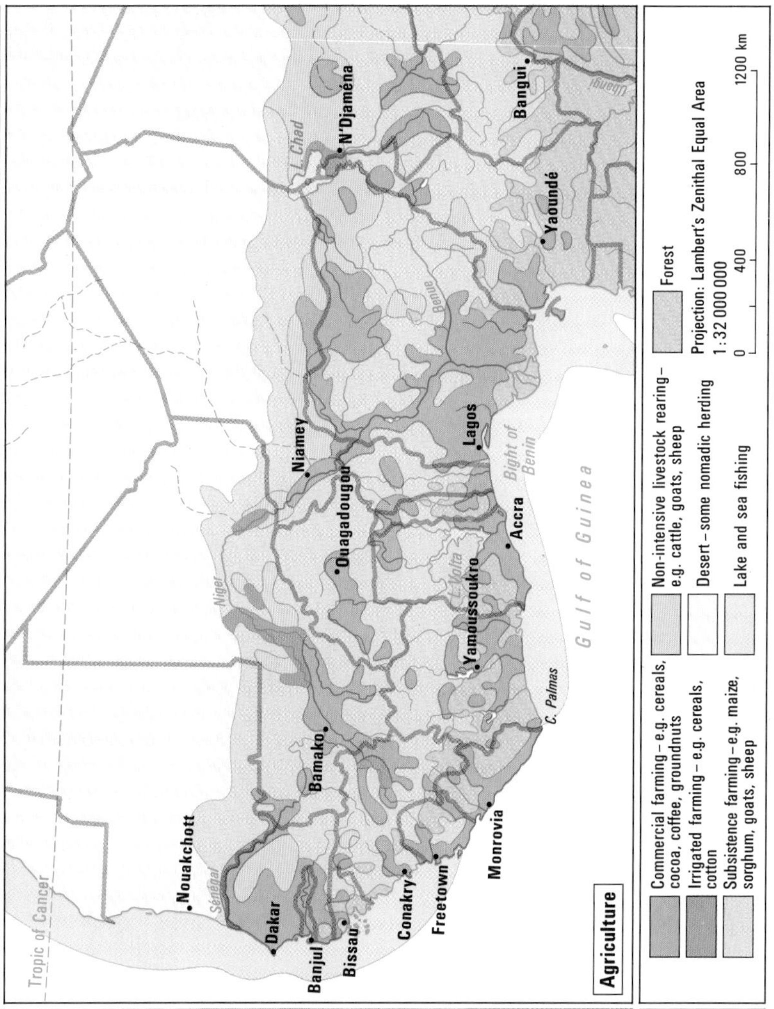

WEST AFRICA – ECONOMIC

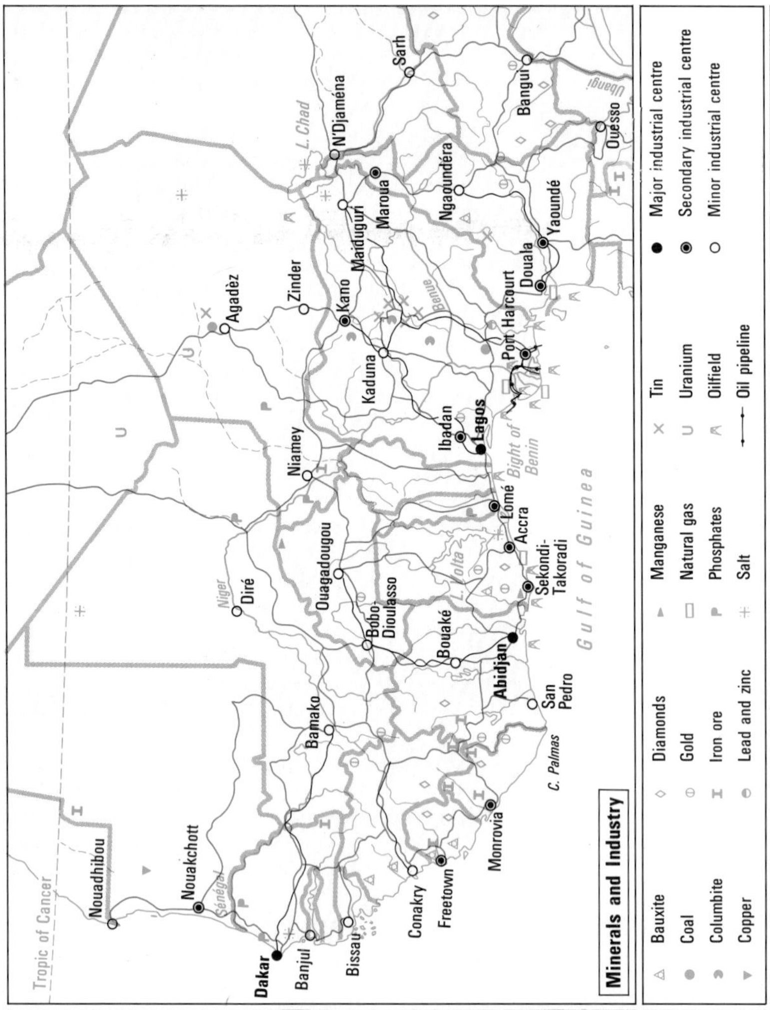
Minerals and Industry
Bauxite
Coal
Columbite
Copper
Diamonds
Gold
Iron ore
Lead and zinc
Manganese
Natural gas
Phosphates
Salt
Tin
Uranium
Oilfield
Oil pipeline
Major industrial centre
Secondary industrial centre
Minor industrial centre
Gulf of Guinea
Bight of Benin

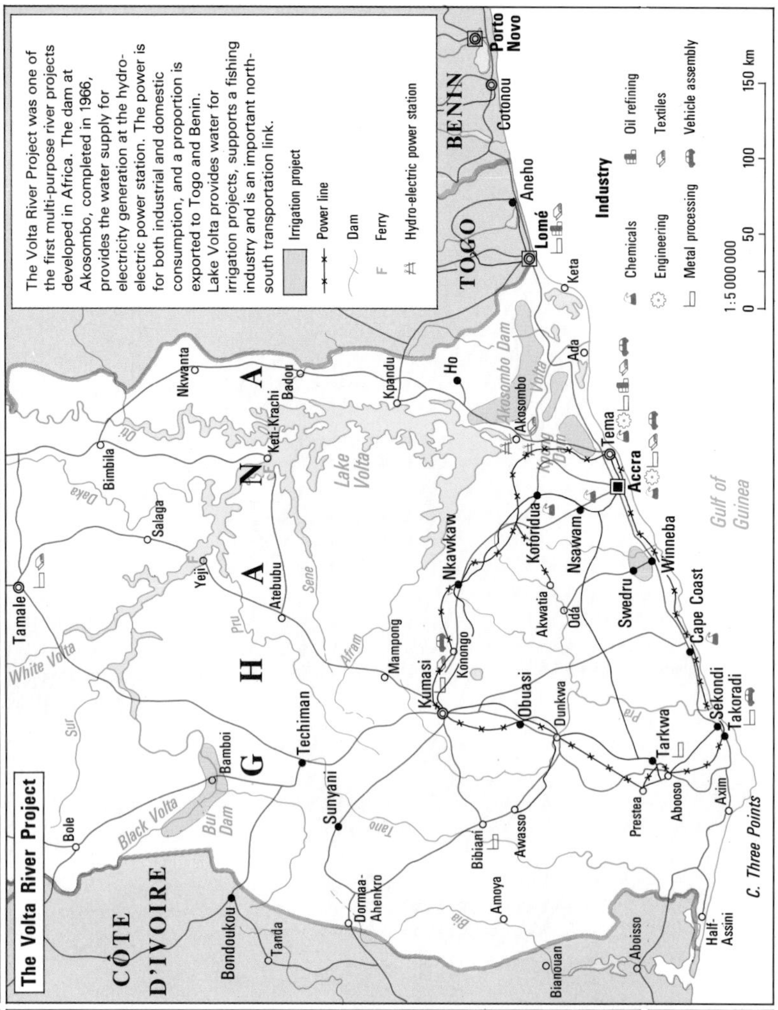
The Volta River Project
The Volta River Project was one of the first multi-purpose river projects developed in Africa. The dam at Akosombo, completed in 1966, provides the water supply for electricity generation at the hydro-electric power station. The power is for both industrial and domestic consumption, and a proportion is exported to Togo and Benin. Lake Volta provides water for irrigation projects, supports a fishing industry and is an important north-south transportation link.
Irrigation project
Power line
Dam
Ferry
Hydro-electric power station
Industry
Chemicals
Engineering
Metal processing
Oil refining
Textiles
Vehicle assembly
1:5 000 000
GHANA
TOGO
BENIN
CÔTE D'IVOIRE
Lake Volta
Gulf of Guinea

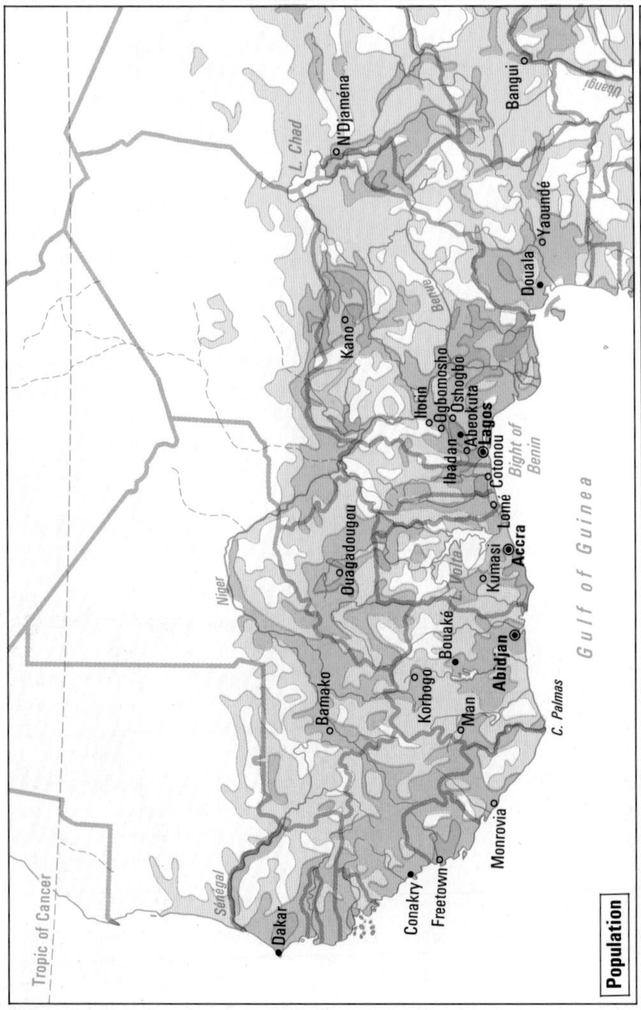
Population
Population Density (persons per sq. km.)
Over 200
100–200
50–100
10–50
1–10
0–1
Population of Cities
1–5 million
500 000–1 million
250 000–500 000
Gulf of Guinea
Bight of Benin

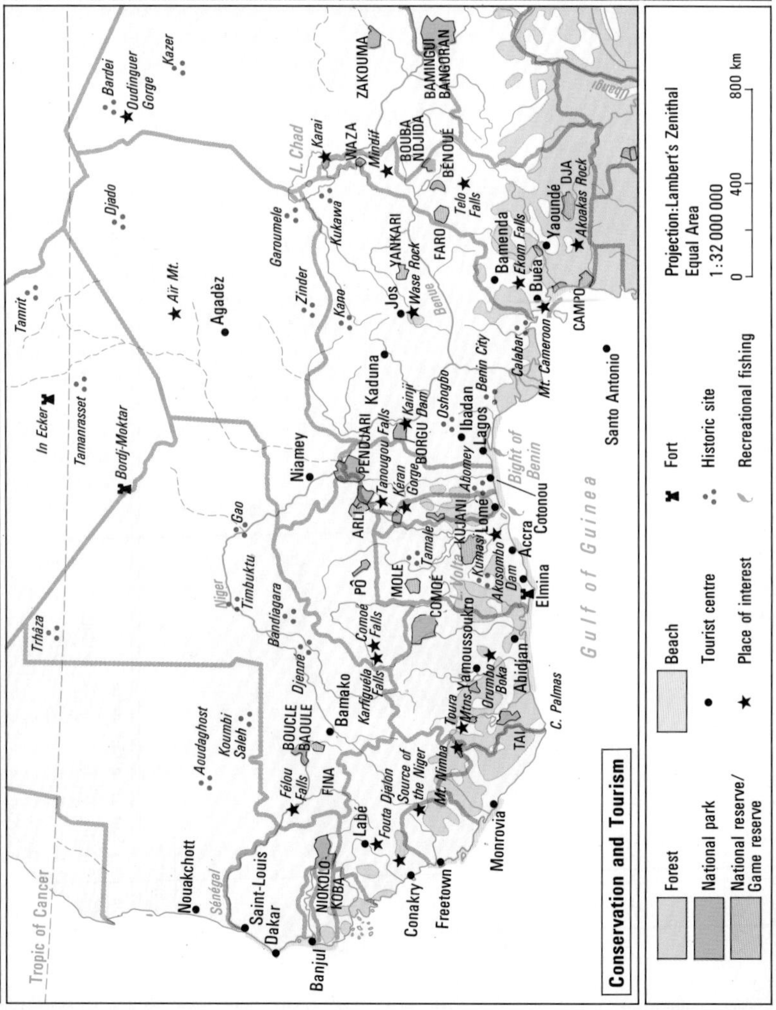
Conservation and Tourism
Forest
National park
National reserve/ Game reserve
Beach
Tourist centre
Place of interest
Fort
Historic site
Recreational fishing
Projection: Lambert's Zenithal Equal Area
1:32 000 000
Gulf of Guinea
Bight of Benin

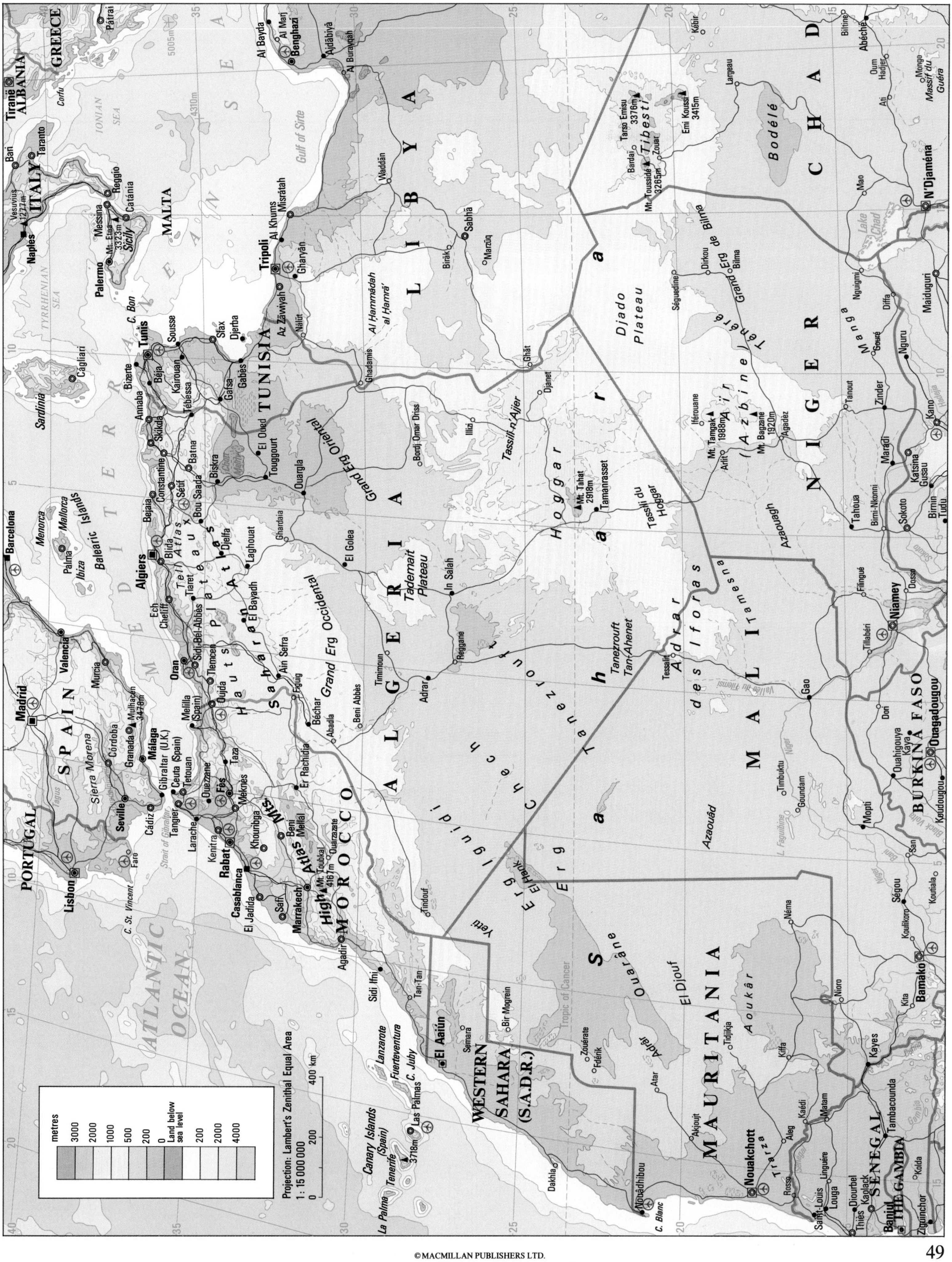
metres
3000
2000
1000
500
200
0
Land below sea level
200
2000
4000
Projection: Lambert's Zenithal Equal Area
1: 15 000 000
0 200 400 km
ATLANTIC OCEAN
MEDITERRANEAN SEA
IONIAN SEA
TYRRHENIAN SEA
Gulf of Sirte
PORTUGAL
SPAIN
ITALY
GREECE
ALBANIA
MALTA
MOROCCO
ALGERIA
TUNISIA
LIBYA
WESTERN SAHARA (S.A.D.R.)
MAURITANIA
MALI
NIGER
CHAD
SENEGAL
THE GAMBIA
BURKINA FASO
Canary Islands (Spain)
Balearic Islands
Sahara
Tropic of Cancer
Lisbon
Madrid
Barcelona
Algiers
Tunis
Tripoli
Rabat
Casablanca
Nouakchott
Niamey
N'Djaména
Bamako
Ouagadougou
El Aaiún
Banjul
Naples
Benghazi
Tiranë
High Atlas
Ahaggar
Tibesti
Grand Erg Occidental
Grand Erg Oriental
Tademait Plateau
Djado Plateau
Lake Chad
©MACMILLAN PUBLISHERS LTD.

NORTH WEST AFRICA – CLIMATE, AGRICULTURE AND ECONOMIC

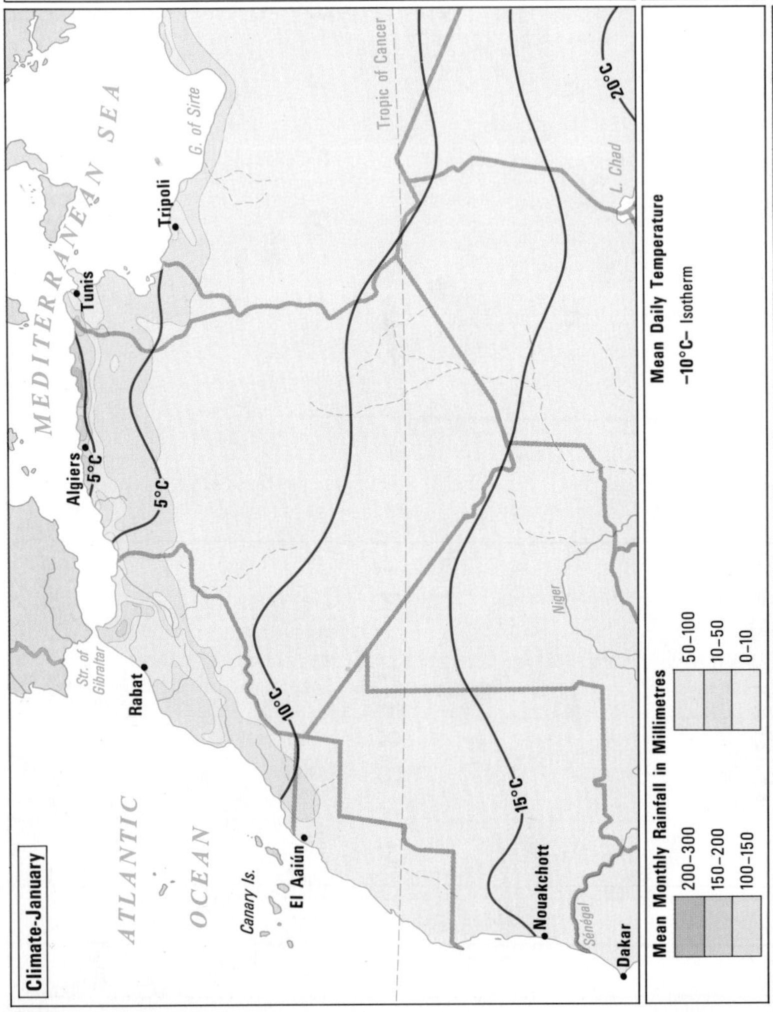

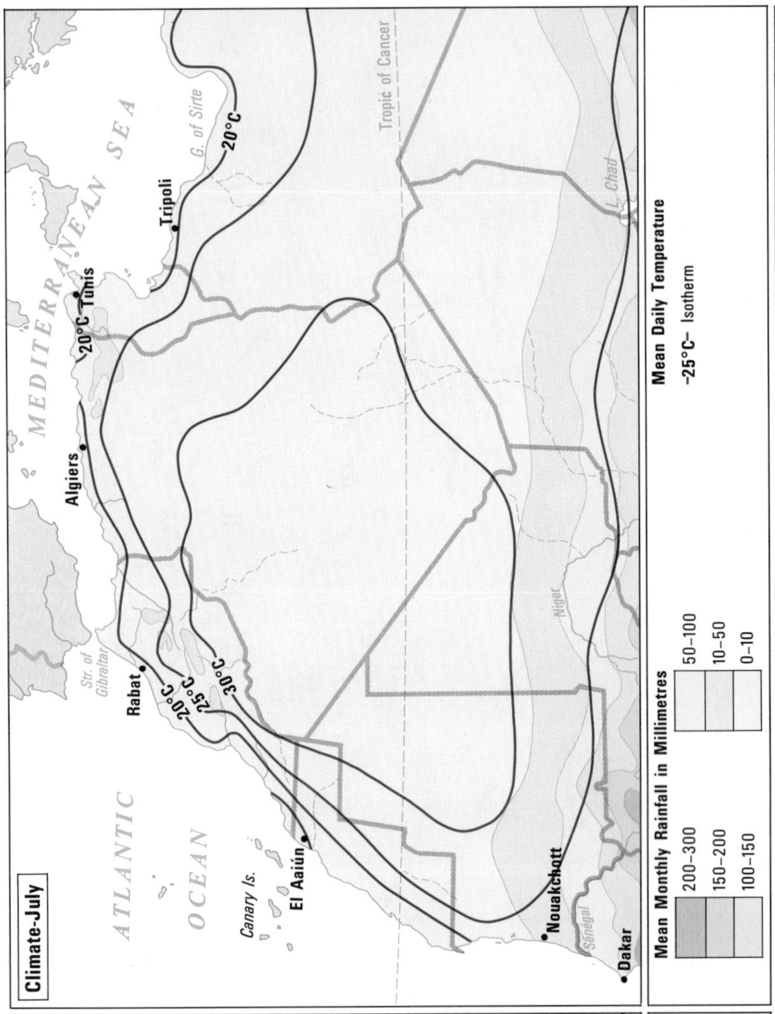

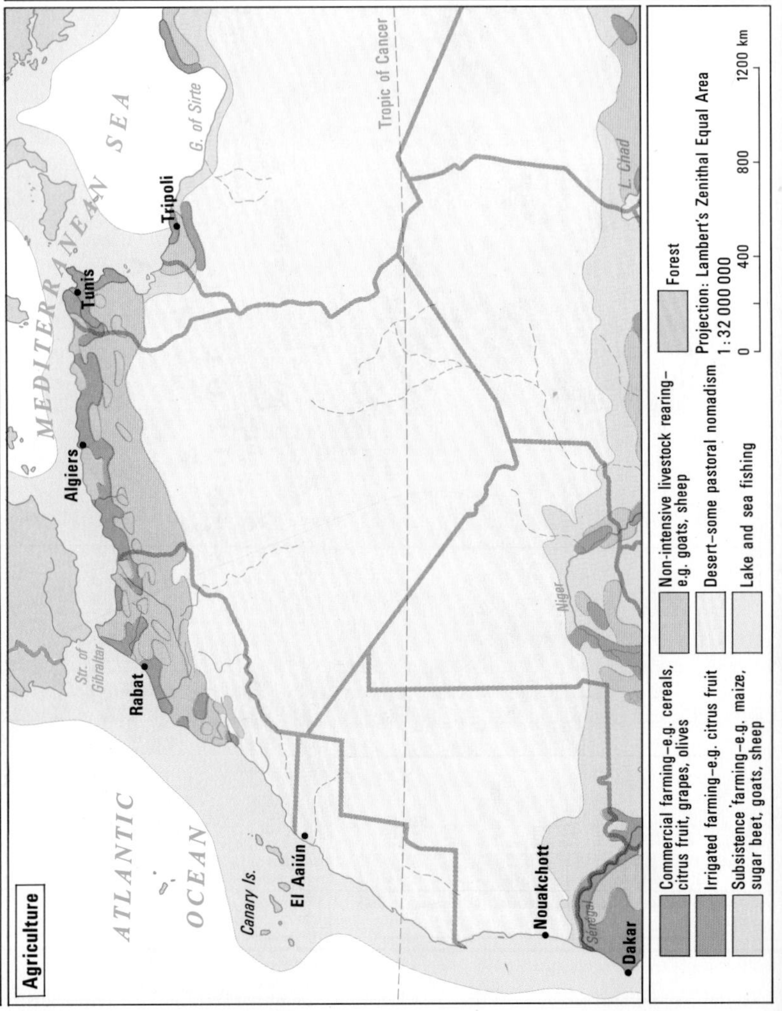

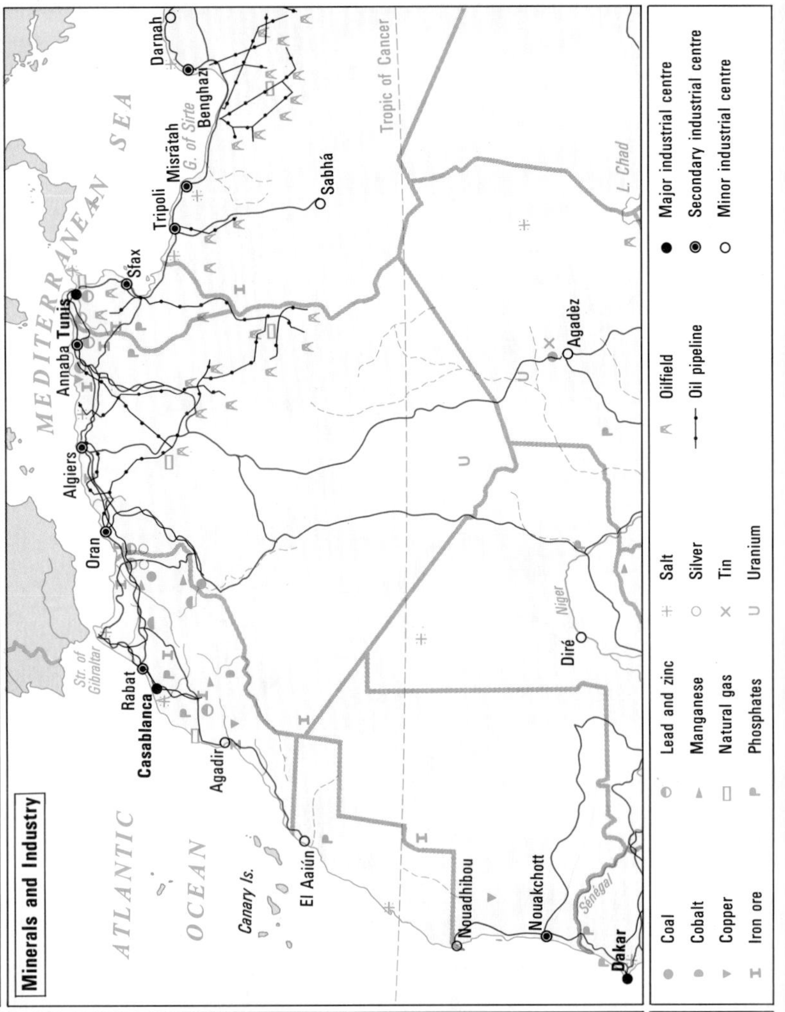

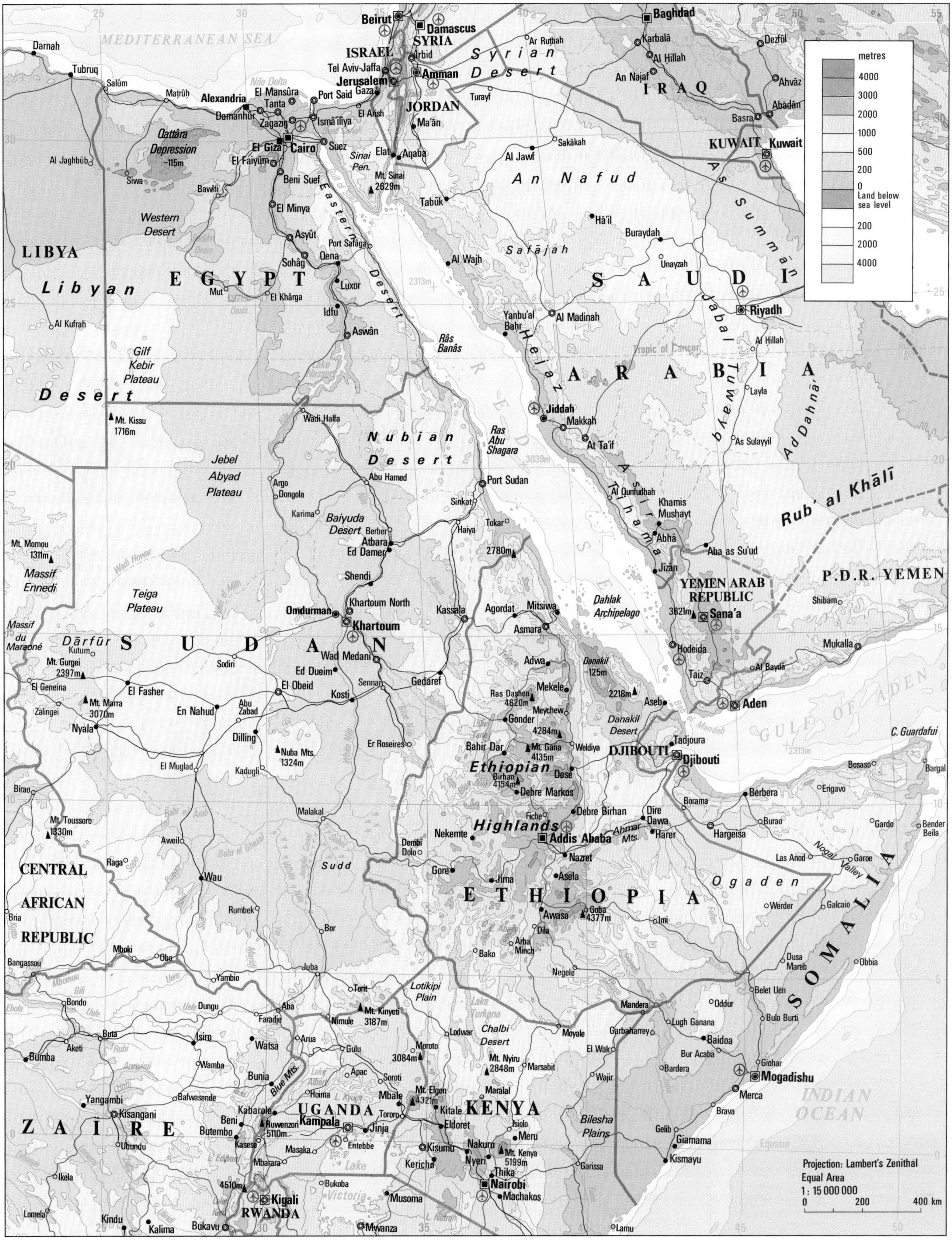
MEDITERRANEAN SEA
metres
4000
3000
2000
1000
500
200
0
Land below sea level
200
2000
4000
LIBYA
EGYPT
SUDAN
ETHIOPIA
SOMALIA
KENYA
UGANDA
ZAÏRE
RWANDA
CENTRAL AFRICAN REPUBLIC
DJIBOUTI
YEMEN ARAB REPUBLIC
P.D.R. YEMEN
SAUDI ARABIA
ISRAEL
SYRIA
JORDAN
IRAQ
KUWAIT
Libyan Desert
Western Desert
Eastern Desert
Nubian Desert
Baiyuda Desert
Gilf Kebir Plateau
Jebel Abyad Plateau
Teiga Plateau
Massif Ennedi
Massif du Maraoné
Dārfūr
Qattâra Depression −115m
Sinai Pen.
Syrian Desert
An Nafud
Safājah
Hejaz
Asir
Tihama
Jabal Tuwayq
As Summān
Ad Dahnā'
Rub' al Khālī
Tropic of Cancer
RED SEA
Dahlak Archipelago
Danakil −125m
Danakil Desert
Ethiopian Highlands
Ahmar Mts.
Ogaden
Nogal Valley
GULF OF ADEN
INDIAN OCEAN
C. Guardafui
Sudd
Lotikipi Plain
Chalbi Desert
Bilesha Plains
Lake Turkana
Lake Victoria
Blue Mts.
Equator
Mt. Kissu 1716m
Mt. Momou 1311m
Mt. Gurgei 2397m
Mt. Marra 3070m
Nuba Mts. 1324m
Mt. Toussoro 1330m
Mt. Sinai 2629m
Ras Dashen 4620m
Mt. Gana 4135m
Birhan 4154m
Goba 4377m
Mt. Kinyeti 3187m
Mt. Elgon 4321m
Mt. Nyiru 2848m
Mt. Kenya 5199m
Ruwenzori 5110m
Beirut
Damascus
Baghdad
Tel Aviv-Jaffa
Jerusalem
Amman
Cairo
Alexandria
Khartoum
Omdurman
Addis Ababa
Asmara
Djibouti
Sana'a
Aden
Riyadh
Jiddah
Makkah
Port Sudan
Mogadishu
Nairobi
Kampala
Kigali
Kuwait
Aswân
Luxor
Wadi Halfa
Atbara
Kassala
Wad Medani
El Obeid
Juba
Wau
Malakal
Kismayu
Mombasa
Berbera
Hargeisa
Harer
Dire Dawa
Gonder
Mekele
Bahir Dar
Kisumu
Eldoret
Jinja
Entebbe
Kisangani
Bukavu
Mwanza
Projection: Lambert's Zenithal Equal Area
1 : 15 000 000
0 200 400 km

NORTH EAST AFRICA – CLIMATE, NATURAL VEGETATION AND AGRICULTURE

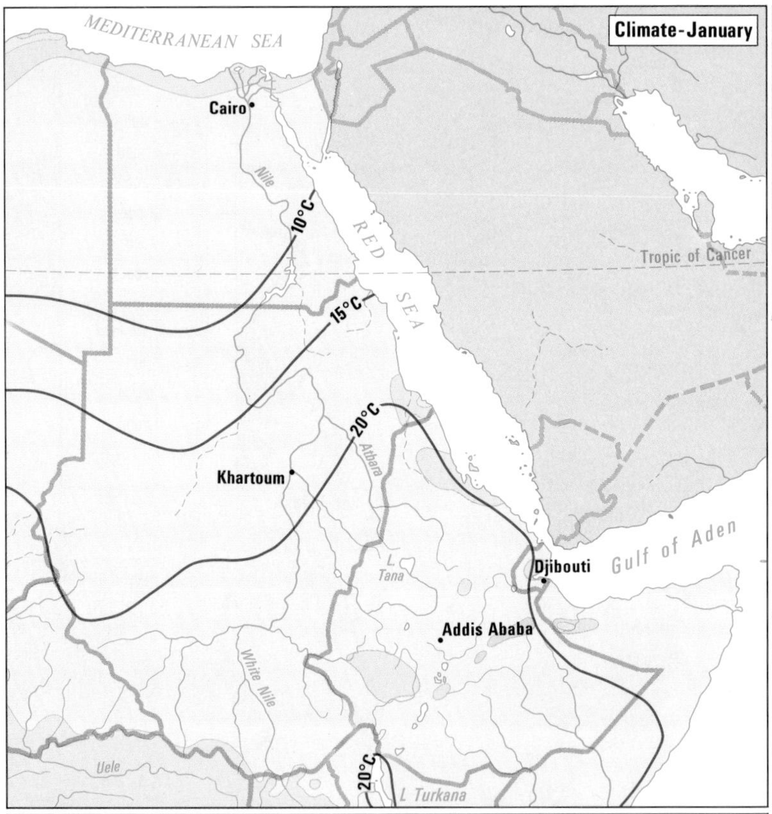

Mean Monthly Rainfall in Millimetres

- 100–150
- 50–100
- 10–50
- 0–10

Mean Daily Temperature

–15°C– Isotherm

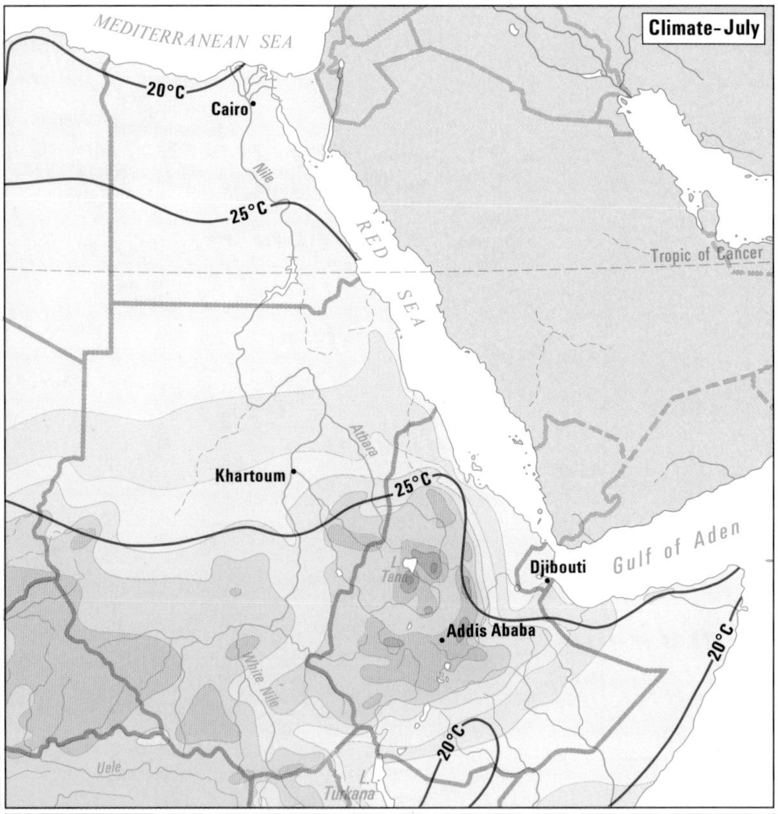

Mean Monthly Rainfall in Millimetres

- over 400
- 300–400
- 200–300
- 150–200
- 100–150
- 50–100
- 10–50
- 0–10

Mean Daily Temperature

–25°C– Isotherm

Projection: Lambert's Zenithal Equal Area
1 : 32 000 000
0 400 800 km

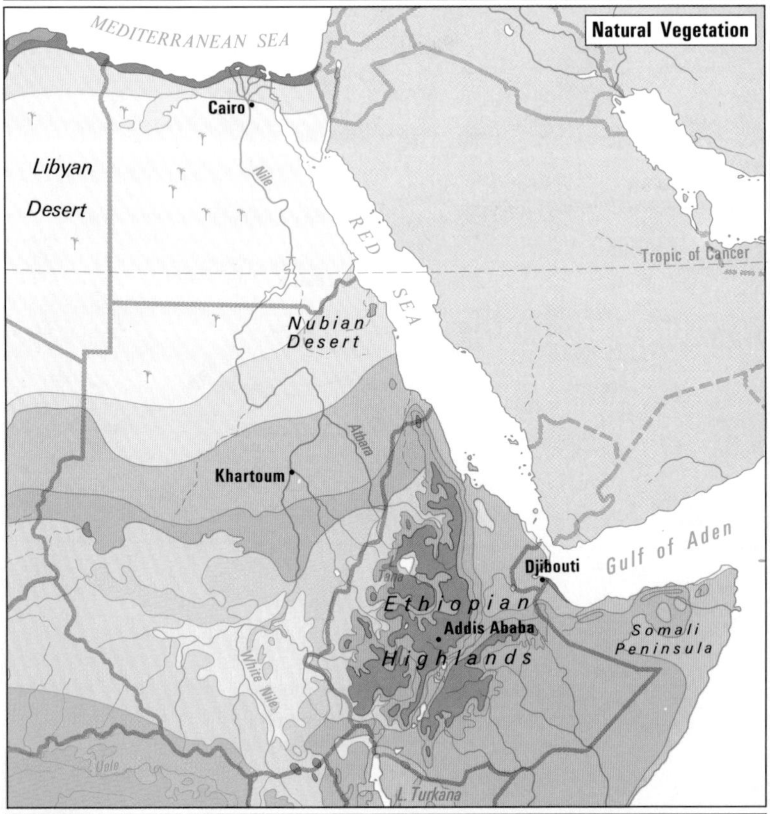

- Rain forest
- Rain forest with grassland
- Mediterranean forest
- Mangrove forest
- Montane forest with scrub
- Woodland
- Wooded grassland
- Grassland
- Evergreen bush-land and thicket
- Deciduous bush-land and thicket
- Semi-desert grassland
- Desert (perennial vegetation)
- Desert (no vegetation)
- Salt-flat vegetation
- Anthropic landscape
- Oasis

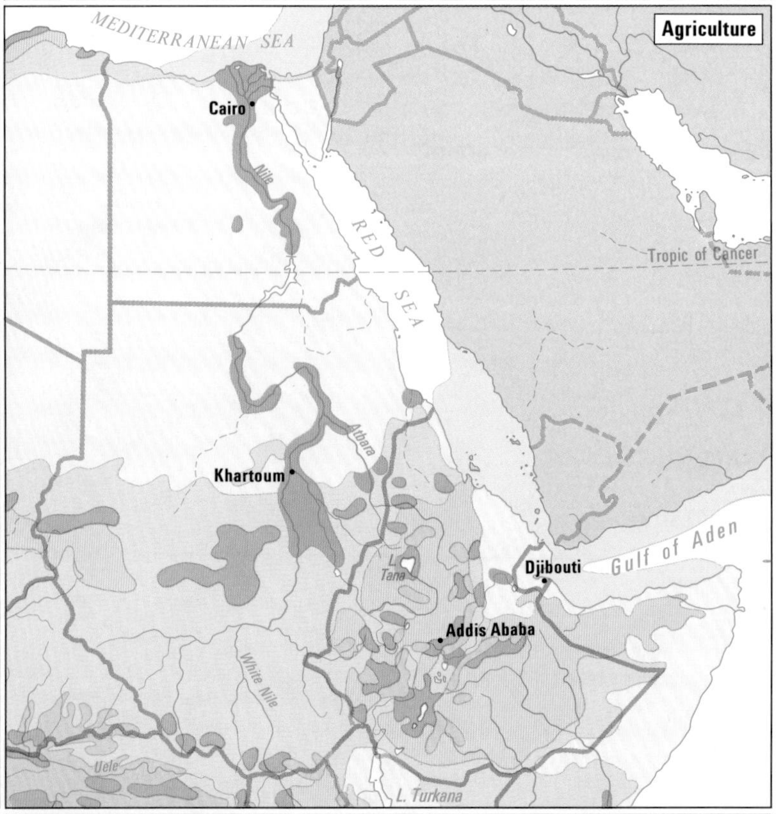

- Commercial farming – e.g. coffee, cotton, groundnuts, tobacco
- Irrigated farming – eg. cereals, cotton, sugar cane
- Subsistence farming – e.g. cassava, maize, millet, cattle, goats
- Non-intensive livestock rearing – e.g. cattle, sheep
- Desert – some nomadic herding
- Lake and sea fishing
- Forest

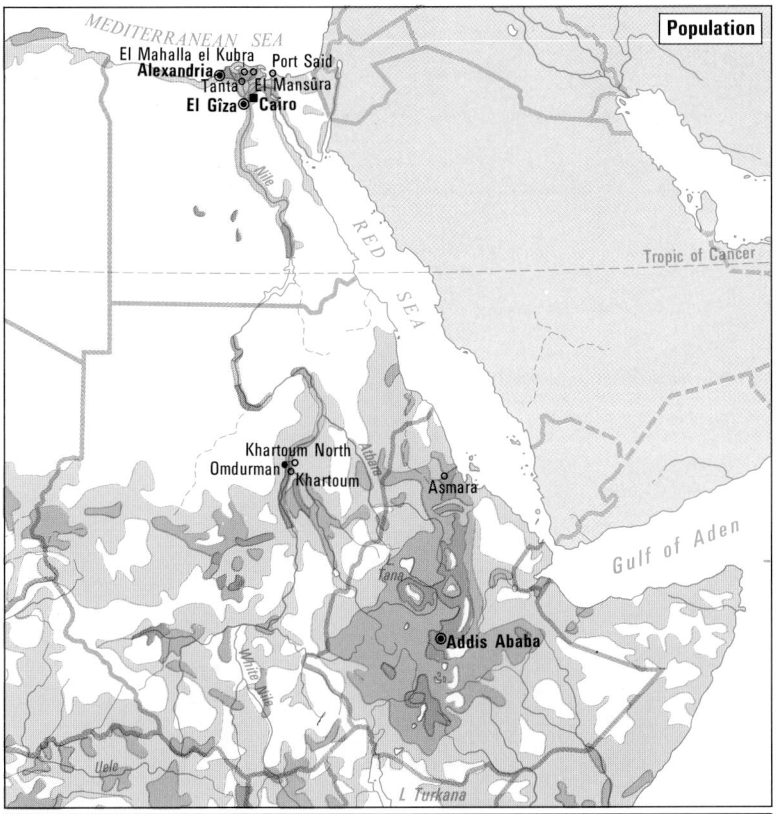

Population Density
(Persons per sq. km.)

- Over 200
- 100–200
- 50–100
- 10–50
- 1–10
- 0–1

Population of Cities

- ■ over 5 Million
- ◉ 1–5 Million
- • 500 000–1 Million
- ○ 250 000–500 000

Projection: Lambert's Zenithal Equal Area
1:32 000 000
0 400 800 km

- Chrome
- Coal
- Copper
- Diamonds
- Gold
- Iron ore
- Lead and zinc
- Manganese
- Natural gas
- Phosphates
- Platinum
- Salt
- Tin
- Uranium
- Oilfield
- Oil pipeline
- Major industrial centre
- Secondary industrial centre
- Minor industrial centre

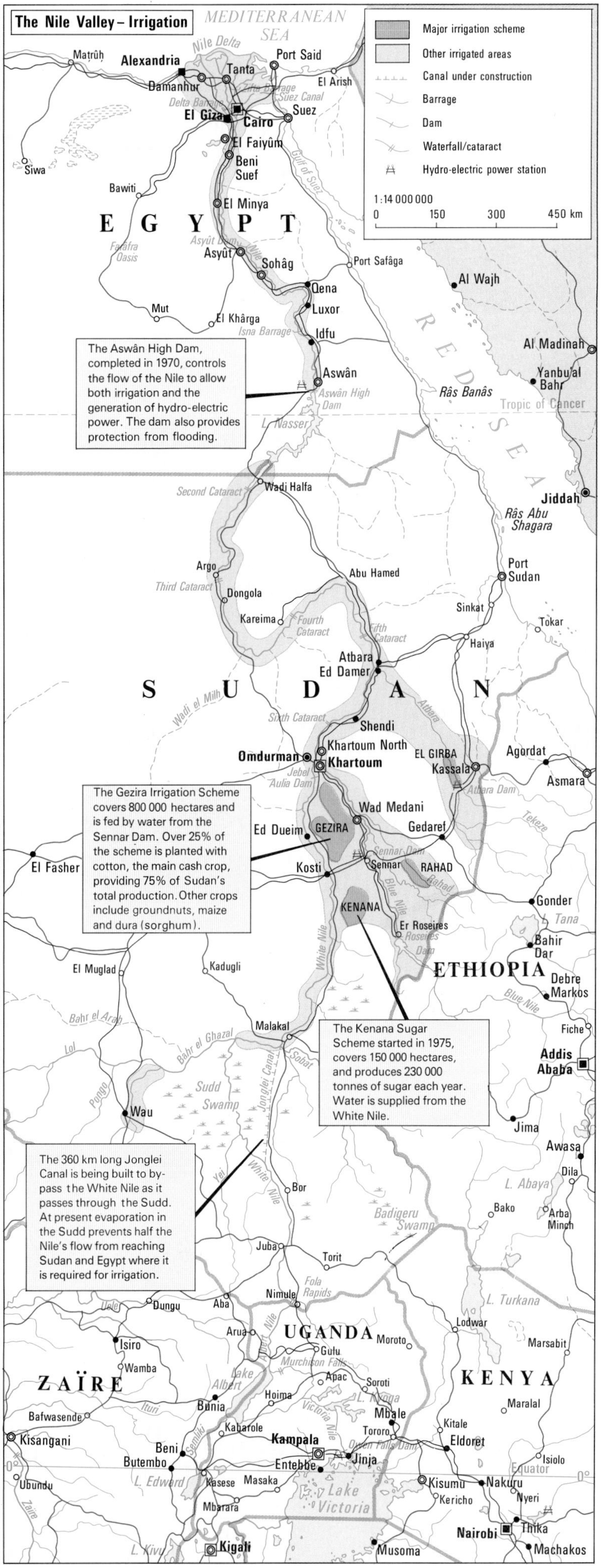

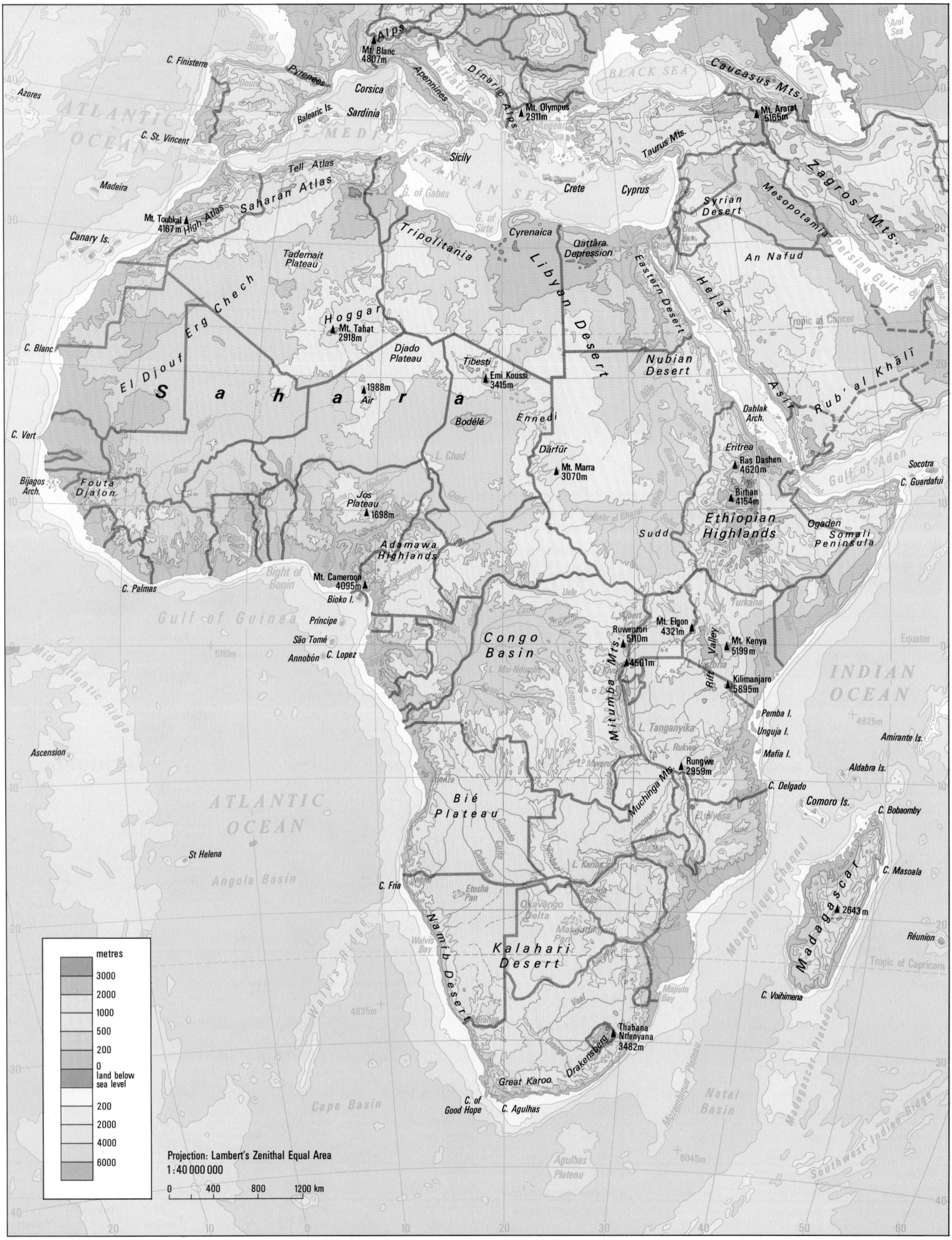
Alps
Mt. Blanc 4807m
C. Finisterre
Pyrenees
Corsica
Apennines
Dinaric Alps
Mt. Olympus 2911m
BLACK SEA
Caucasus Mts.
Mt. Ararat 5165m
Azores
ATLANTIC OCEAN
Balearic Is.
Sardinia
C. St. Vincent
MEDITERRANEAN SEA
Sicily
Taurus Mts.
Zagros Mts.
Tell Atlas
Madeira
Saharan Atlas
G. of Gabes
Crete
Cyprus
Syrian Desert
Mesopotamia
Mt. Toubkal 4167m
High Atlas
G. of Sirte
Canary Is.
Tripolitania
Cyrenaica
Qattâra Depression
Tademait Plateau
Libyan Desert
Eastern Desert
An Nafud
Hejaz
Persian Gulf
Erg Chech
Hoggar
Mt. Tahat 2918m
Tropic of Cancer
C. Blanc
Djado Plateau
Tibesti
Emi Koussi 3415m
Nubian Desert
El Djouf
Sahara
1988m Aïr
Bodélé
Ennedi
Asir
Rub' al Khali
Dahlak Arch.
C. Vert
L. Chad
Dârfûr
Mt. Marra 3070m
Eritrea
Ras Dashen 4620m
Gulf of Aden
Socotra
C. Guardafui
Bijagos Arch.
Fouta Djalon
Jos Plateau 1698m
Birhan 4154m
Ethiopian Highlands
Sudd
Ogaden
Somali Peninsula
Adamawa Highlands
Mt. Cameroon 4095m
Bight of Benin
C. Palmas
Bioko I.
Gulf of Guinea
Príncipe
São Tomé
Annobón
C. Lopez
Congo Basin
Ruwenzori 5110m
Mt. Elgon 4321m
Mt. Kenya 5199m
Rift Valley
4501m
Kilimanjaro 5895m
Mitumba Mts.
Equator
INDIAN OCEAN
Mid-Atlantic Ridge
Pemba I.
Unguja I.
Mafia I.
Amirante Is.
L. Tanganyika
L. Rukwa
Ascension
Rungwe 2959m
Aldabra Is.
Muchinga Mts.
C. Delgado
Comoro Is.
C. Bobaomby
Bié Plateau
ATLANTIC OCEAN
St Helena
Angola Basin
L. Kariba
Mozambique Channel
C. Masoala
Madagascar
2643m
C. Fria
Etosha Pan
Okavango Delta
Namib Desert
Walvis Bay
Walvis Ridge
Kalahari Desert
Réunion
Tropic of Capricorn
C. Voihimena
Maputo Bay
4835m
Thabana Ntlenyana 3482m
Drakensberg
Great Karoo
Natal Basin
C. of Good Hope
C. Agulhas
Cape Basin
Agulhas Plateau
6045m
Southwest Indian Ridge
Madagascar Plateau
Mozambique Plateau
metres
3000
2000
1000
500
200
0
land below sea level
200
2000
4000
6000
Projection: Lambert's Zenithal Equal Area
1:40 000 000
0 400 800 1200 km

FRANCE
SWITZ.
AUSTRIA
HUNGARY
ROMANIA
YUGOSLAVIA
U.S.S.R.
BULGARIA
BLACK SEA
CASPIAN SEA
ITALY
Corsica
Sardinia
ALBANIA
Balearic Is.
PORTUGAL
SPAIN
Azores (Port.)
ATLANTIC OCEAN
GREECE
TURKEY
MEDITERRANEAN SEA
Algiers
Annaba
Tunis
Sicily
Tangier
Rabat
Oran
Constantine
Sfax
CYPRUS
SYRIA
Casablanca
Fès
Madeira (Port.)
Marrakech
MOROCCO
Agadir
TUNISIA
Tripoli
Crete
LEBANON
IRAQ
IRAN
ISRAEL
Port Said
JORDAN
Misratah
Benghazi
Alexandria
Canary Is. (Sp.)
El Gîza
Cairo
Suez
KUWAIT
El Aaiún
ALGERIA
LIBYA
WESTERN SAHARA (S.A.D.R.)
Asyût
QATAR
EGYPT
SAUDI ARABIA
U.A.E.
RED SEA
Tropic of Cancer
Aswân
L. Nasser
MAURITANIA
Port Sudan
OMAN
Nouakchott
MALI
Nile
NIGER
Atbara
P.D.R. YEMEN
St-Louis
Gao
Agadèz
CHAD
Khartoum
Kassala
Asmara
YEMEN ARAB REPUBLIC
Dakar
Sénégal
SENEGAL
Kayes
Niger
THE GAMBIA
Banjul
Bamako
Ouagadougou
Niamey
El Obeid
Wad Medani
Gulf of Aden
Bissau
GUINEA-BISSAU
BURKINA FASO
L. Chad
Sokoto
Kano
N'Djaména
DJIBOUTI
Djibouti
Socotra (P.D.R.Y.)
GUINEA
Maiduguri
SUDAN
White Nile
Conakry
SIERRA LEONE
CÔTE D'IVOIRE
Tamale
Kaduna
NIGERIA
Addis Ababa
Dire Dawa
Hargeisa
Freetown
Bouaké
GHANA
TOGO
BENIN
Abuja
Garoua
Benue
CENTRAL AFRICAN REPUBLIC
SOMALIA
Monrovia
Kumasi
Ogbomosho
Ibadan
Wau
ETHIOPIA
LIBERIA
Yamoussoukro
Accra
Lomé
Porto Novo
Lagos
Enugu
Abidjan
Sekondi-Takoradi
CAMEROON
Port Harcourt
Douala
Bangui
Juba
Malabo
Yaoundé
Uele
L. Turkana
Gulf of Guinea
SÃO TOMÉ & PRÍNCIPE
EQ. GUINEA
Ubangi
Zaïre
UGANDA
KENYA
Mogadishu
Kisangani
São Tomé
Libreville
Mbandaka
Kampala
Kisumu
Equator
GABON
CONGO
ZAÏRE
L. Victoria
Kismayu
RWANDA
Kigali
Nairobi
Bukavu
Mwanza
INDIAN OCEAN
Brazzaville
Bujumbura
BURUNDI
Mombasa
Kasai
Kinshasa
TANZANIA
Cabinda (Angola)
Kananga
Mbuji-Mayi
L. Tanganyika
Zanzibar
Dodoma
Dar es Salaam
Luanda
Mbeya
SEYCHELLES
CAPE VERDE
MAURITANIA
Praia
SENEGAL
Dakar
THE GAMBIA
Banjul
ATLANTIC OCEAN
ANGOLA
Lubumbashi
MALAWI
Moroni
COMOROS
Lobito
Kitwe
Ndola
L. Nyasa
Mayotte (Fr.)
Huambo
ZAMBIA
Lilongwe
Nampula
St. Helena (U.K.)
Lubango
Cubango
Lusaka
Blantyre
Mahajanga
Zambezi
Mozambique Channel
Harare
Kariba
NAMIBIA
ZIMBABWE
Antananarivo
Victoria
SEYCHELLES
Bulawayo
Beira
MOZAMBIQUE
Francistown
Walvis Bay
Windhoek
BOTSWANA
Limpopo
MADAGASCAR
Réunion (Fr.)
Toliara
Gaborone
Pretoria
Tropic of Capricorn
Lüderitz
Johannesburg
Maputo
SWAZILAND
Mbabane
Orange
Kimberley
Vaal
LESOTHO
Maseru
REPUBLIC OF SOUTH AFRICA
Durban
Mahajanga
MADAGASCAR
Cape Town
East London
Port Elizabeth
Antananarivo
MAURITIUS
Port Louis
Réunion (Fr.)
Projection: Lambert's Zenithal Equal Area
1:40 000 000
0
400
800
1200 km

AFRICA – CLIMATE

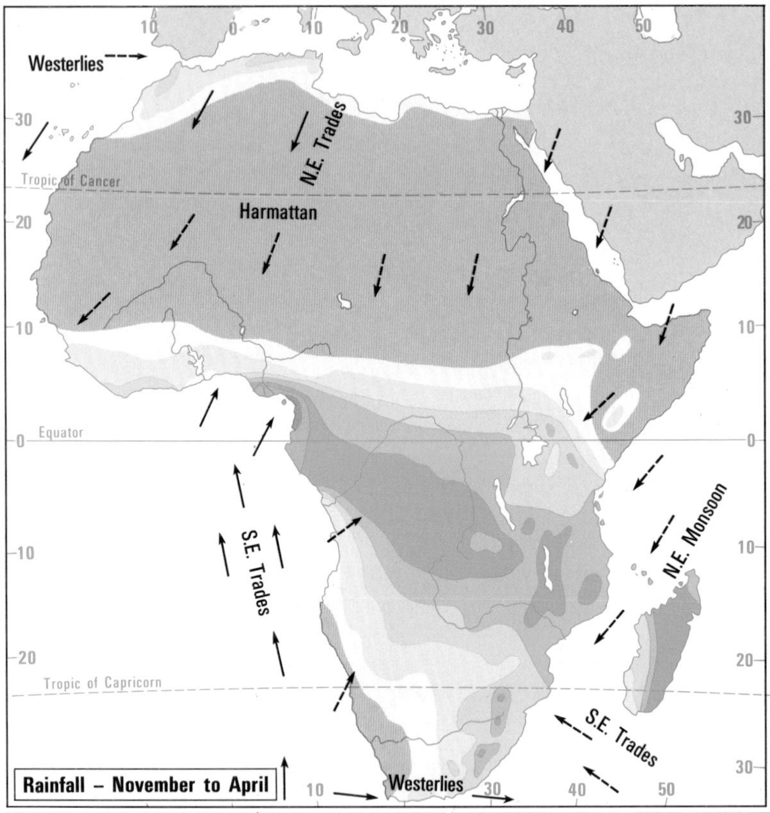

Rainfall – November to April

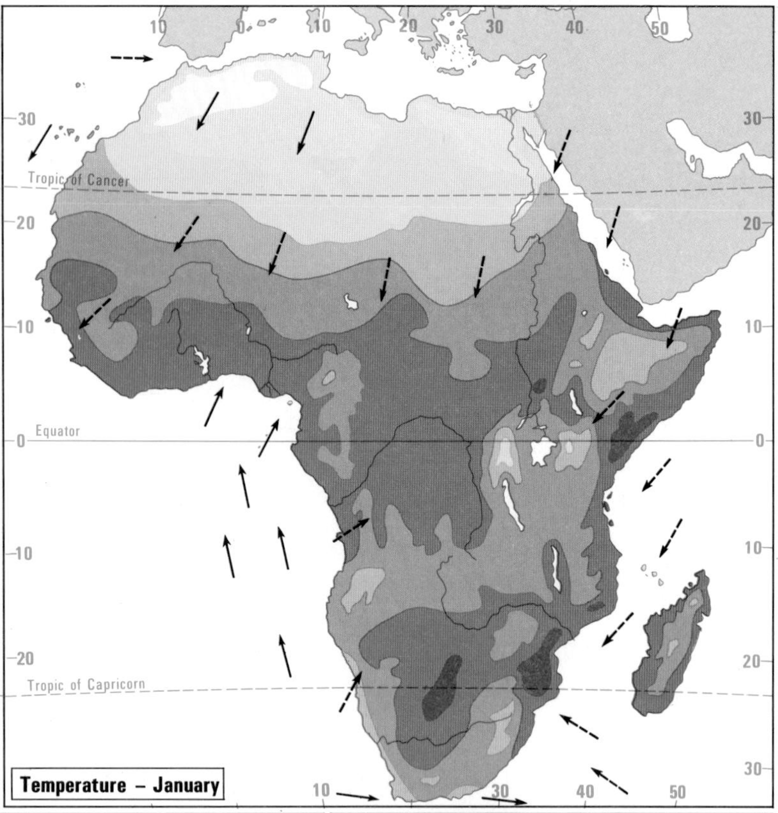

Temperature – January

Rainfall in millimetres

- Over 2000
- 1500–2000
- 1000–1500
- 750–1000
- 500–750
- 250–500
- 125–250
- 0–125

0 1000 2000 3000 kilometres

Prevailing winds

- Blowing all year
- November to April
- May to October

Surface temperature °C

36
32
28
24
20
16
12
8
4

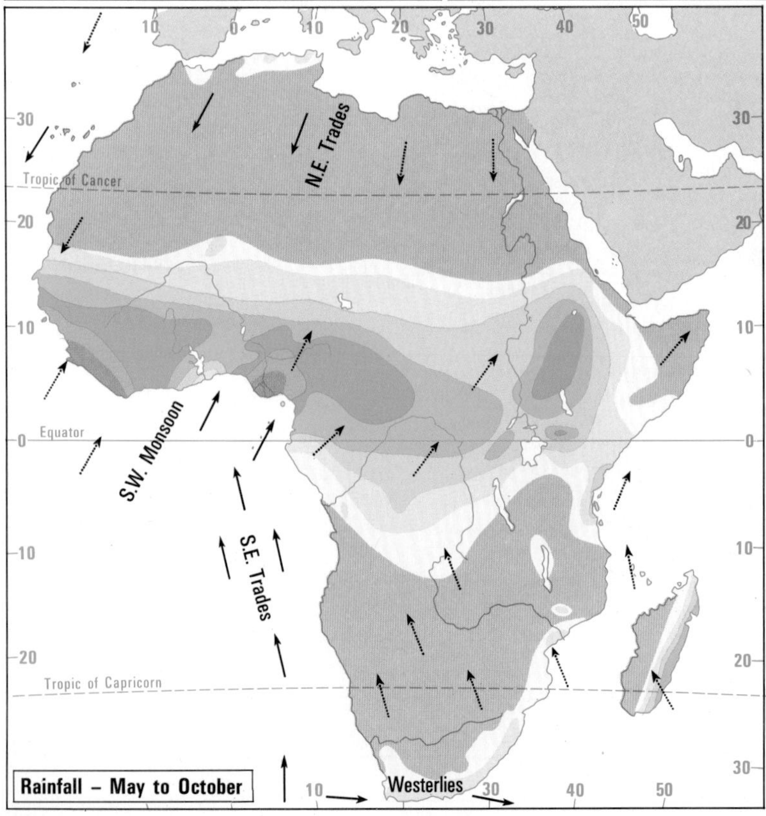

Rainfall – May to October

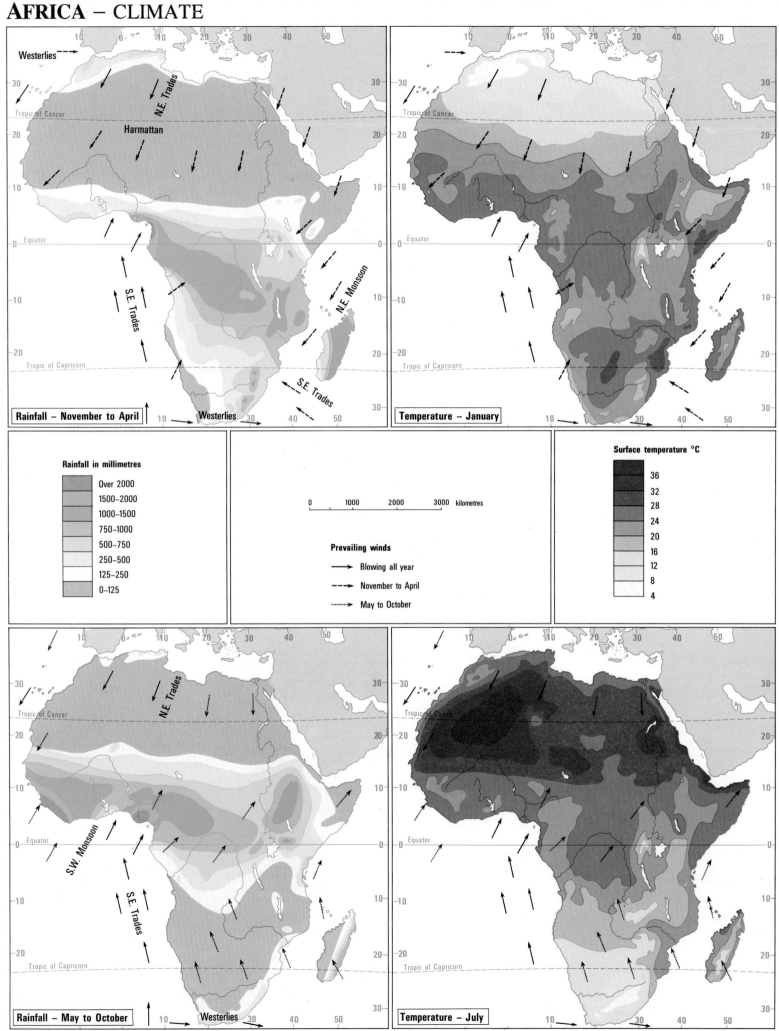

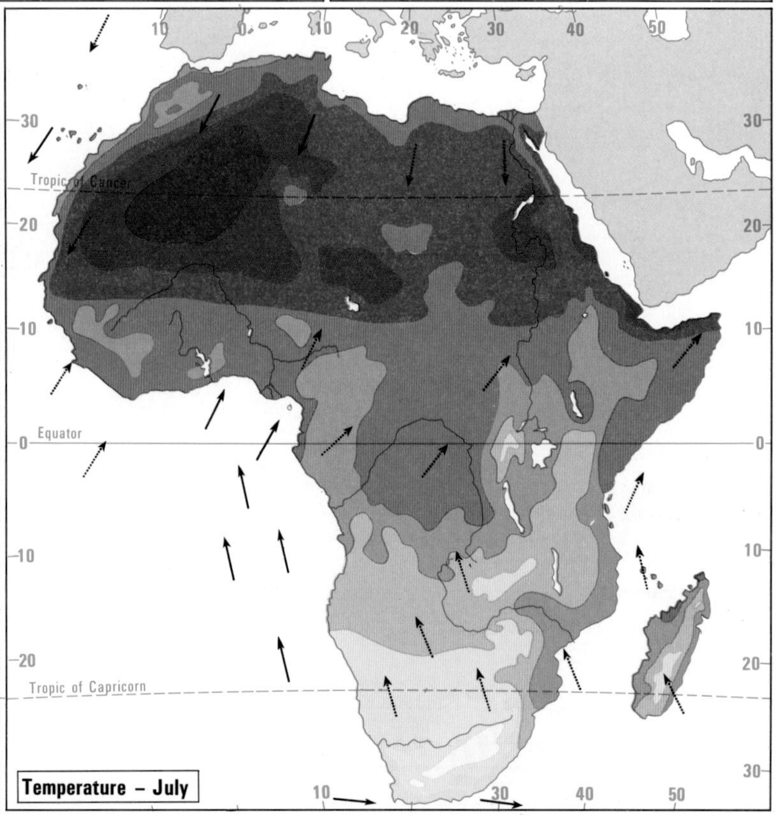

Temperature – July

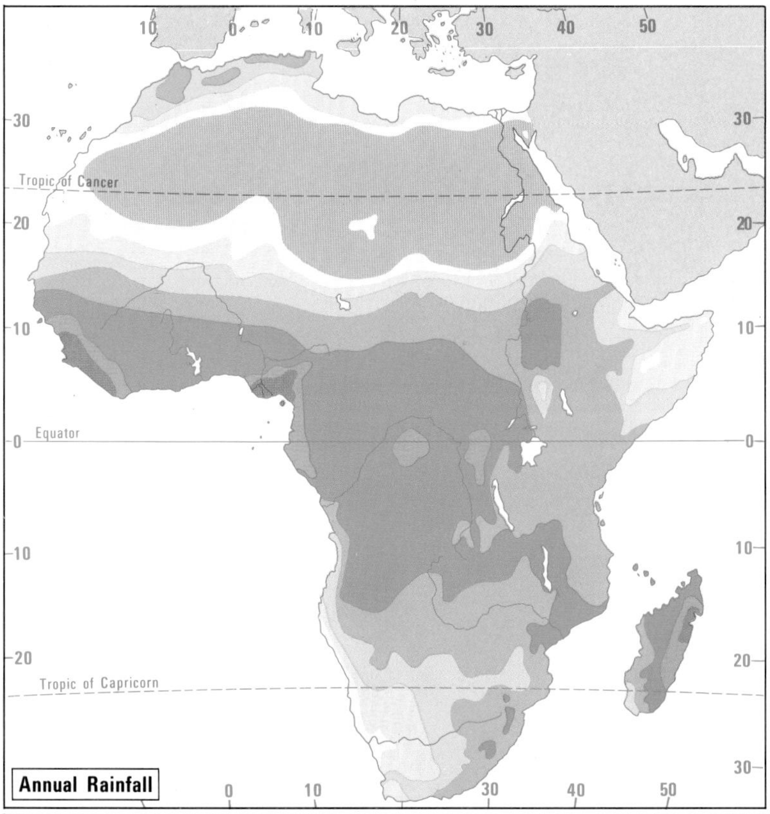

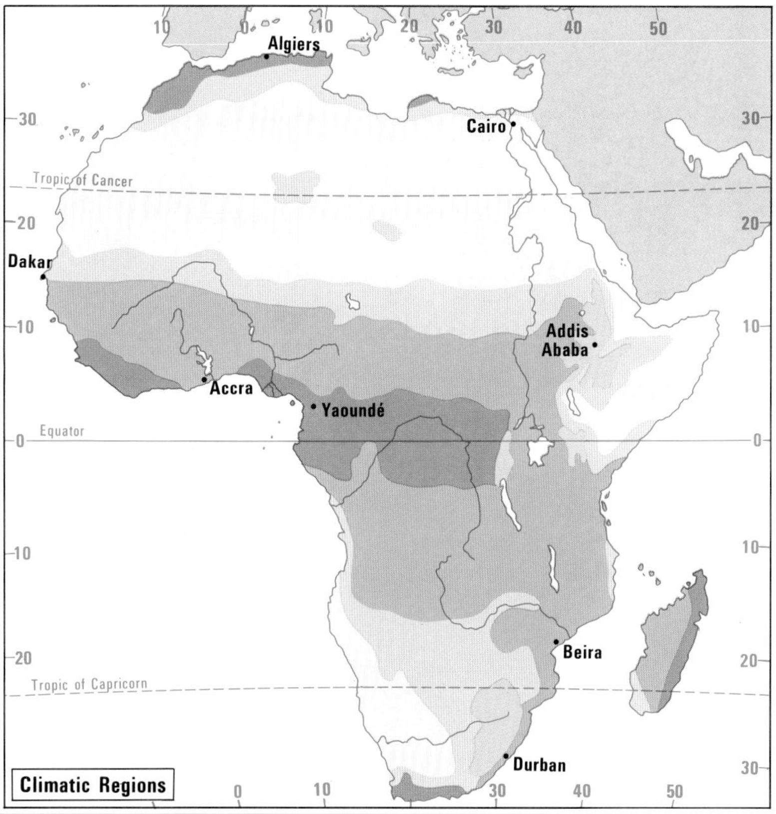

Rainfall in millimetres

- Over 3000
- 2000–3000
- 1000–2000
- 500–1000
- 250–500
- 100–250
- 50–100
- 0–50

0 1000 2000 3000 kilometres

- Tropical rain forest
- Tropical savanna
- Steppe
- Desert
- Mediterranean
- Maritime west coast

For further information on Köppen's Climatic Classification see page 106

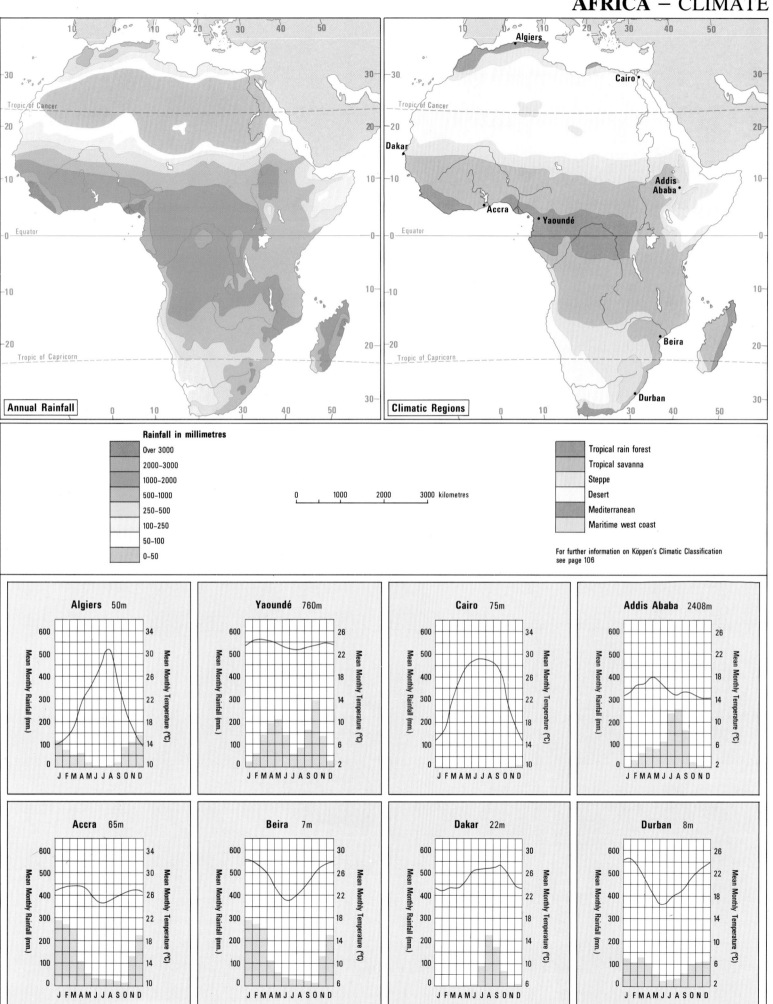

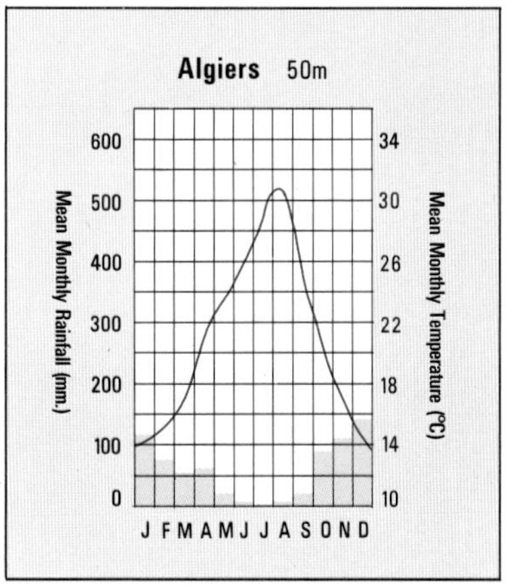

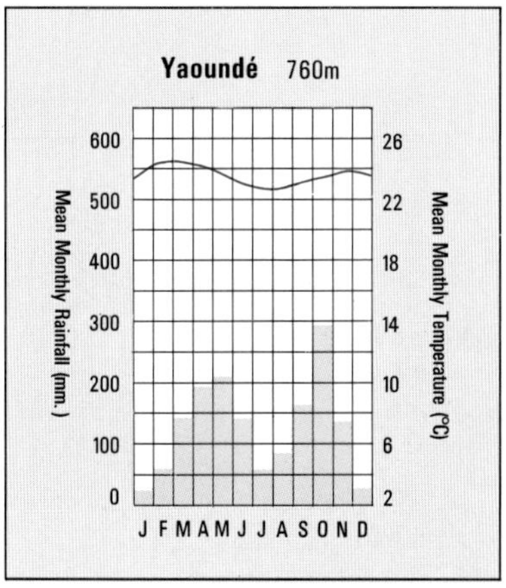

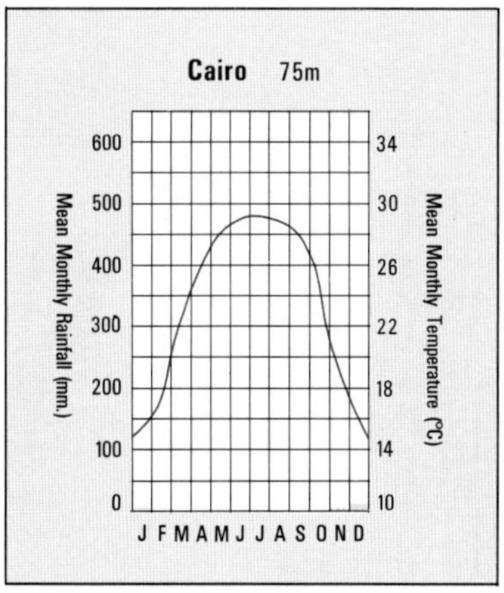

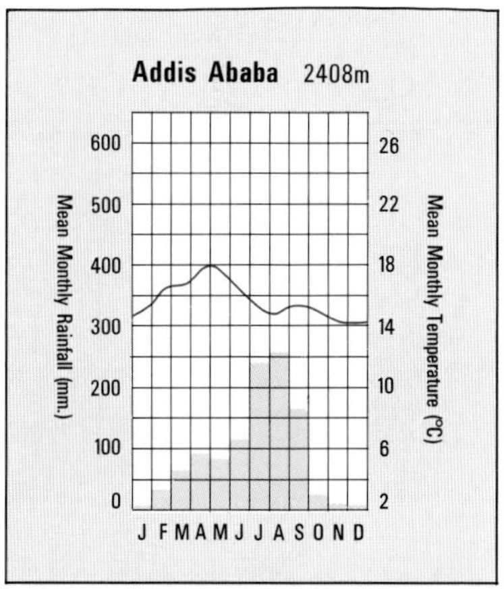

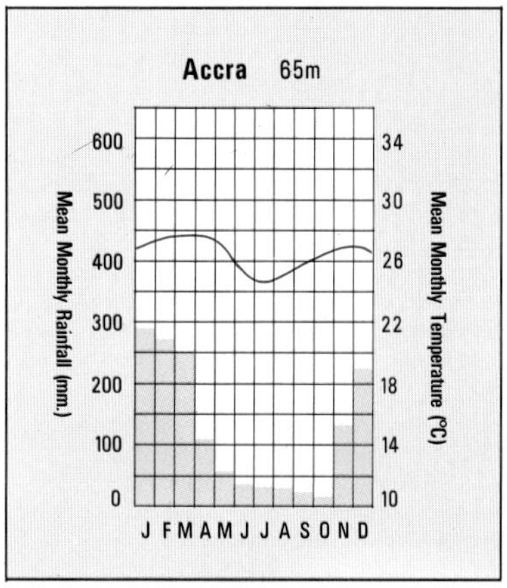

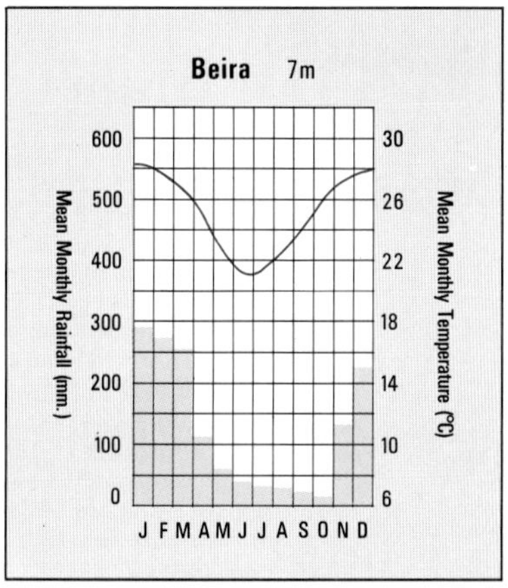

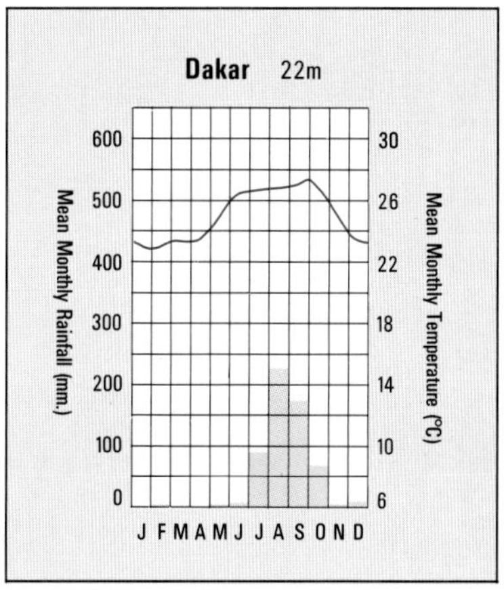

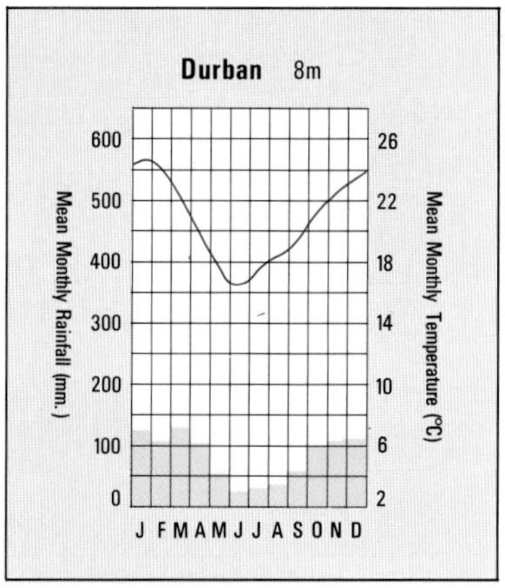

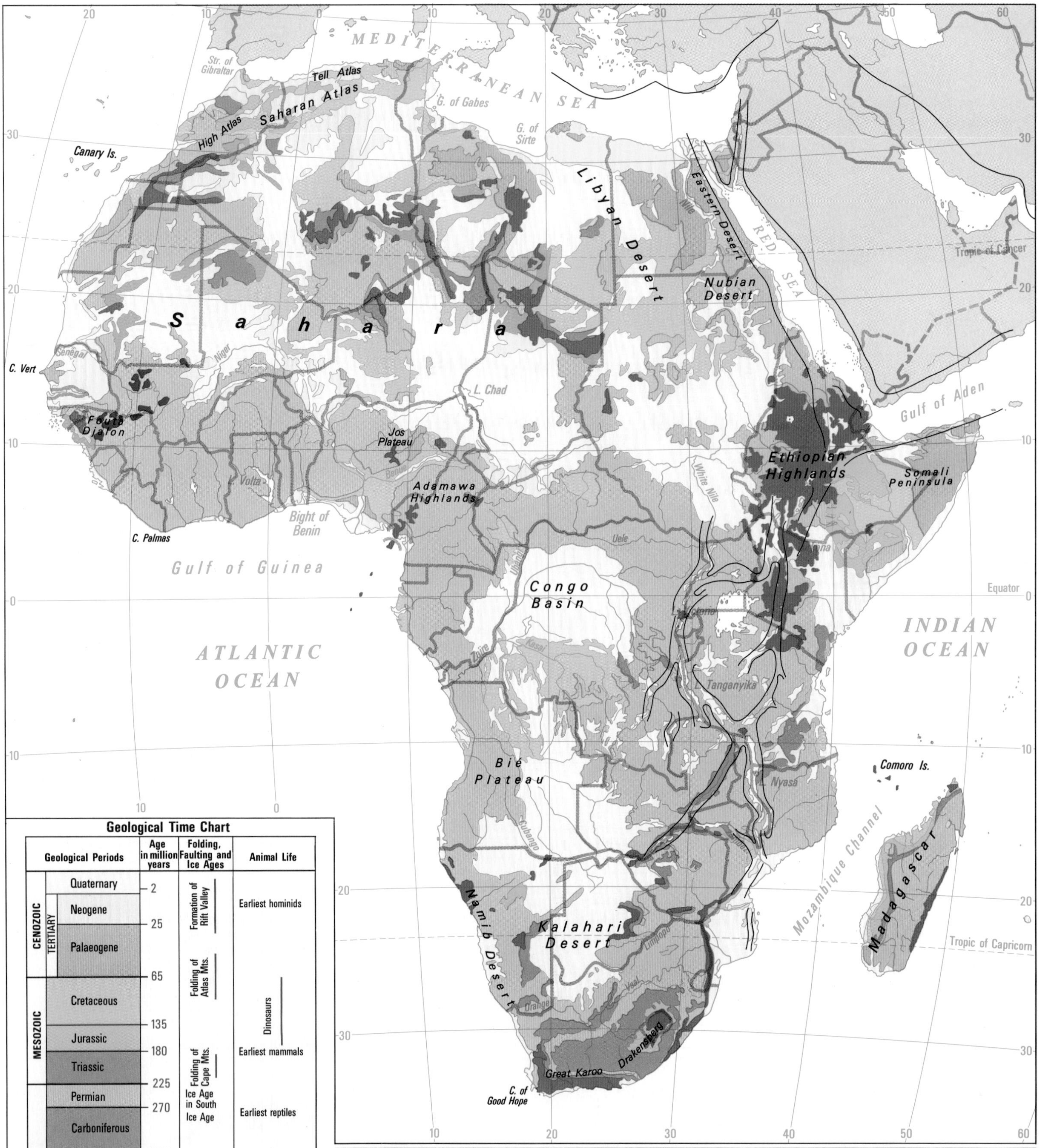

Geological Time Chart

Era	Geological Periods	Age in million years	Folding, Faulting and Ice Ages	Animal Life
CENOZOIC	Quaternary	2	Formation of Rift Valley	
CENOZOIC – TERTIARY	Neogene	25		Earliest hominids
CENOZOIC – TERTIARY	Palaeogene	65	Folding of Atlas Mts.	
MESOZOIC	Cretaceous	135		Dinosaurs
MESOZOIC	Jurassic	180		Earliest mammals
MESOZOIC	Triassic	225	Folding of Cape Mts.	
PALAEOZOIC	Permian	270	Ice Age in South Ice Age	Earliest reptiles
PALAEOZOIC	Carboniferous	350		
PALAEOZOIC	Devonian	395		Earliest amphibians
PALAEOZOIC	Silurian	440	Ice Age in Sahara	Earliest fish
PALAEOZOIC	Ordovician	500		
PALAEOZOIC	Cambrian	570	Pan African folding	
PRE-CAMBRIAN	Pre-Cambrian (undifferentiated)	3 500		Earliest micro-organisms (eg algae)

Vertical scale is not proportional to elapsed time

Extrusive igneous rocks

Intrusive igneous rocks

Major fault line

Source: UNESCO Geological Map of the World

Projection: Lambert's Zenithal Equal Area
1 : 40 000 000

0 400 800 1200 km

Africa is a stable continental landmass composed of metamorphic Pre-Cambrian rocks partly covered by younger sedimentary rocks. African Pre-Cambrian rocks include some of the oldest rocks known, and were folded in a series of mountain-building events. Since the Pre-Cambrian period Africa has been mostly emergent, therefore sediments created by erosion are usually continental. These include the thin covering of Quaternary river and dune sands of the Sahara and Kalahari deserts, and the sediments of Lake Chad and the East African lakes.

Africa, together with South America, India, Antarctica and Arabia, was originally part of a larger landmass (Gondwanaland), which was broken up by rifting and continental drift. Sedimentary basins formed during this process include the Triassic basins of East Africa, basins along the continental margins and the East African rift valleys. Extensive lavas are associated with the movement of Arabia away from Africa and the formation of the Red Sea.

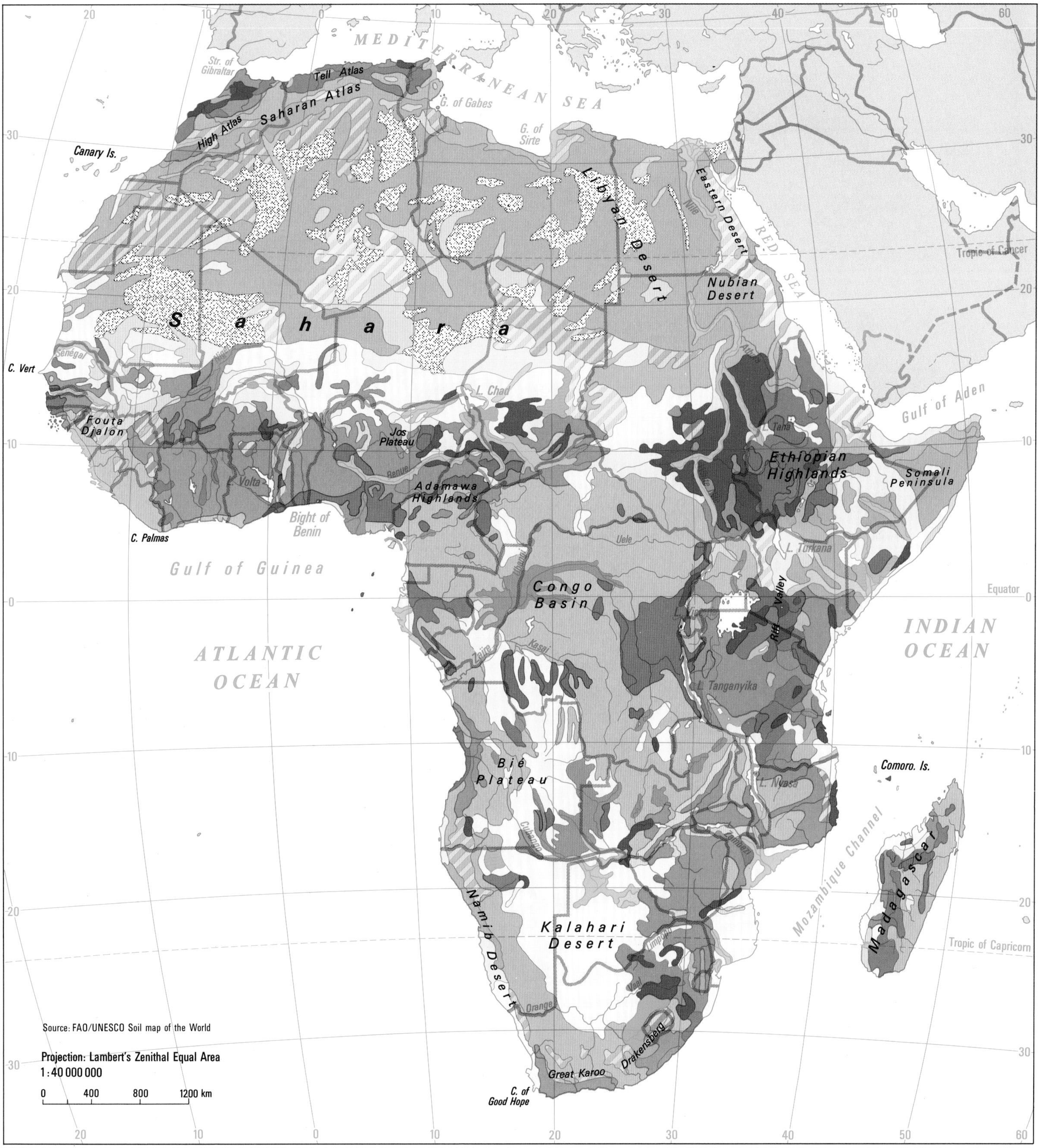

- Fluvisols–soils developed from recent alluvial deposits
- Gleysols–soils in poorly drained sites subject to water logging, often with a high clay content
- Regosols–soils developed from loose material such as sand with low water and plant nutrient content
- Lithosols–shallow soils, generally too shallow and stony for agriculture
- Arenosols–weakly developed, coarse textured soils with a low clay content
- Andosols–soils derived from volcanic deposits
- Vertisols–dark, cracking clays, sticky when wet and cracked when dry; where well managed can be agriculturally productive
- Solonchaks–saline soils
- Solonetz –alkaline soils, often develop under irrigation
- Yermosols–desert soils with very low organic matter
- Xerosols–semi-desert soils with low organic matter
- Kastanozems–brown soils typical of temperate steppe environments
- Cambisols–brown soils with no major accumulations or depositions within the soil
- Luvisols– soils with clay accumulation beneath the surface
- Podzols–soils with accumulations of iron, aluminium and/or organic matter beneath the surface
- Planosols–soils with a bleached horizon near the surface showing signs of water logging
- Acrisols–soils with clay accumulations beneath the surface being more strongly leached than luvisols
- Nitosols–fertile, tropical soils with clay accumulation beneath the surface
- Ferralsols–strongly weathered soils of the humid tropics, generally low fertility
- No soil present–wind-blown sand or rock

Note: striped areas indicate mixed soil types

AFRICA – NATURAL VEGETATION

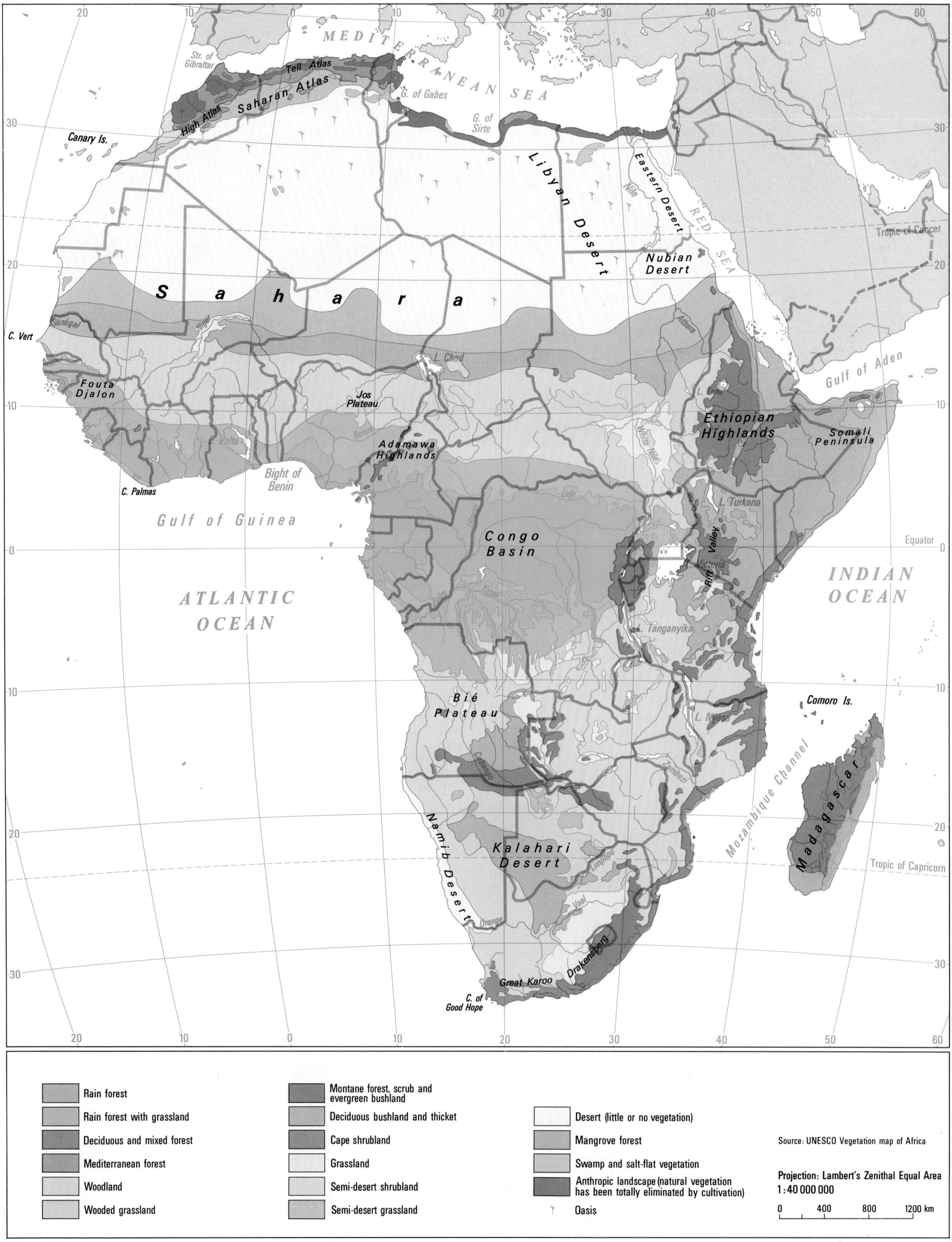

- Rain forest
- Rain forest with grassland
- Deciduous and mixed forest
- Mediterranean forest
- Woodland
- Wooded grassland
- Montane forest, scrub and evergreen bushland
- Deciduous bushland and thicket
- Cape shrubland
- Grassland
- Semi-desert shrubland
- Semi-desert grassland
- Desert (little or no vegetation)
- Mangrove forest
- Swamp and salt-flat vegetation
- Anthropic landscape (natural vegetation has been totally eliminated by cultivation)
- Oasis

Source: UNESCO Vegetation map of Africa

Projection: Lambert's Zenithal Equal Area
1 : 40 000 000

0 400 800 1200 km

- Cultivated land
- Cultivated land and grazing land
- Grassland and grazing land
- Forest and woodland
- Swamp and marsh
- Semi-desert and desert

Projection: Lambert's Zenithal Equal Area
1 : 40 000 000

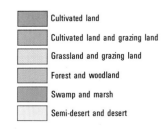

0 400 800 1200 km

The above map shows the present day natural environment of the African continent. This map differs from the vegetation map on page 60 in that it shows how man has modified the natural landscape through intensive agriculture, grazing and urban development.
However, vast areas of the continent's desert and tropical rain forest environment remain relatively unaffected by man. The table on the right indicates the percentage distribution of land use in Africa, divided into four main categories.

Land Use	1964–1966	1981–1983
Cultivated land	5%	6%
Grassland and grazing land	26%	26%
Forest and woodland	25%	23%
Other land	43%	44%

AFRICA – DESERTIFICATION AND ENVIRONMENTAL PROBLEMS

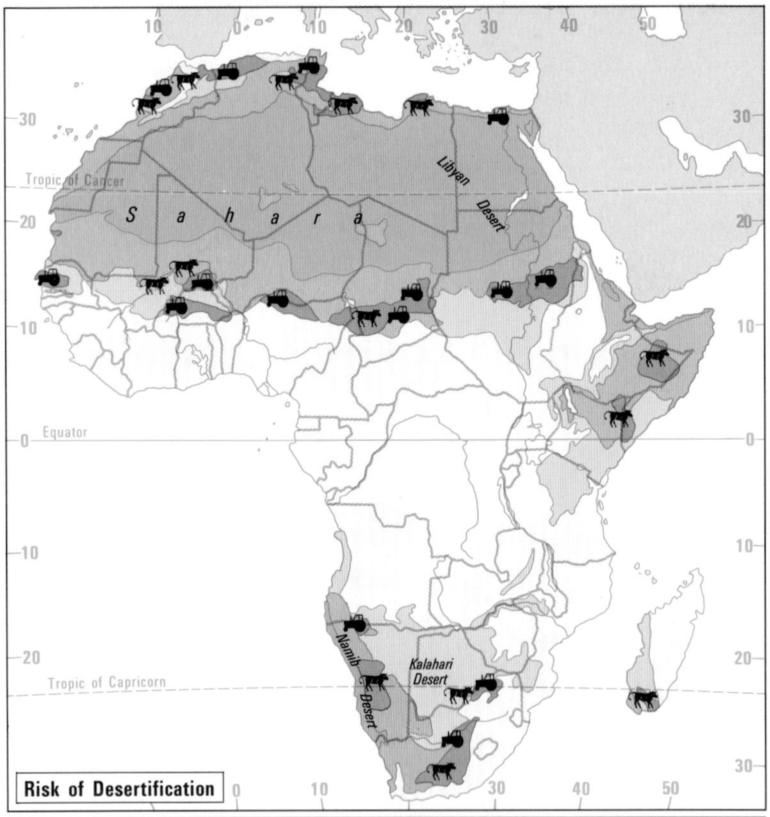

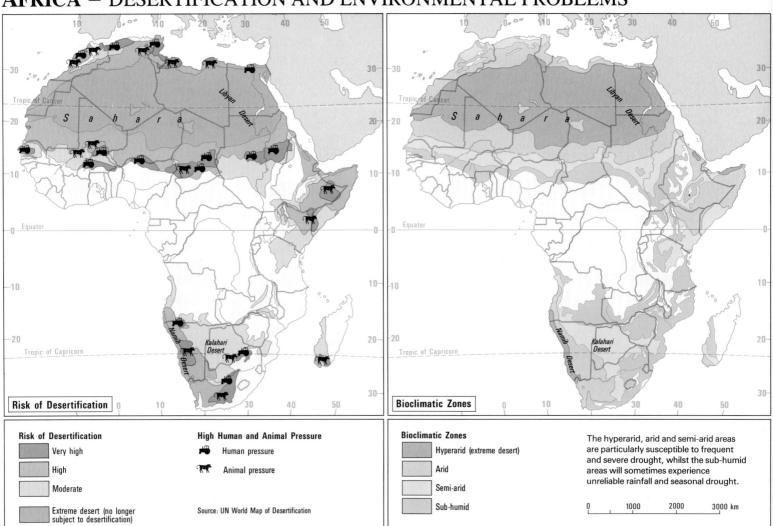

Risk of Desertification

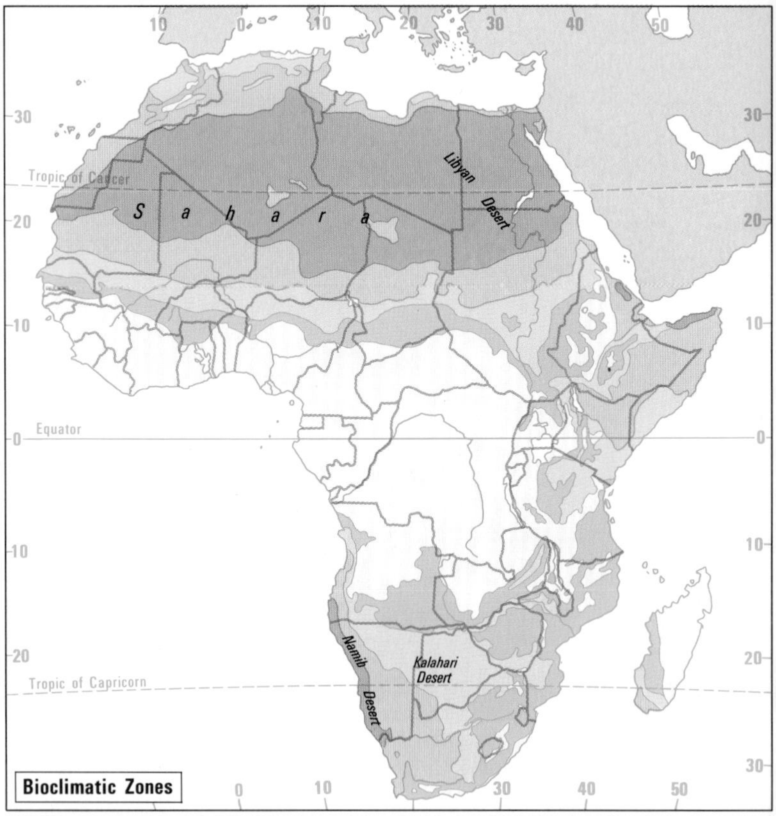

Bioclimatic Zones

Risk of Desertification

- Very high
- High
- Moderate
- Extreme desert (no longer subject to desertification)

High Human and Animal Pressure

- Human pressure
- Animal pressure

Source: UN World Map of Desertification

Bioclimatic Zones

- Hyperarid (extreme desert)
- Arid
- Semi-arid
- Sub-humid

The hyperarid, arid and semi-arid areas are particularly susceptible to frequent and severe drought, whilst the sub-humid areas will sometimes experience unreliable rainfall and seasonal drought.

0 1000 2000 3000 km

Desertification

Desertification is the extension of desert conditions to the arid and semi-arid zones bordering extreme deserts. It is becoming more commonly accepted that the major causes of desertification are a combination of increasing human and animal pressure on the land, in conjunction with occasional periods of severe drought. Man's activities of overcultivation, overgrazing and large scale destruction of woodland for fuel in these vulnerable areas has led to the removal of the natural vegetation cover. Surface run-off and the resultant soil erosion has become an increasing problem leading to the inability of the environment to support ever increasing population levels.

During the dry season the ground becomes hard, developing wide cracks, and cannot hold enough moisture to permit plant growth. The unreliability of the rainfall can lead to severe droughts causing famine and starvation.

Deforestation

Deforestation is the term used to describe the removal and clearance of forested areas, by man. It is estimated that some 12 million hectares of forest are being removed annually throughout the world, and consequently such severe deforestation is having an extremely serious effect on the environment.

In Africa, both tropical forest and open woodland are being rapidly depleted to provide the expanding population with fuelwood, timber and additional land for cultivation. The poor tropical soils left after forest clearance are generally unsuitable for cultivation, only remaining fertile for a short period of time, and they are particularly susceptible to soil erosion.

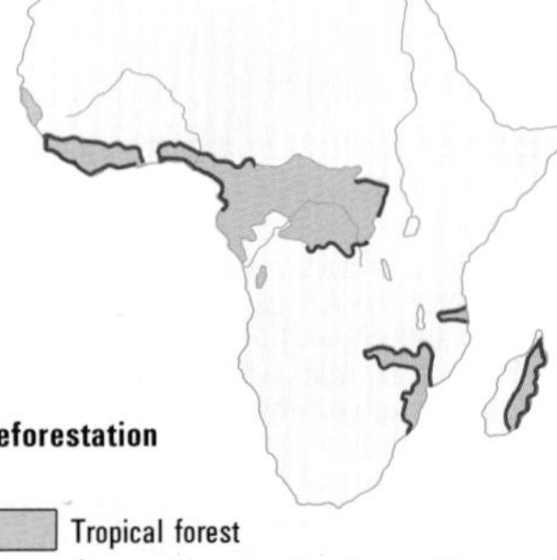

Deforestation

- Tropical forest
- Areas of forest undergoing rapid removal

0 2000 4000 km

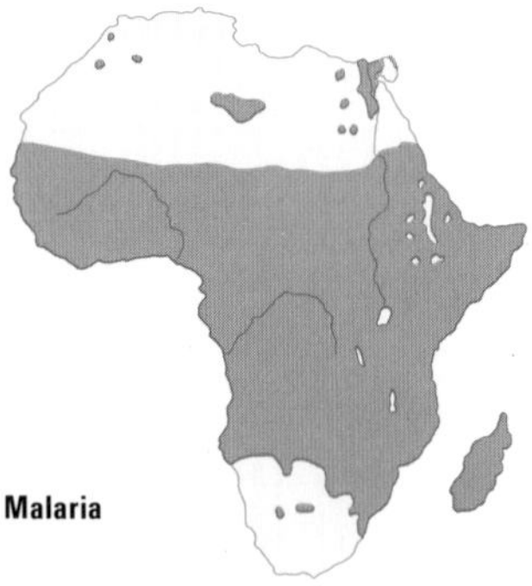

Malaria

Schistosomiasis

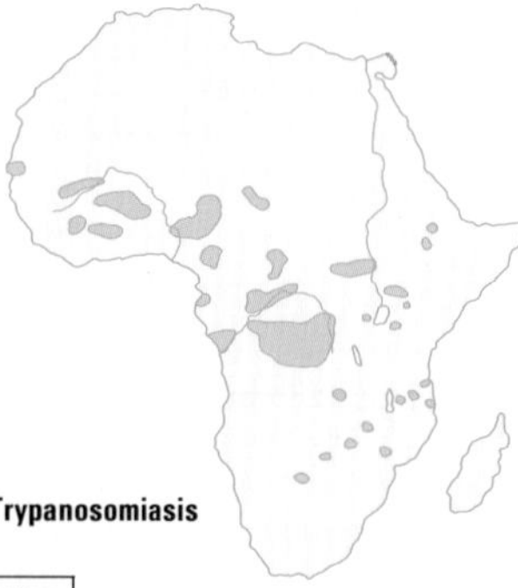

Trypanosomiasis

Yellow Fever

Disease

- Areas where malarial infection may occur
- Areas where trypanosomiasis (sleeping sickness) may occur
- Areas where schistosomiasis (bilharzia) may occur
- Areas where yellow fever may occur

AFRICA – AGRICULTURE AND IRRIGATION

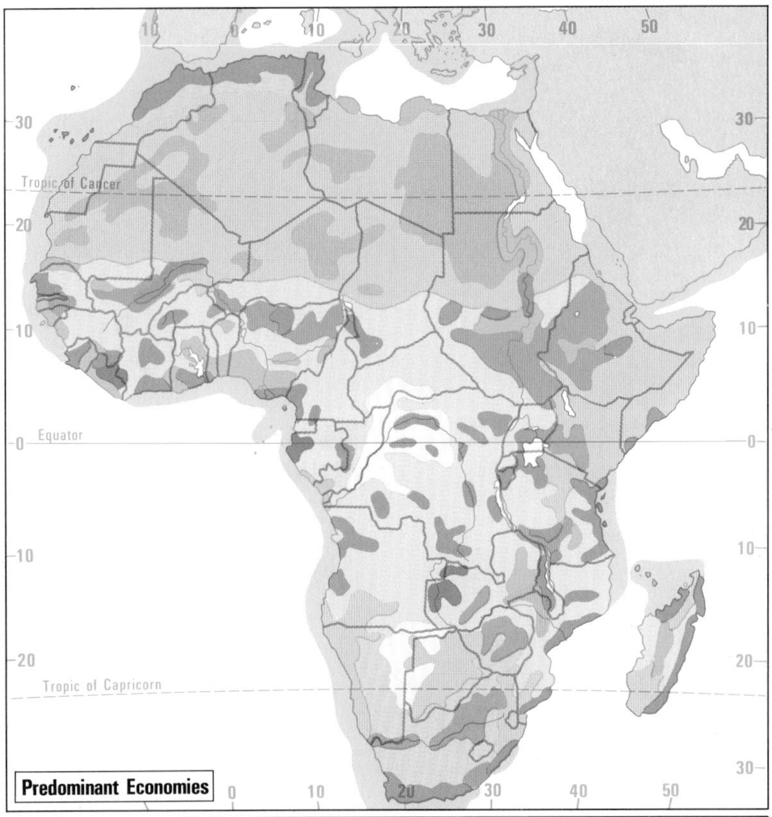

Predominant Economies

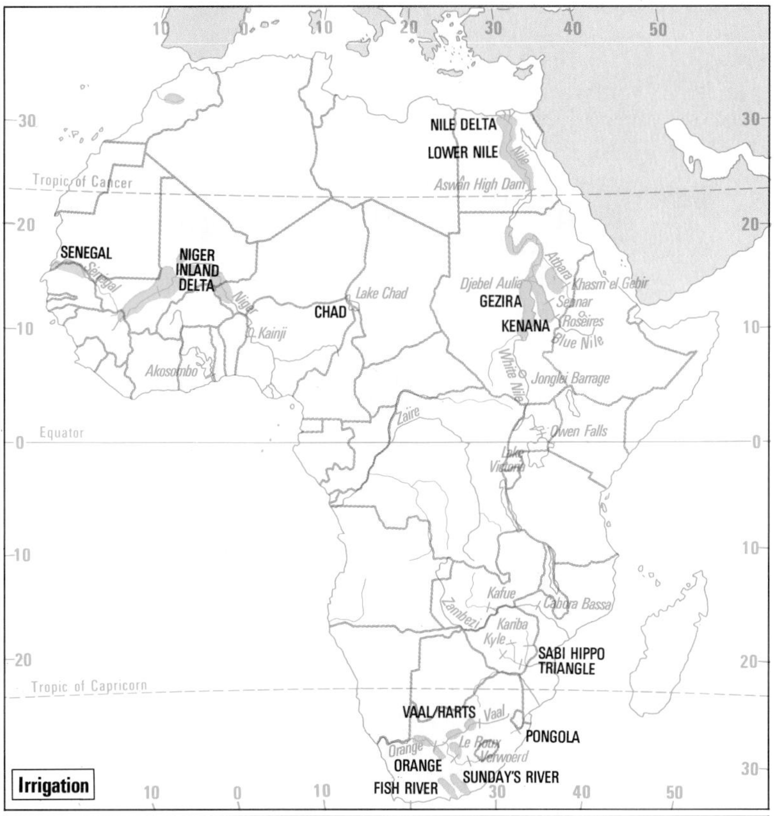

Irrigation

Commercial Economies

- Specialised cultivation – plantation and market gardening
- Mixed crop and livestock
- Extensive grazing and stock raising
- Timber production
- Fishing

Subsistence Economies

- Gathering, hunting, fishing and primitive cultivation
- Shifting and marginal cultivation
- Intensive crop cultivation
- Mixed crop and livestock
- Nomadic herding
- Little or no activity

- Major areas of irrigated land
- ORANGE Large scale irrigation schemes
- Dam
- Barrage

0 1000 2000 3000 kilometres

Total irrigated land area in Africa: 10 390 000 hectares

Egypt	2 474 000 hectares
Sudan	1 700 000 hectares
Nigeria	1 200 000 hectares
South Africa	1 128 000 hectares

Land Use (in millions of hectares)
Total land area – 2 966.5 million hectares

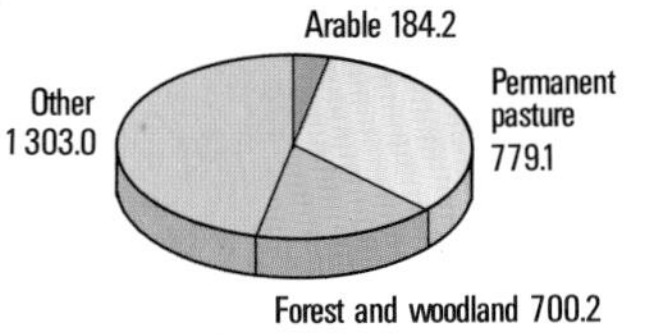

Livestock 1985 (in millions of head)

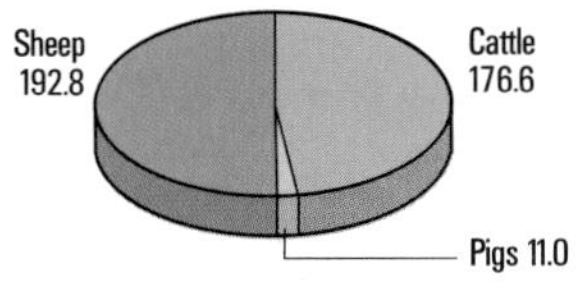

Crop Production 1985 (in millions of tonnes)

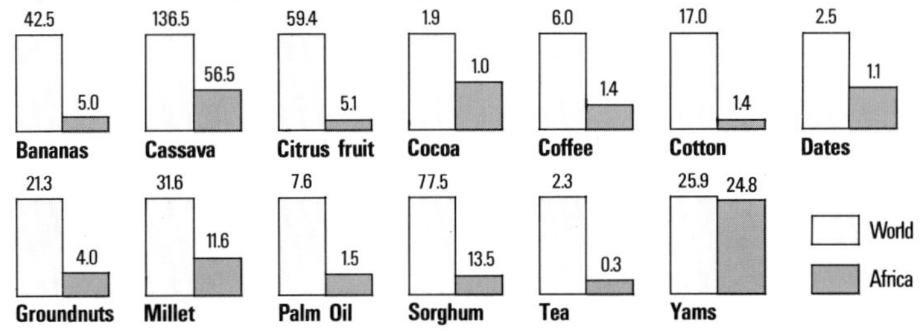

Selected Crops-Production by Country 1985

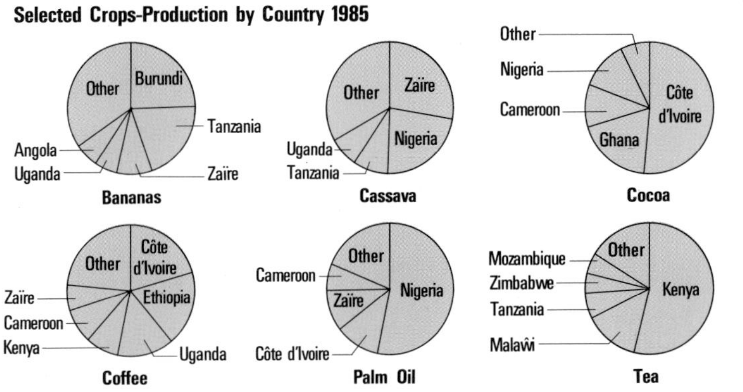

Selected Countries – Agriculture as a percentage of G.D.P. 1984

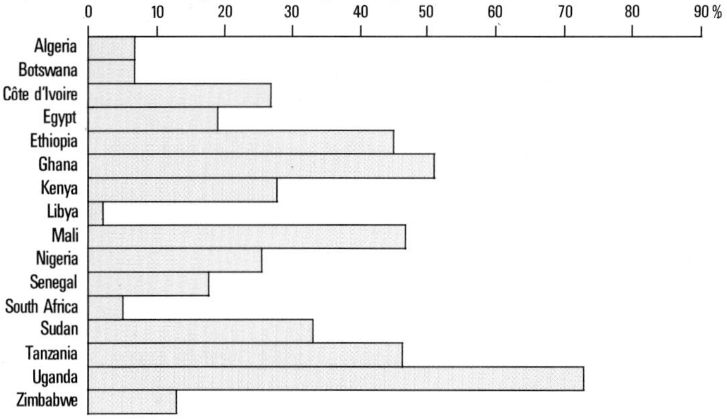

Selected Countries – Percentage of the economically active population employed in agriculture 1985

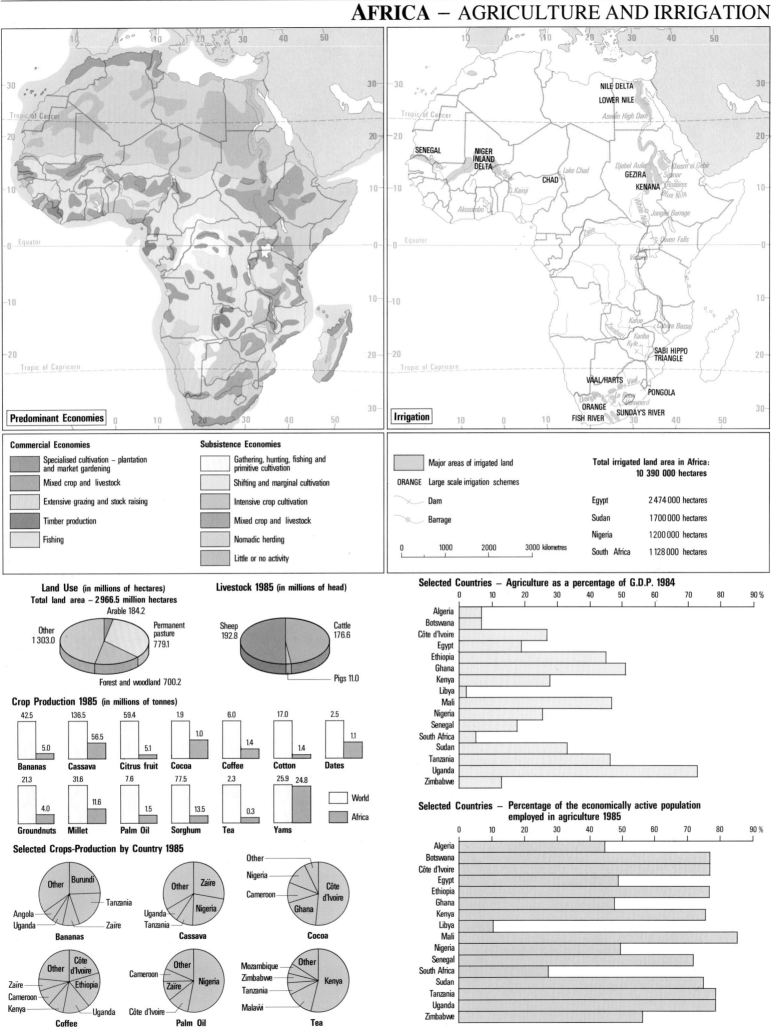

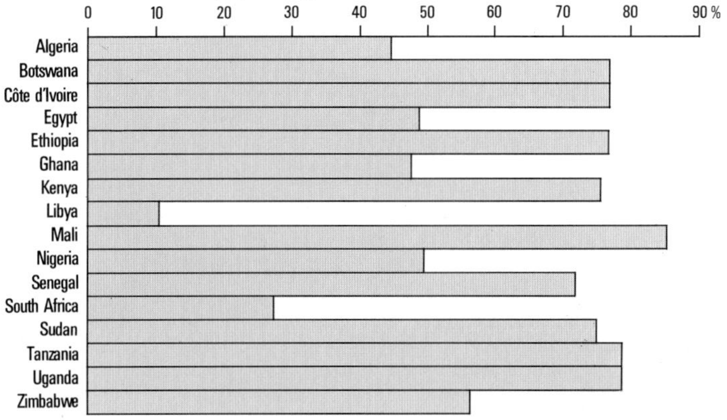

AFRICA – AGRICULTURE

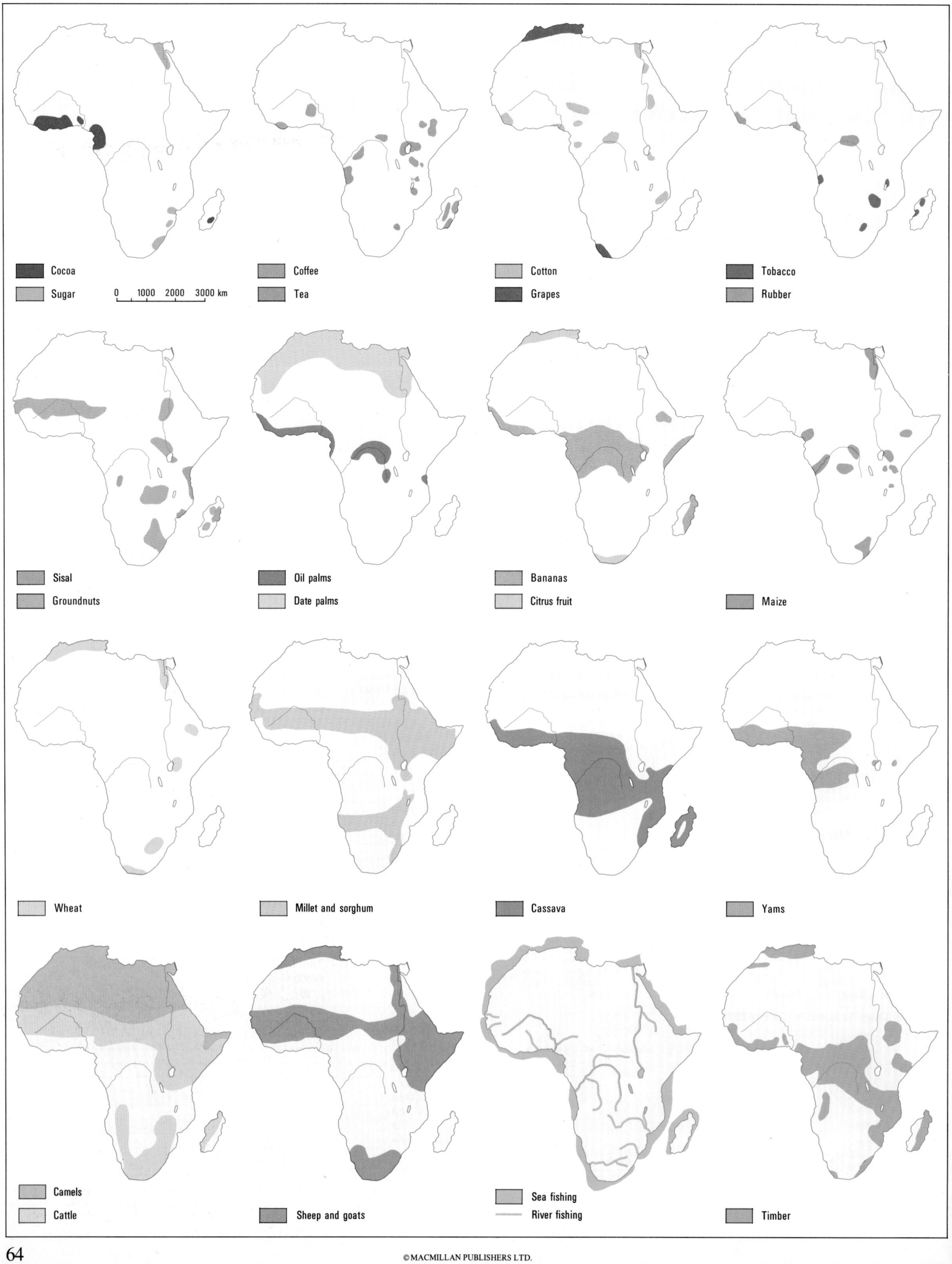
Cocoa
Sugar
0 1000 2000 3000 km
Coffee
Tea
Cotton
Grapes
Tobacco
Rubber
Sisal
Groundnuts
Oil palms
Date palms
Bananas
Citrus fruit
Maize
Wheat
Millet and sorghum
Cassava
Yams
Camels
Cattle
Sheep and goats
Sea fishing
River fishing
Timber

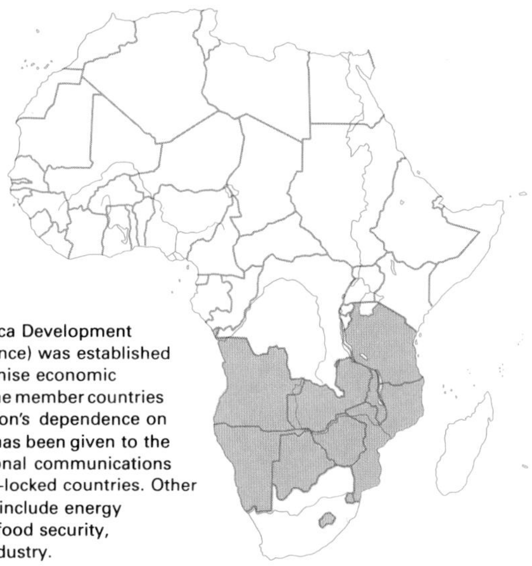

SADCC (Southern Africa Development Co-ordination Conference) was established in July 1979 to harmonise economic development among the member countries and to reduce the region's dependence on South Africa. Priority has been given to the improvement of a regional communications system to aid the land-locked countries. Other development projects include energy planning, agriculture, food security, disease control and industry.

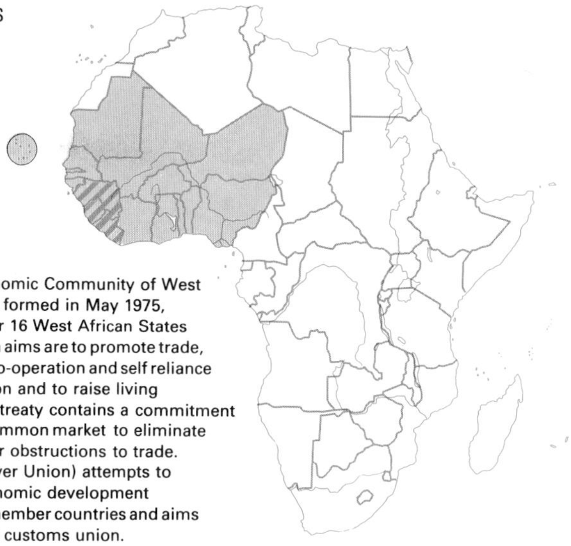

ECOWAS (Economic Community of West African States), formed in May 1975, groups together 16 West African States whose common aims are to promote trade, development, co-operation and self reliance within the region and to raise living standards. The treaty contains a commitment to establish a common market to eliminate tariffs and other obstructions to trade.
MRU (Mano River Union) attempts to harmonise economic development between the 3 member countries and aims to form its own customs union.

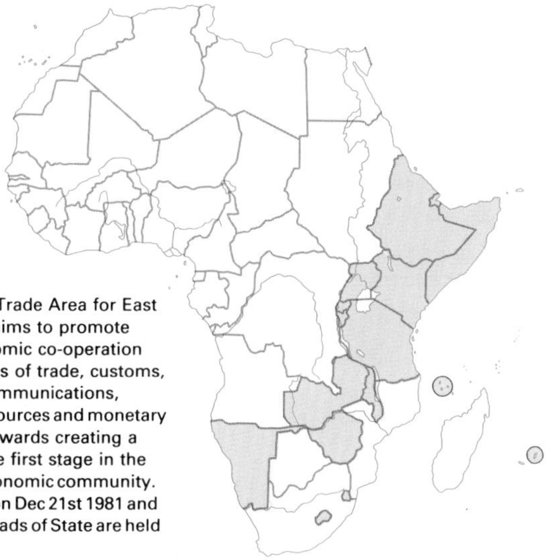

The PTA (Preferential Trade Area for East and Southern Africa) aims to promote commercial and economic co-operation particularly in the fields of trade, customs, industry, transport, communications, agriculture, natural resources and monetary affairs. It is directed towards creating a common market as the first stage in the establishment of an economic community. The treaty was signed on Dec 21st 1981 and summit meetings of Heads of State are held annually.

CEAO (Economic Community of West Africa) is an economic grouping of Francophone West African States, formed in January 1974, to replace the West African Customs Union (UDEAO). The community promotes trade, regional economic co-operation and integration, and aims to lower customs duties between members. A community development fund is financed by member states according to their respective share in manufacturing trade within the community.

UDEAC (Customs and Economic Union of Central Africa) was established in 1964 between countries in former French Equatorial Africa. Duty-free trade takes place between the member countries and a common external tariff is applied to imports from non member countries. The eventual aim is to create a common market with free movement of goods, people, services and capital.
CEPGL (Economic Community of the Great Lake Countries) groups Zaïre, Rwanda and Burundi together for common economic objectives.

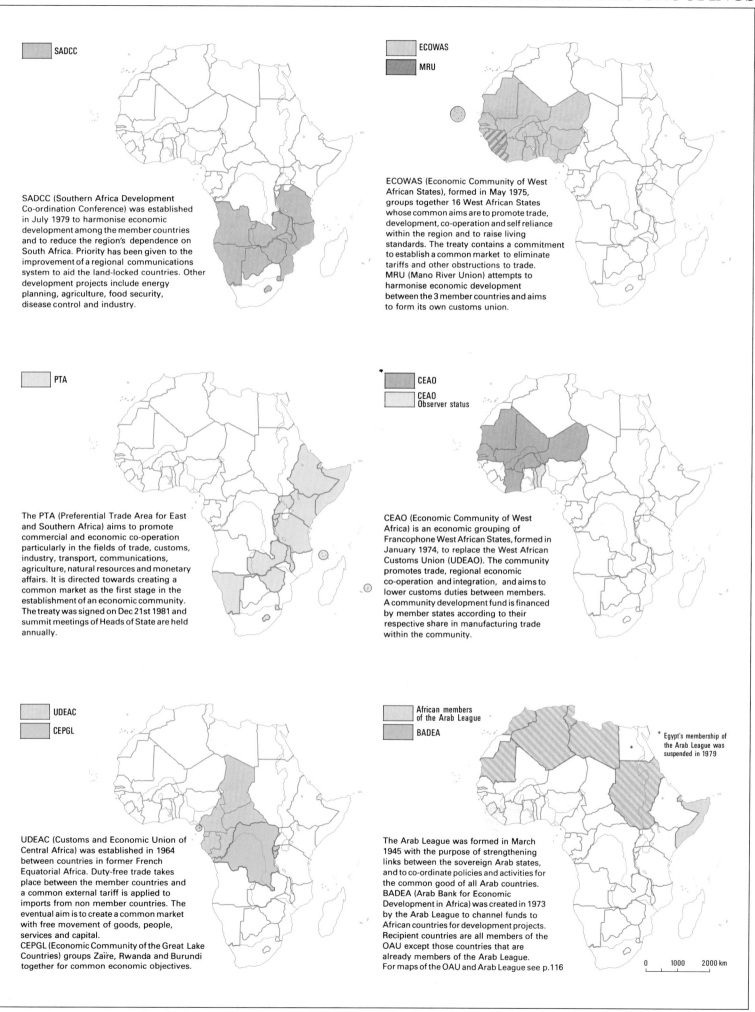

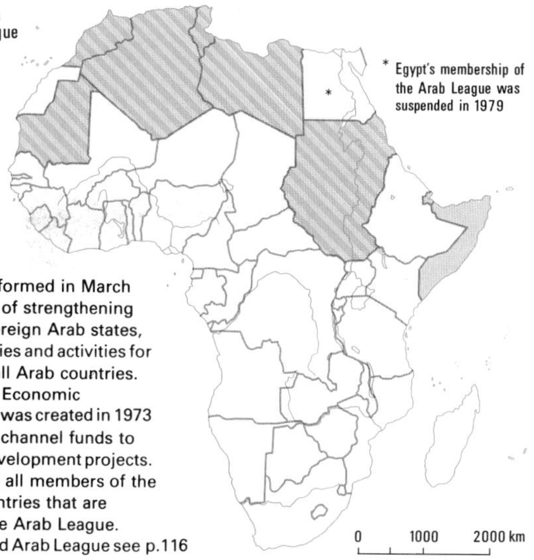

The Arab League was formed in March 1945 with the purpose of strengthening links between the sovereign Arab states, and to co-ordinate policies and activities for the common good of all Arab countries.
BADEA (Arab Bank for Economic Development in Africa) was created in 1973 by the Arab League to channel funds to African countries for development projects. Recipient countries are all members of the OAU except those countries that are already members of the Arab League.
For maps of the OAU and Arab League see p.116

AFRICA – MINERALS AND ENERGY

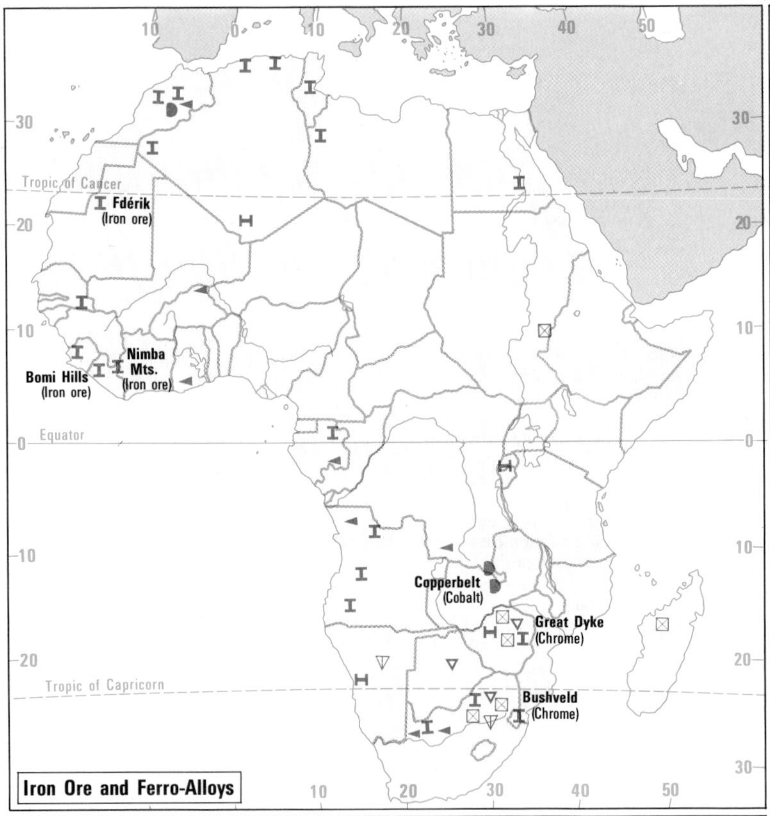

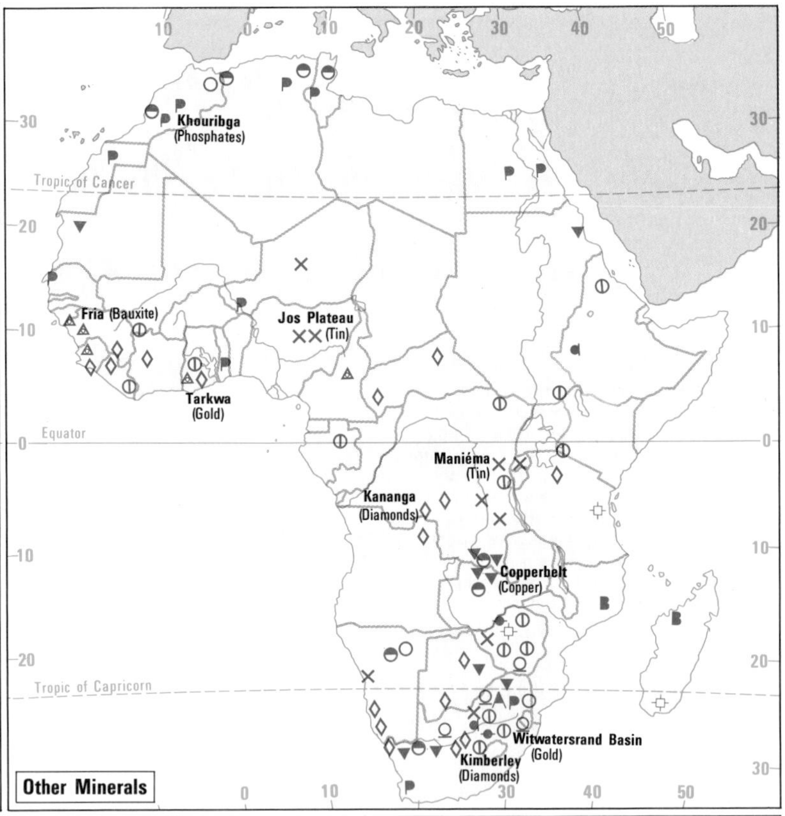

Chrome
Cobalt
Iron ore
Manganese
Nickel
Tungsten
Vanadium

0 1000 2000 3000 kilometres

Antimony
Asbestos
Bauxite
Beryllium
Copper
Diamonds
Gold
Lead and zinc
Mica
Phosphates
Platinum
Pyrites
Silver
Tin

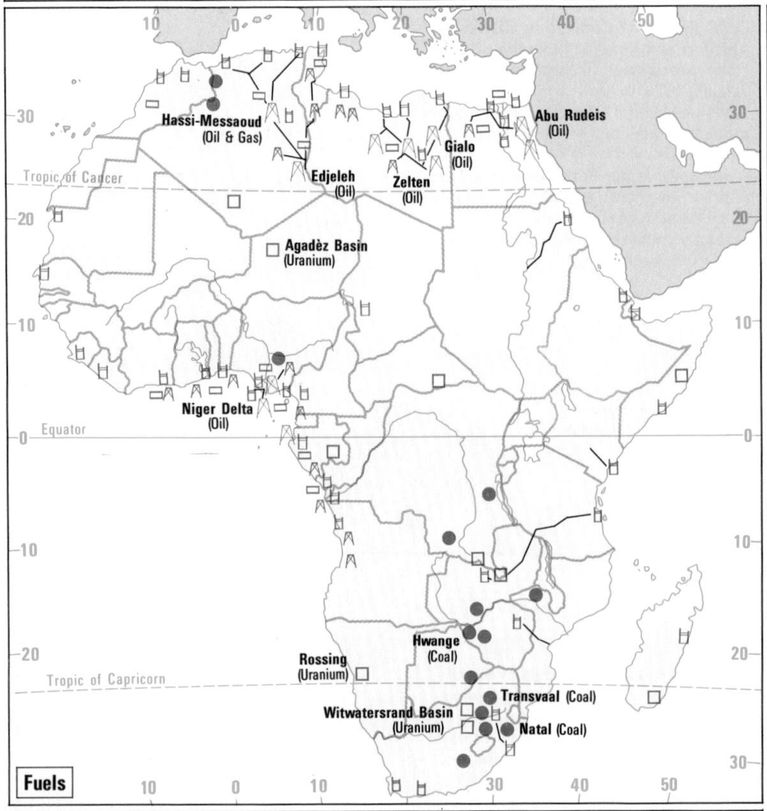

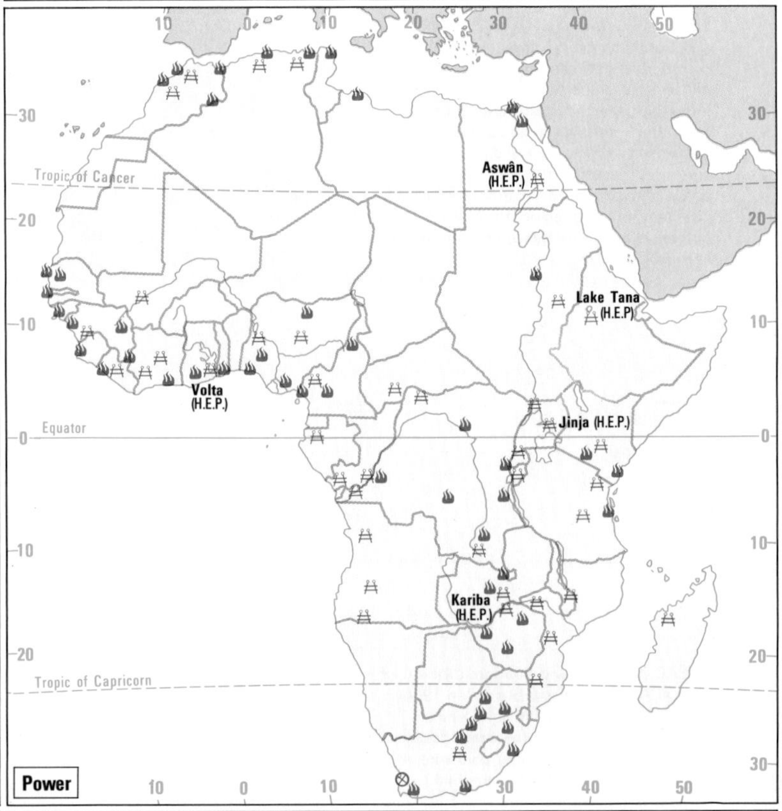

Oilfield
Oil refinery
Oil pipeline
Natural Gas
Coal
Uranium

Thermal power station
Hydro-electric power station
Nuclear power station

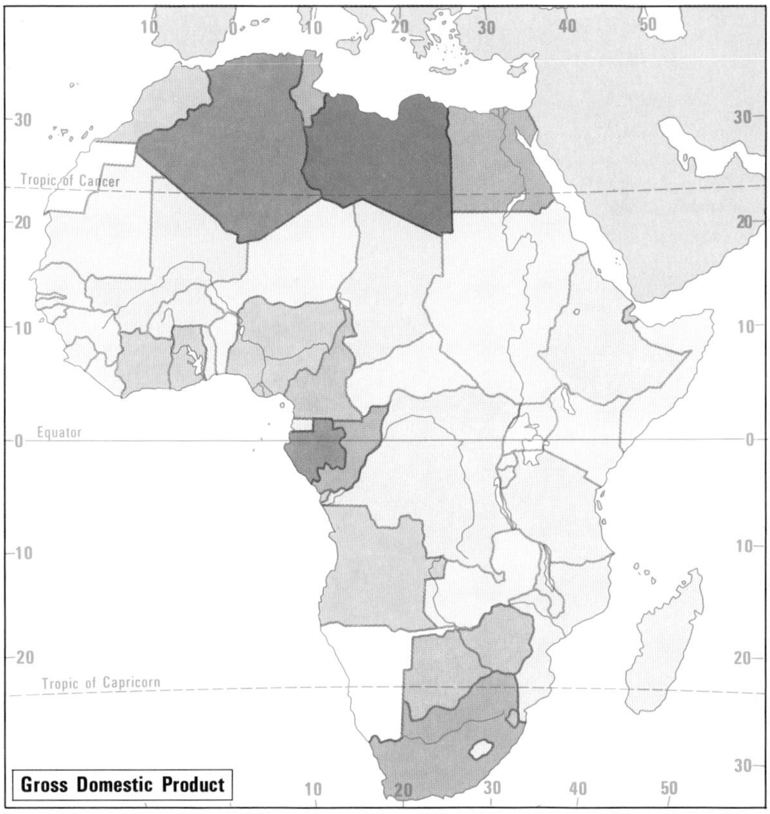

G.D.P. per capita 1984/85 (US dollars)

- More than 5000
- 2000–4999
- 1000–1999
- 750–999
- 500–749
- 250–499
- 0–249
- No data available

NB. Guinea-Bissau, Morocco, Uganda and Zimbabwe show 1983 statistics

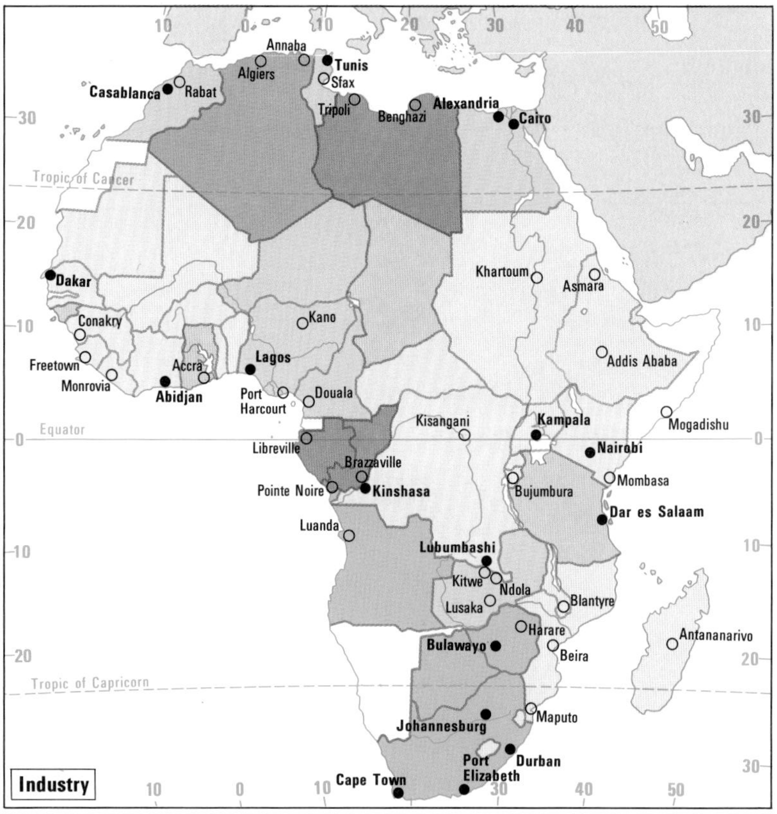

Industrial Activity (including mineral production)
as a percentage of G.D.P.

- More than 60%
- 50%–59%
- 40%–49%
- 30%–39%
- 20%–29%
- 10%–19%
- 0%–9%
- No data available

- ● Major industrial centre
- ○ Other industrial centre

Latest available statistics

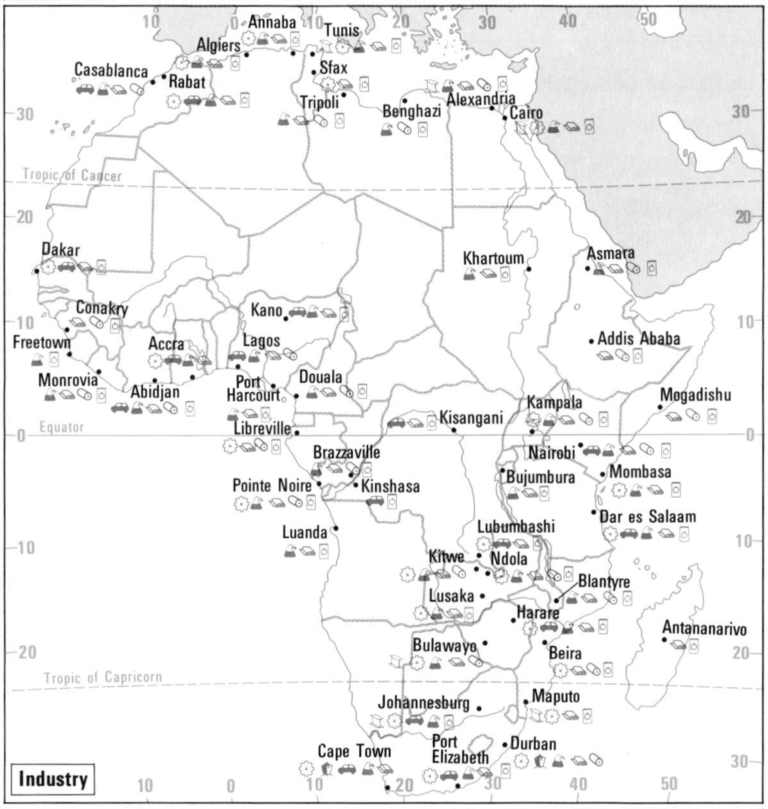

- Iron and steel
- Engineering
- Ship building
- Vehicles
- Chemicals
- Textiles
- Timber and paper
- Food processing

0 1000 2000 3000 kilometres

Exports by Commodity Group for Selected Countries 1982/83

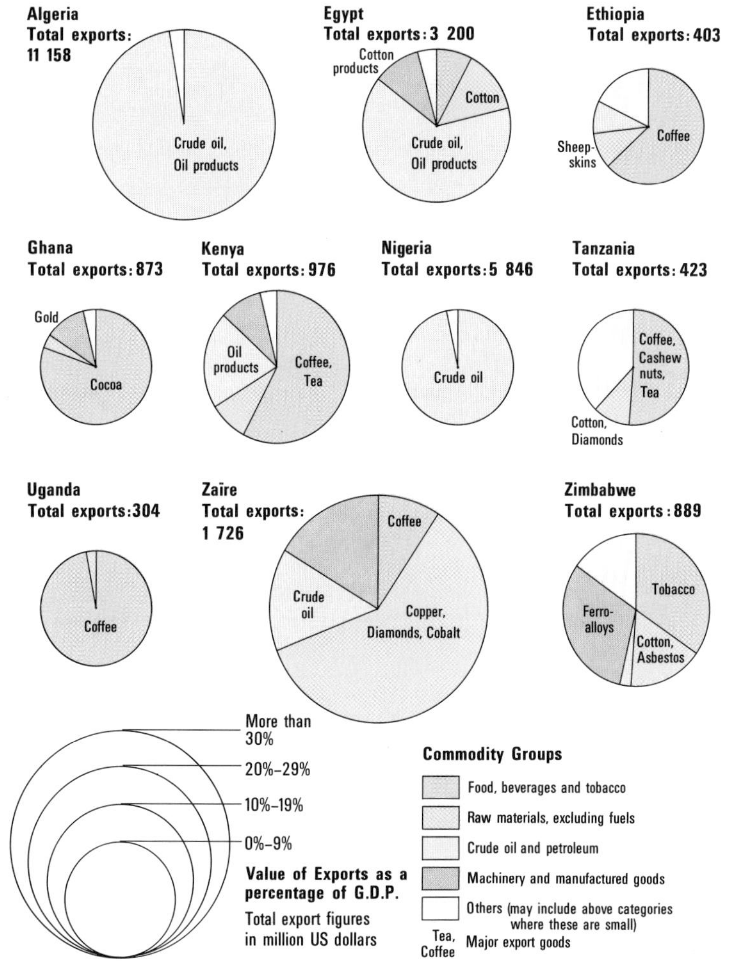

Commodity Groups

- Food, beverages and tobacco
- Raw materials, excluding fuels
- Crude oil and petroleum
- Machinery and manufactured goods
- Others (may include above categories where these are small)
- Tea, Coffee: Major export goods

AFRICA – TRANSPORT AND COMMUNICATIONS

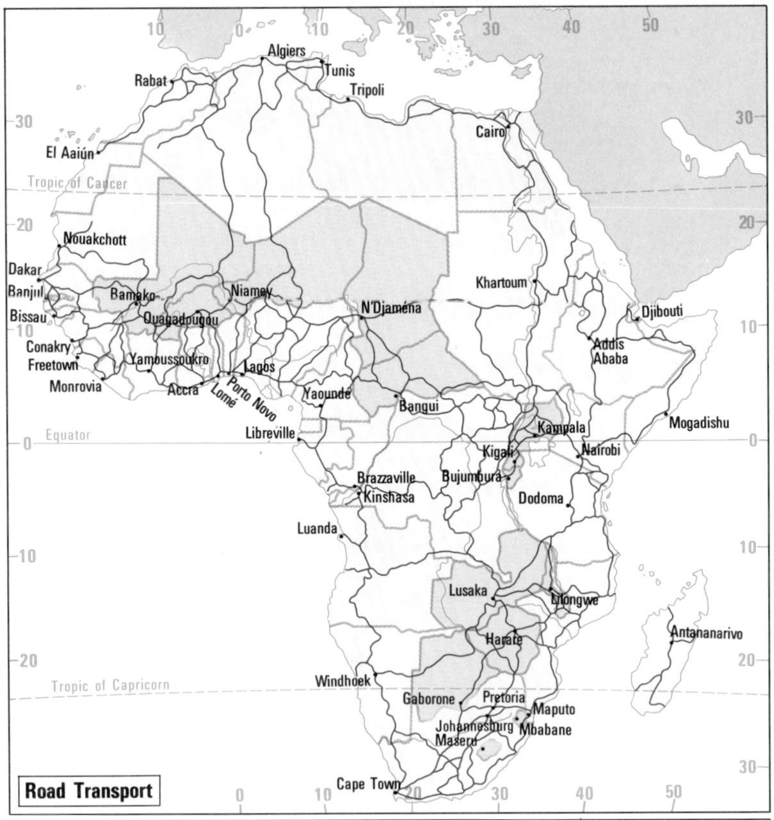

- Major road
- Landlocked country

0 1000 2000 3000 kilometres

- Railway
- Landlocked country
- Landlocked country without railways
- Major sea port
- Major sea route
- Navigable river (permanent)
- Navigable river (seasonal)
- Canal

- Major international airport
- Other international airport
- Landlocked country

International Seaborne Trade Passing Through Selected African Countries in 1984

Cargo in millions of tonnes

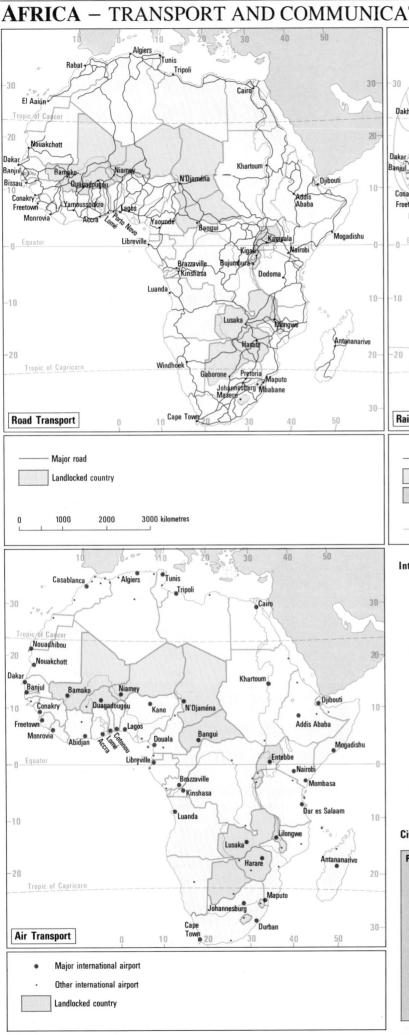

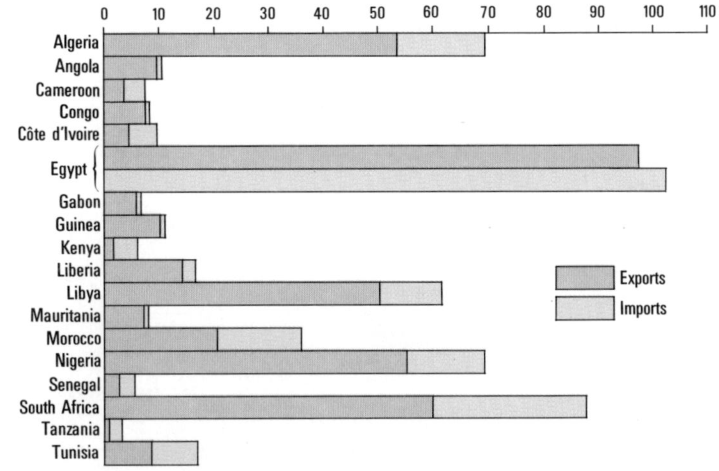

Civil Aviation – Scheduled Services in 1984

Passengers carried (in thousands)	
South Africa	4 461
Algeria	3 781
Egypt	2 786
Nigeria	2 375
Kenya	2 058
Libya	1 584
Tunisia	1 260
Tanzania	1 161
Morocco	1 046
Angola	690
Cameroon	677
Sudan	485
Zimbabwe	441

Freight carried (in million tonnes/km)	
South Africa	442.9
Egypt	88.2
Cameroon	55.5
Ethiopia	43.8
Kenya	42.1
Nigeria	40.1
Morocco	38.1
Zaire	32.0
Gabon	31.8
Zambia	25.3
Angola	25.1
Madagascar	22.1
Congo	19.0

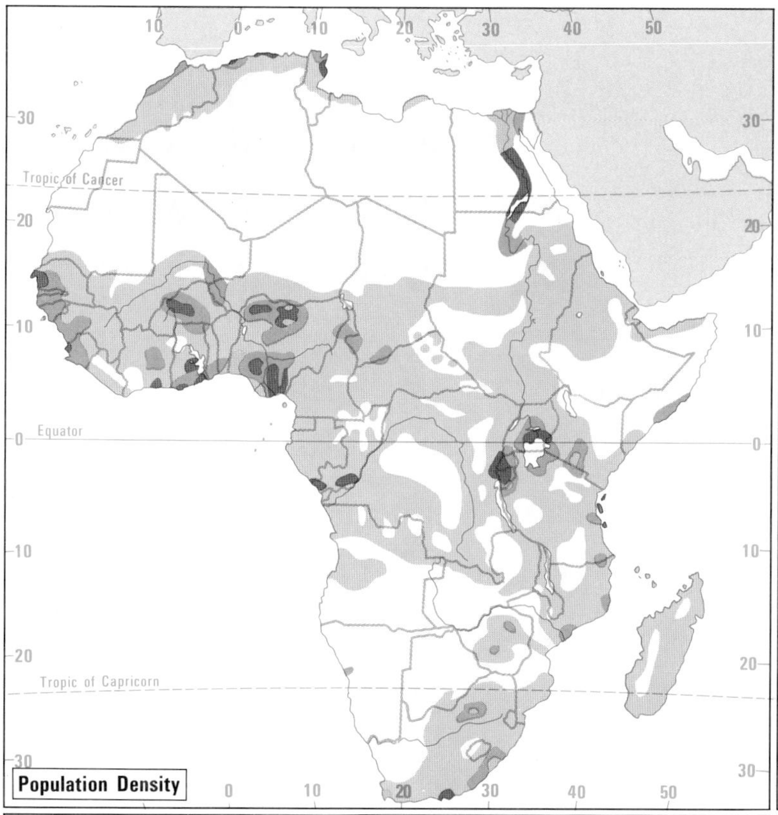

Density (persons per sq. km)
- Over 100
- 50–100
- 25–50
- 5–25
- Under 5

0 1000 2000 3000 kilometres

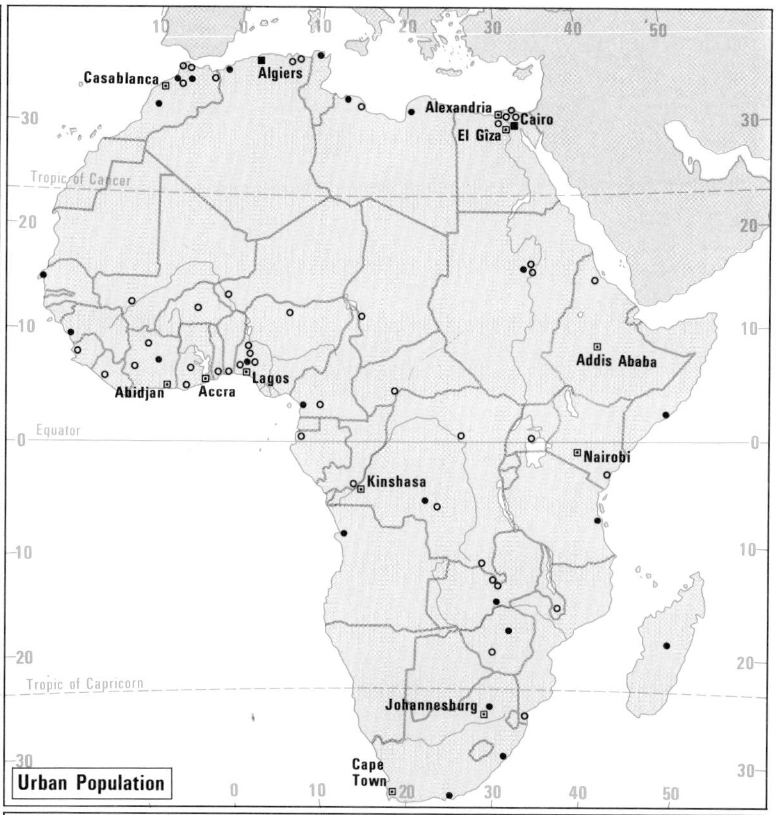

Population of Cities
- ■ Over 5 million
- ⊡ 1–5 million
- ● 500 000 – 1 million
- ○ 250 000 – 500 000

0 1000 2000 3000 kilometres

Population by Selected Country

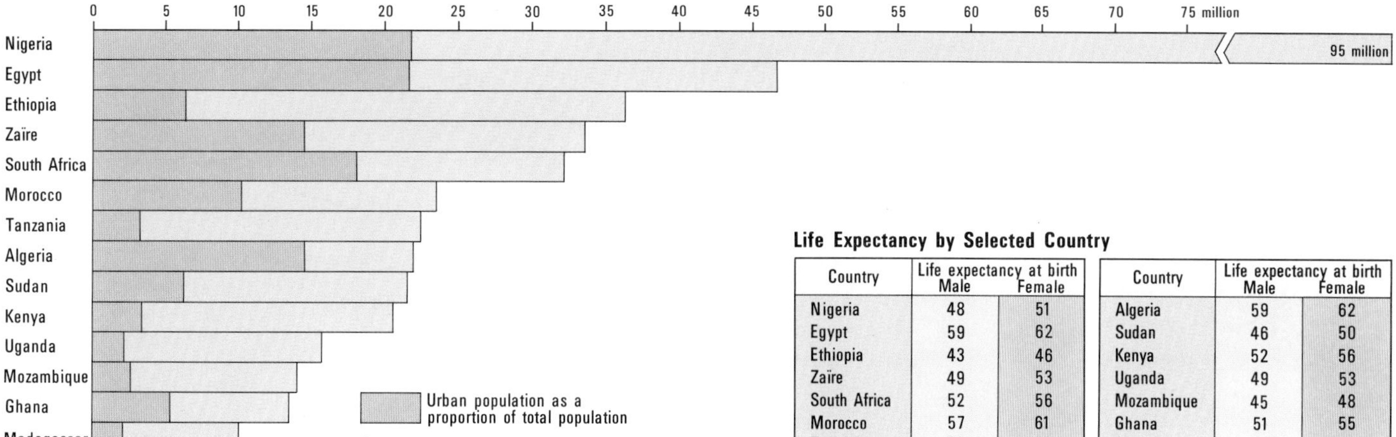

Source: U.N. Statistics 1985 (estimates)

Life Expectancy by Selected Country

Country	Life expectancy at birth Male	Female
Nigeria	48	51
Egypt	59	62
Ethiopia	43	46
Zaïre	49	53
South Africa	52	56
Morocco	57	61
Tanzania	50	53

Country	Life expectancy at birth Male	Female
Algeria	59	62
Sudan	46	50
Kenya	52	56
Uganda	49	53
Mozambique	45	48
Ghana	51	55
Madagascar	51	54

Percentage Population Breakdown by Age and Sex

Source: UN Demographic Yearbook

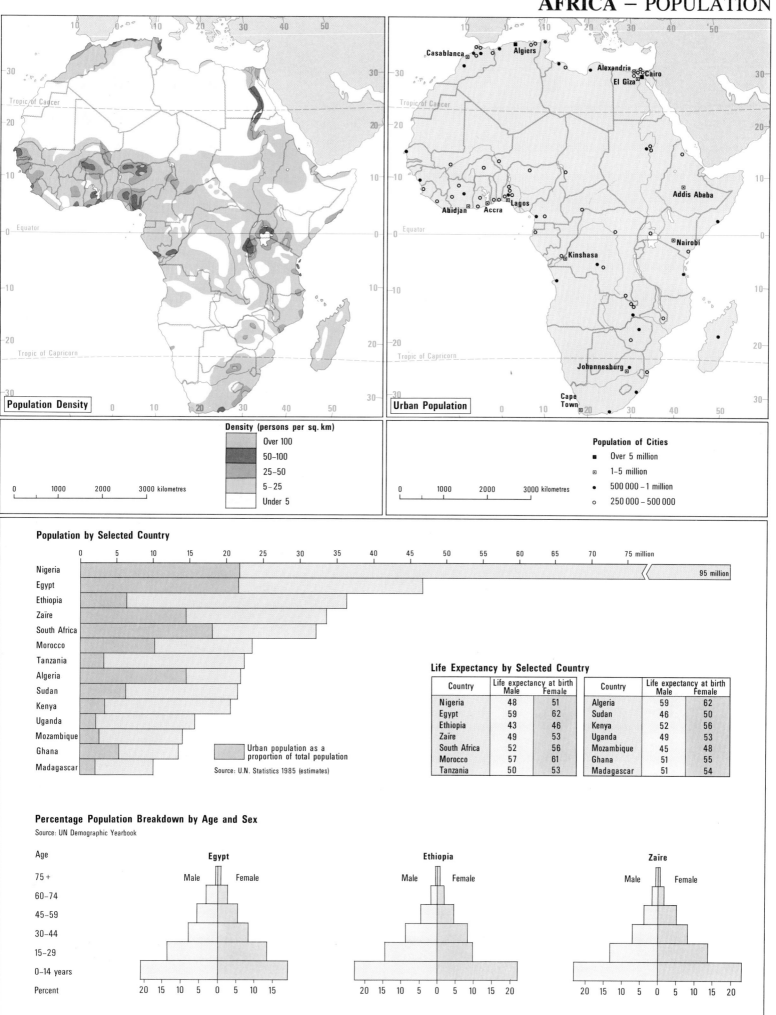

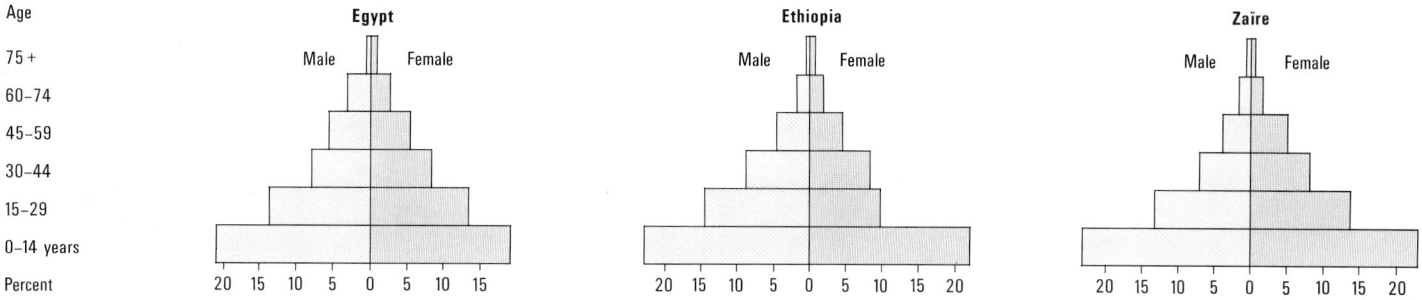

AFRICA – HISTORY

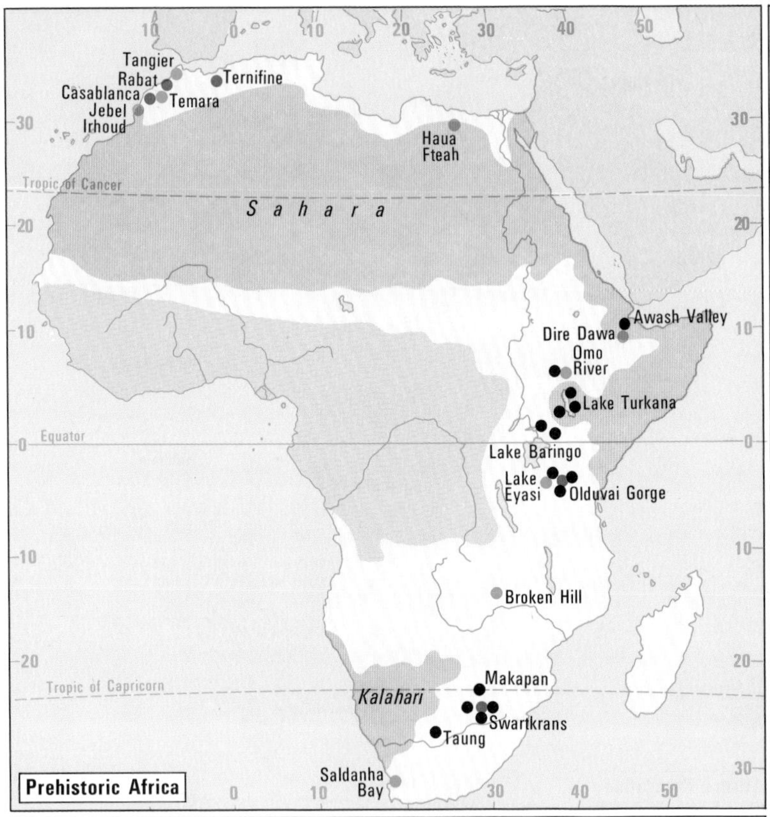

Prehistoric Africa

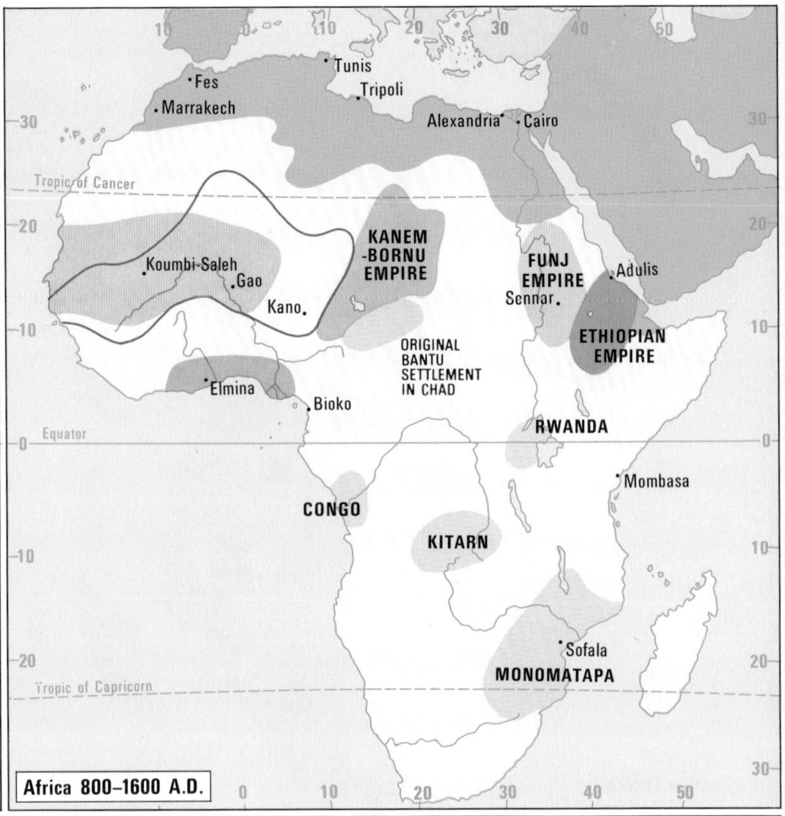

Africa 800–1600 A.D.

0 1000 2000 3000 kilometres

- Desert
- Forest
- ● Australopithecine sites
- ● Homo-erectus sites
- ● Homo-sapiens sites

- Kingdom of Ghana 800–1240 A.D.
- Kingdom of Mali 1240–1550 A.D.
- Songhai Kingdom 1460–1600 A.D.
- Kanem-Bornu Empire 11th–16th centuries
- Oyo and Benin in 13th century
- Extent of the Islamic World in 800 A.D.
- • Important town
- Bantu states
- Extent of the Funj Empire
- Extent of the Ethiopian Empire

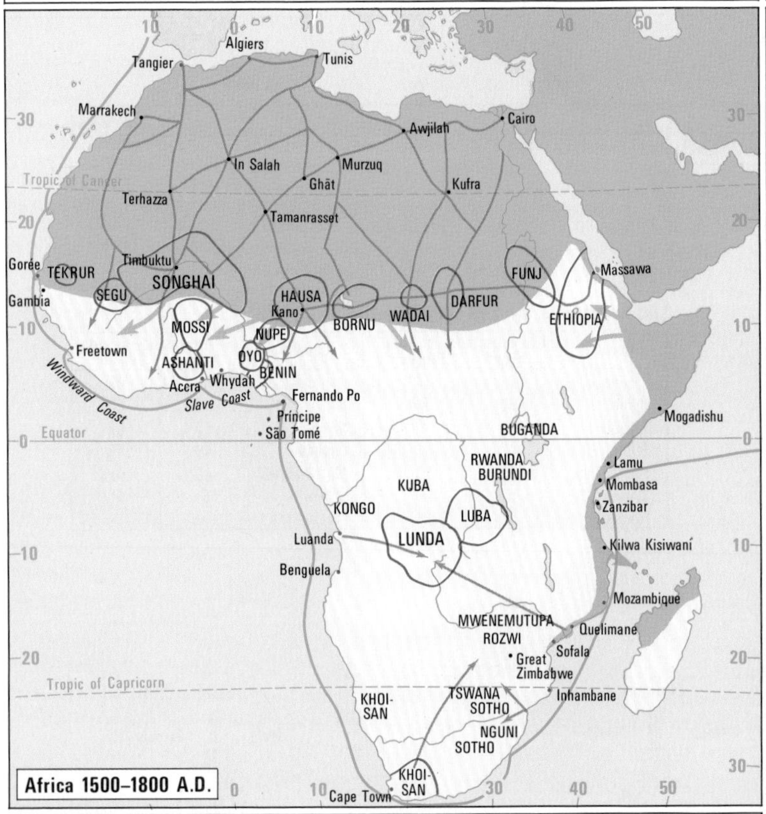

Africa 1500–1800 A.D.

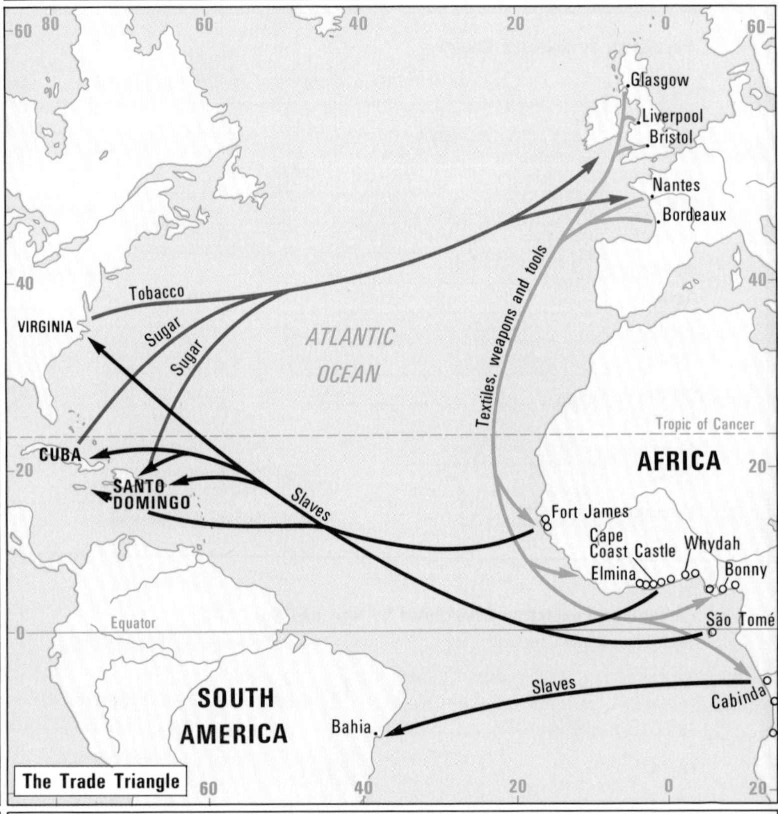

The Trade Triangle

- African Kingdom or State
- Extent of the Islamic world in 1800 A.D.
- Spread of Islam
- Major trade route
- Portuguese sea route
- • European coastal settlement
- • Other settlement

0 1000 2000 kilometres

- ○ Principal Slave trading port

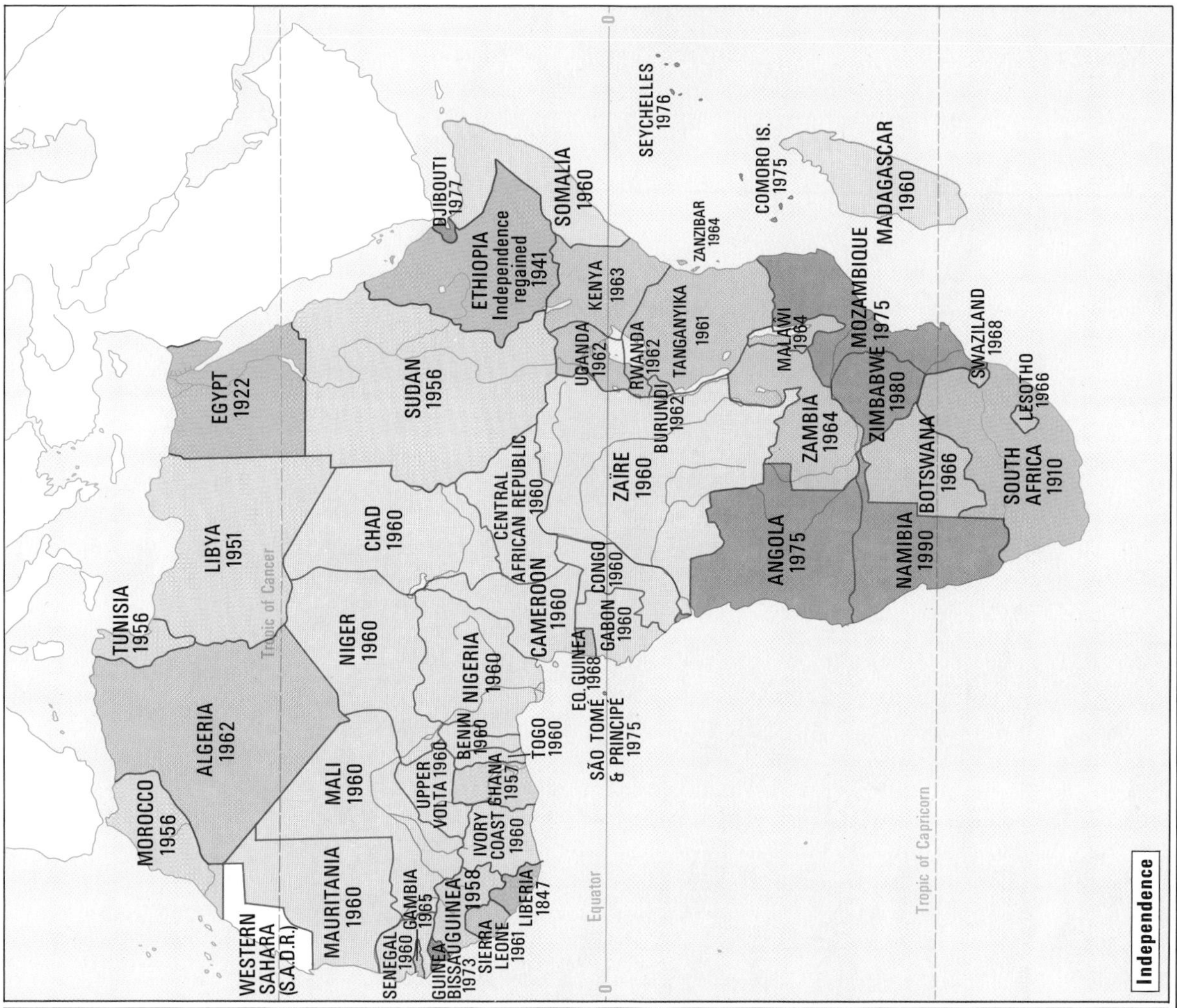
Independence
MOROCCO 1956
ALGERIA 1962
TUNISIA 1956
LIBYA 1951
EGYPT 1922
WESTERN SAHARA (S.A.D.R.)
MAURITANIA 1960
MALI 1960
NIGER 1960
CHAD 1960
SUDAN 1956
ETHIOPIA Independence regained 1941
DJIBOUTI 1977
SOMALIA 1960
SENEGAL 1960
GAMBIA 1965
GUINEA BISSAU 1973
GUINEA 1958
SIERRA LEONE 1961
LIBERIA 1847
IVORY COAST 1960
UPPER VOLTA 1960
GHANA 1957
TOGO 1960
BENIN 1960
NIGERIA 1960
CAMEROON 1960
EQ. GUINEA 1968
SÃO TOMÉ & PRINCIPE 1975
GABON 1960
CONGO 1960
CENTRAL AFRICAN REPUBLIC 1960
ZAÏRE 1960
UGANDA 1962
KENYA 1963
RWANDA 1962
BURUNDI 1962
TANGANYIKA 1961
ZANZIBAR 1964
SEYCHELLES 1976
COMORO IS. 1975
MADAGASCAR 1960
MALAWI 1964
MOZAMBIQUE 1975
ZAMBIA 1964
ZIMBABWE 1980
ANGOLA 1975
NAMIBIA 1990
BOTSWANA 1966
SWAZILAND 1968
LESOTHO 1966
SOUTH AFRICA 1910
Tropic of Cancer
Equator
Tropic of Capricorn

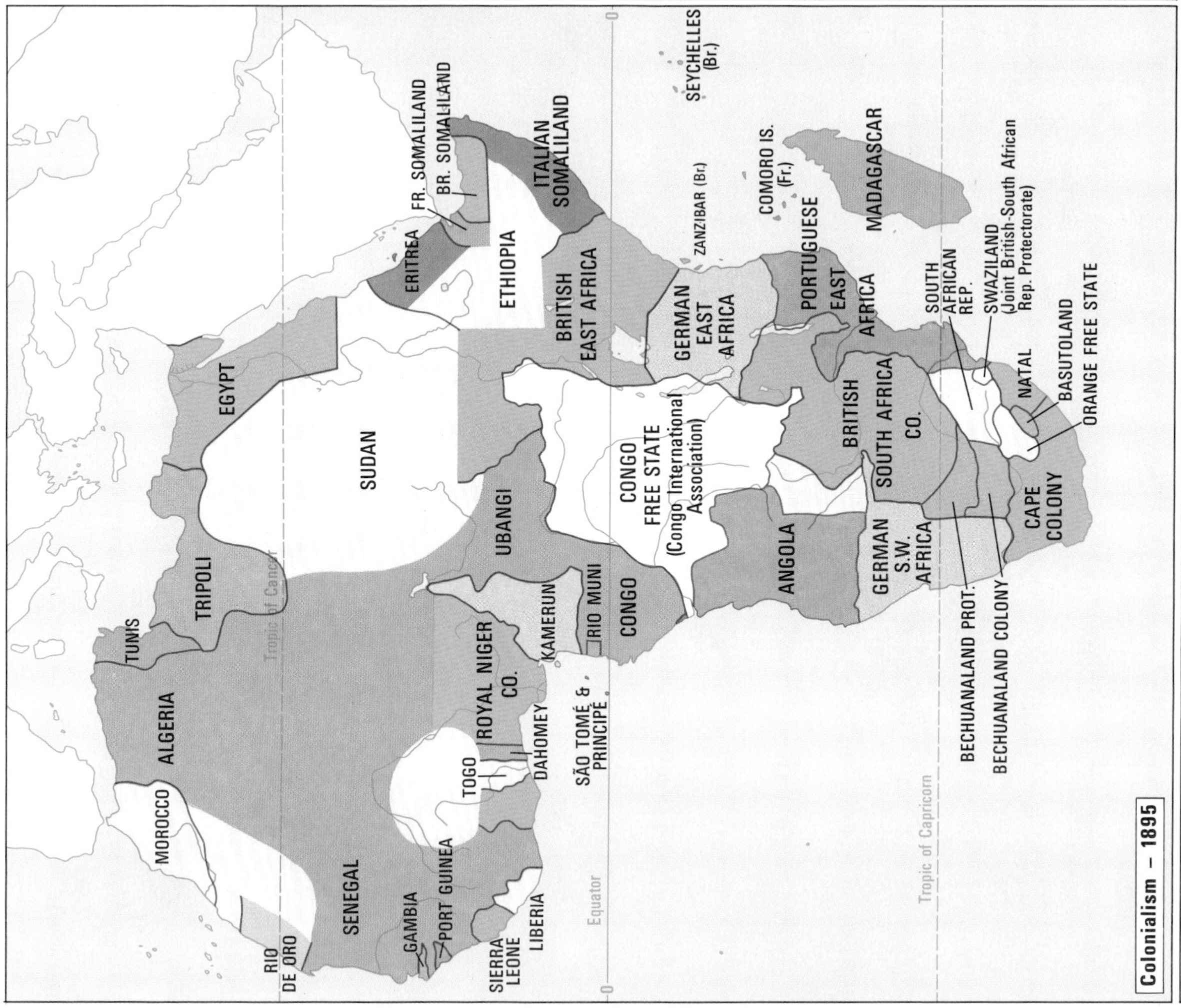
Colonialism – 1895
RIO DE ORO
MOROCCO
ALGERIA
TUNIS
TRIPOLI
EGYPT
SUDAN
ERITREA
FR. SOMALILAND
BR. SOMALILAND
ITALIAN SOMALILAND
ETHIOPIA
SENEGAL
GAMBIA
PORT GUINEA
SIERRA LEONE
LIBERIA
TOGO
DAHOMEY
ROYAL NIGER CO.
KAMERUN
RIO MUNI
SÃO TOMÉ & PRINCIPE
CONGO
UBANGI
CONGO FREE STATE (Congo International Association)
BRITISH EAST AFRICA
GERMAN EAST AFRICA
ZANZIBAR (Br.)
SEYCHELLES (Br.)
COMORO IS. (Fr.)
MADAGASCAR
PORTUGUESE EAST AFRICA
BRITISH SOUTH AFRICA CO.
ANGOLA
GERMAN S.W. AFRICA
BECHUANALAND PROT.
BECHUANALAND COLONY
SOUTH AFRICAN REP.
SWAZILAND (Joint British-South African Rep. Protectorate)
NATAL
BASUTOLAND
ORANGE FREE STATE
CAPE COLONY
Tropic of Cancer
Equator
Tropic of Capricorn

Colonial Powers

- Great Britain
- France
- Spain
- Portugal
- Germany
- Turkey
- Italy
- Independent

Independence

- Before 1945
- 1950 – 59
- 1960
- 1961 – 70
- After 1970

0 — 1000 — 2000 km

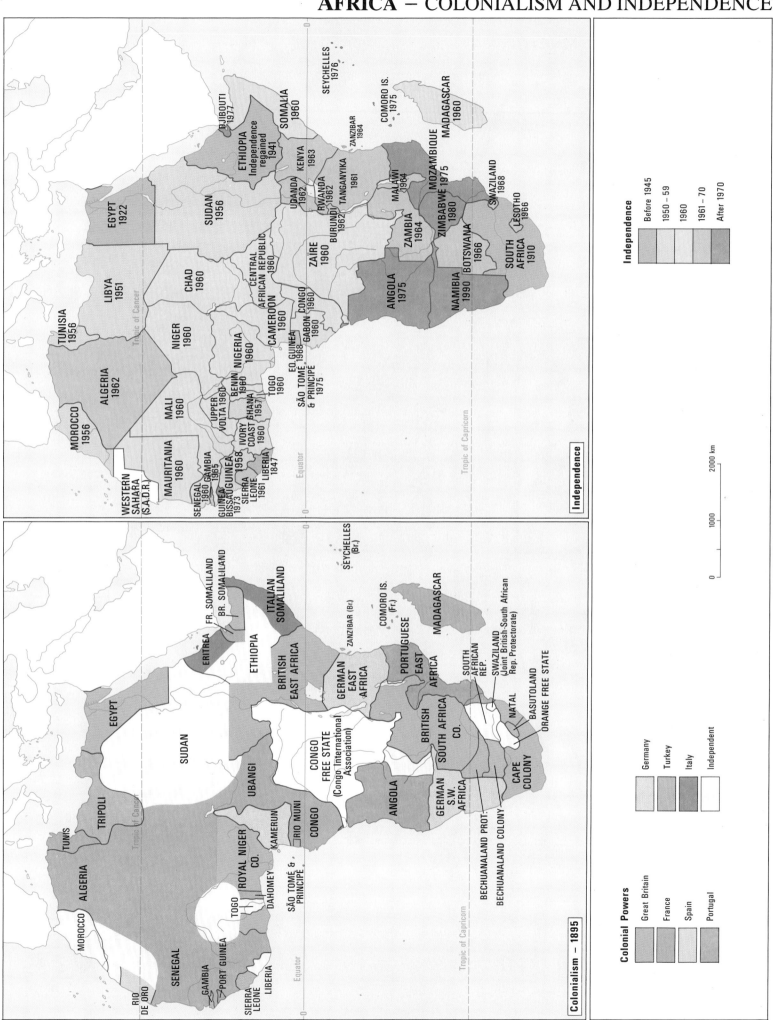

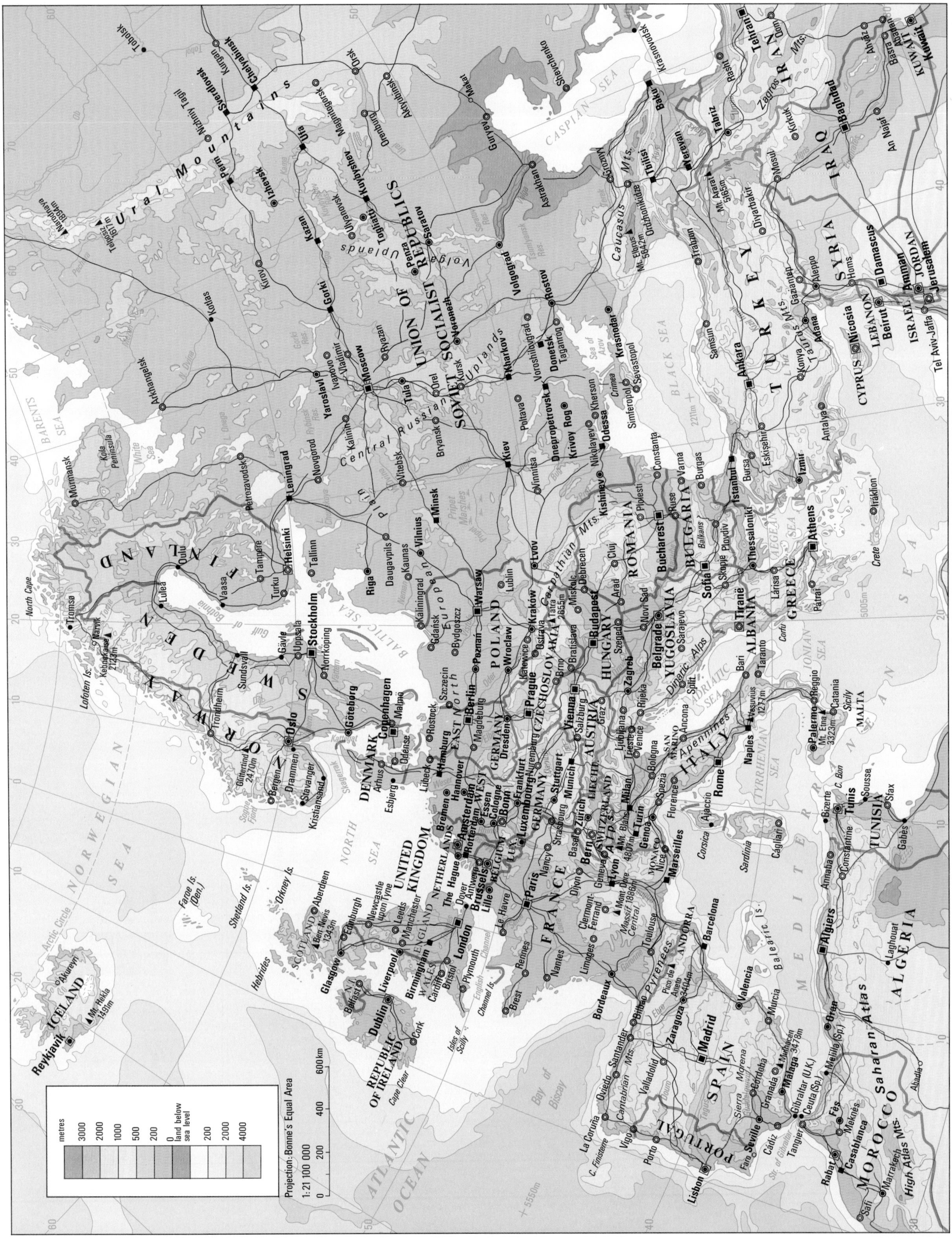
metres
3000
2000
1000
500
200
0
land below sea level
200
2000
4000
Projection: Bonne's Equal Area
1:21 100 000
0 200 400 600 km
UNION OF SOVIET SOCIALIST REPUBLICS
FINLAND
SWEDEN
NORWAY
DENMARK
UNITED KINGDOM
REPUBLIC OF IRELAND
ICELAND
NETHERLANDS
BELGIUM
LUX.
FRANCE
WEST GERMANY
EAST GERMANY
POLAND
CZECHOSLOVAKIA
AUSTRIA
SWITZERLAND
LIECHT.
HUNGARY
ROMANIA
BULGARIA
YUGOSLAVIA
ALBANIA
GREECE
ITALY
SAN MARINO
MONACO
ANDORRA
SPAIN
PORTUGAL
MOROCCO
ALGERIA
TUNISIA
MALTA
TURKEY
CYPRUS
SYRIA
LEBANON
ISRAEL
JORDAN
IRAQ
IRAN
KUWAIT
NORWEGIAN SEA
NORTH SEA
BALTIC SEA
BARENTS SEA
White Sea
BLACK SEA
CASPIAN SEA
Sea of Azov
AEGEAN SEA
IONIAN SEA
ADRIATIC SEA
TYRRHENIAN SEA
MEDITERRANEAN SEA
ATLANTIC OCEAN
Bay of Biscay
Ural Mountains
Caucasus Mts.
Carpathian Mts.
Alps
Pyrenees
Apennines
Balkans
Dinaric Alps

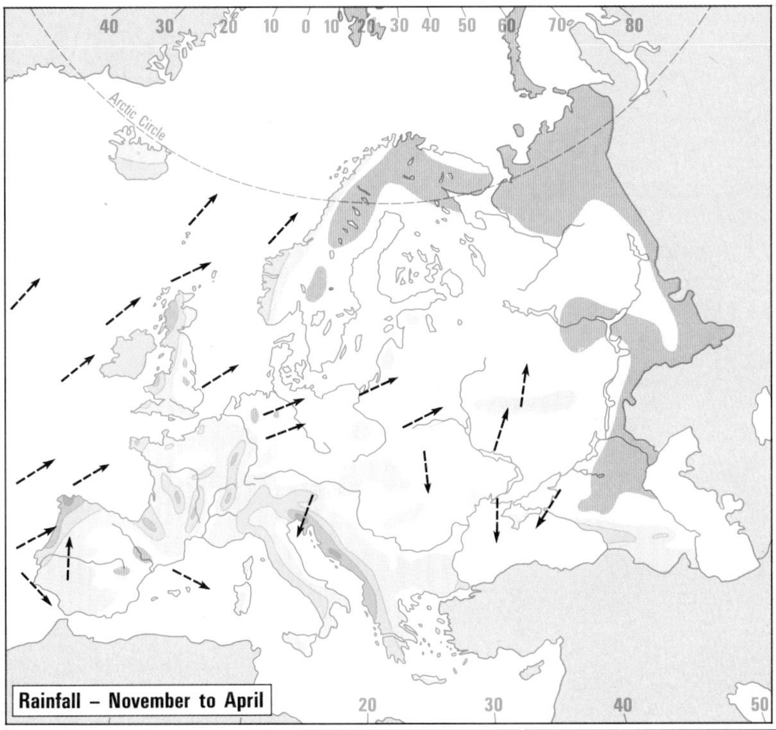

Rainfall – November to April

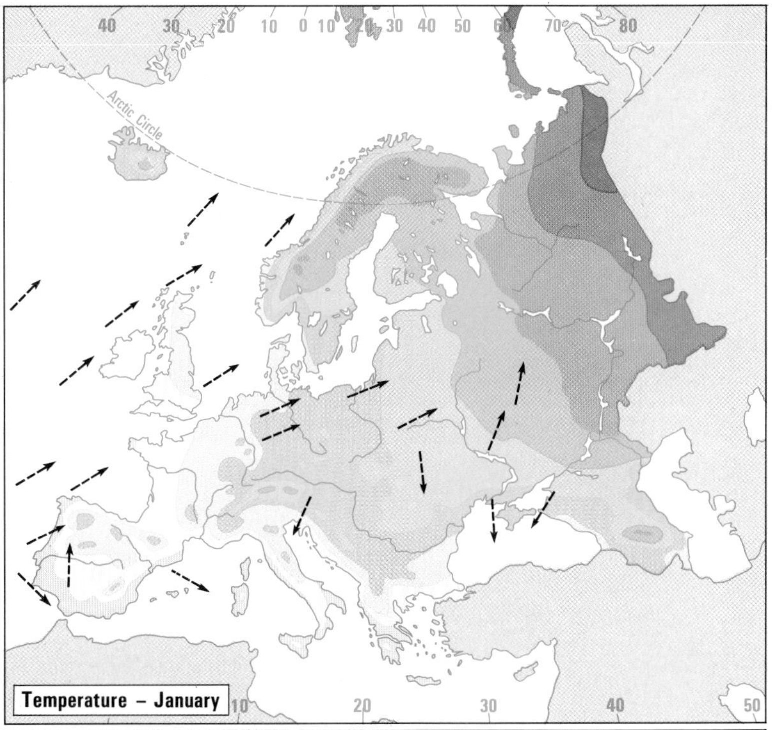

Temperature – January

Rainfall in millimetres

- 1000–1500
- 750–1000
- 500–750
- 250–500
- 125–250
- 0–125

0 500 1000 1500 kilometres

Prevailing winds

- November to April
- May to October

Surface temperature °C

28
24
20
16
12
8
4
0
−4
−8
−12
−16
below −20

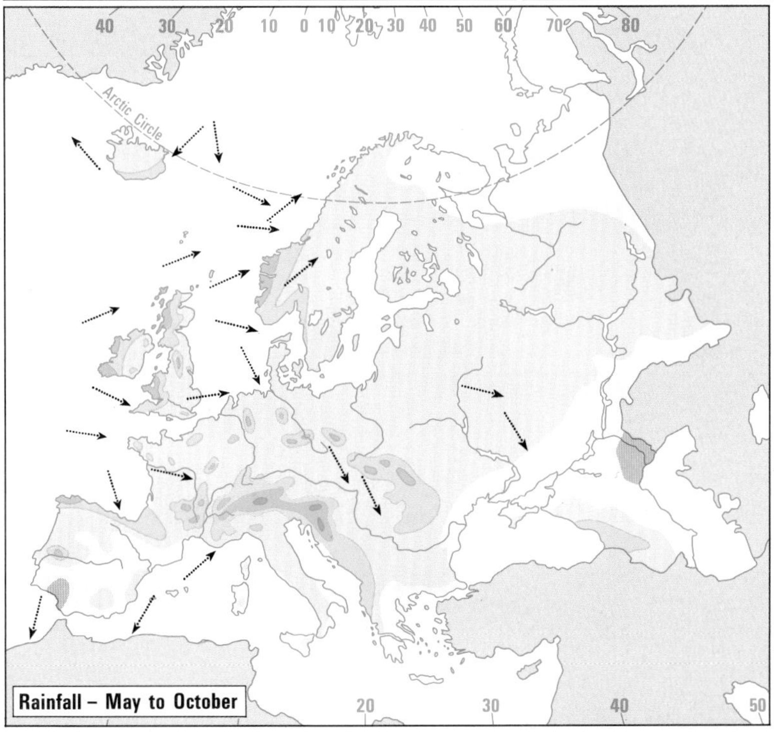

Rainfall – May to October

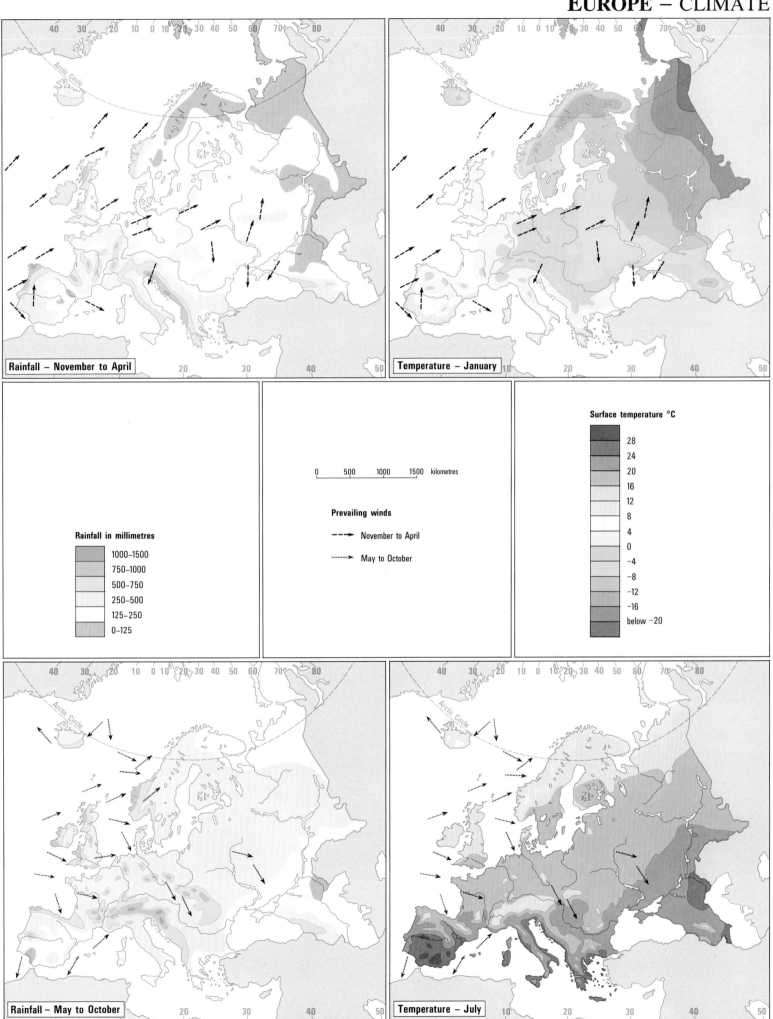

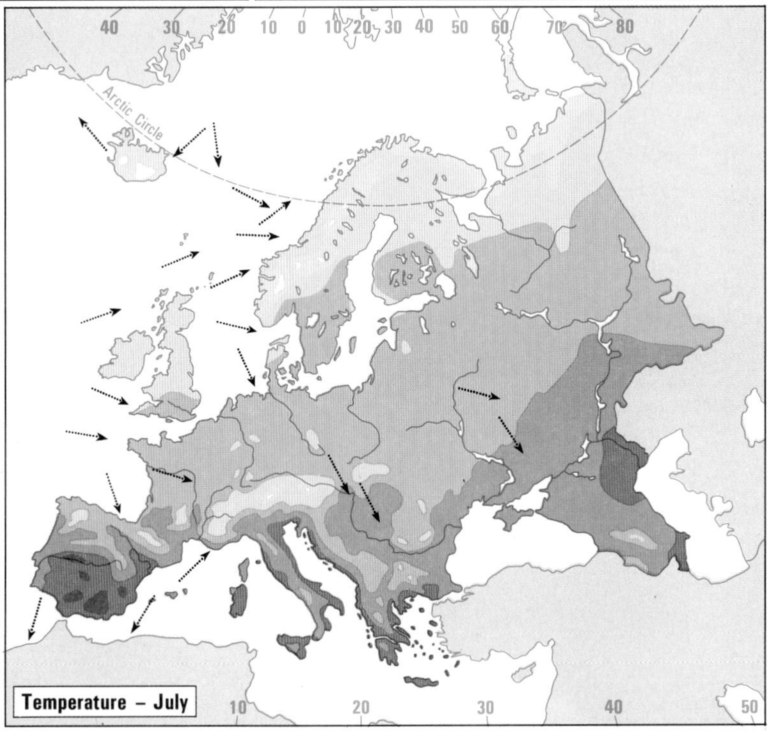

Temperature – July

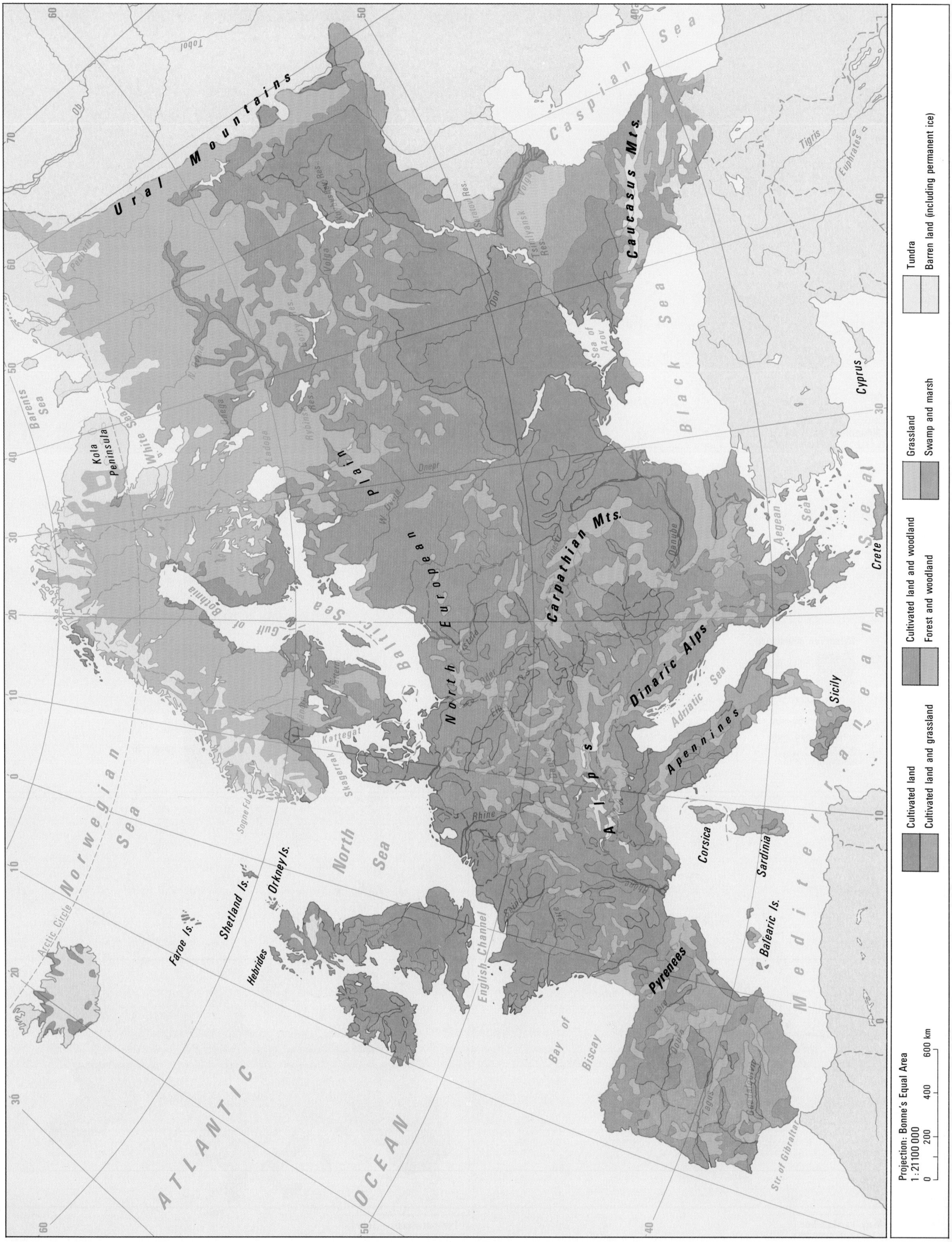
Ural Mountains
Caucasus Mts.
Caspian Sea
Black Sea
Sea of Azov
Cyprus
Crete
Aegean Sea
Carpathian Mts.
North European Plain
Dinaric Alps
Adriatic Sea
Sicily
Apennines
Alps
Corsica
Sardinia
Balearic Is.
Mediterranean Sea
Pyrenees
Bay of Biscay
English Channel
North Sea
Orkney Is.
Shetland Is.
Faroe Is.
Hebrides
Norwegian Sea
Arctic Circle
ATLANTIC OCEAN
Baltic Sea
Gulf of Bothnia
Kattegat
Skagerrak
Sogne Fd.
Kola Peninsula
White Sea
Barents Sea
Ladoga
Onega
Rybinsk Res.
Gorky Res.
Kuybyshev Res.
Saratov Res.
Tsimlyansk Res.
Volga
Don
Dnepr
W. Dvina
Danube
Rhine
Elbe
Oder
Vistula
Loire
Seine
Rhône
Ebro
Tagus
Guadalquivir
Str. of Gibraltar
Pechora
Ob
Tobol
Tigris
Euphrates
Cultivated land
Cultivated land and grassland
Cultivated land and woodland
Forest and woodland
Grassland
Swamp and marsh
Tundra
Barren land (including permanent ice)
Projection: Bonne's Equal Area
1: 21 100 000
0 200 400 600 km

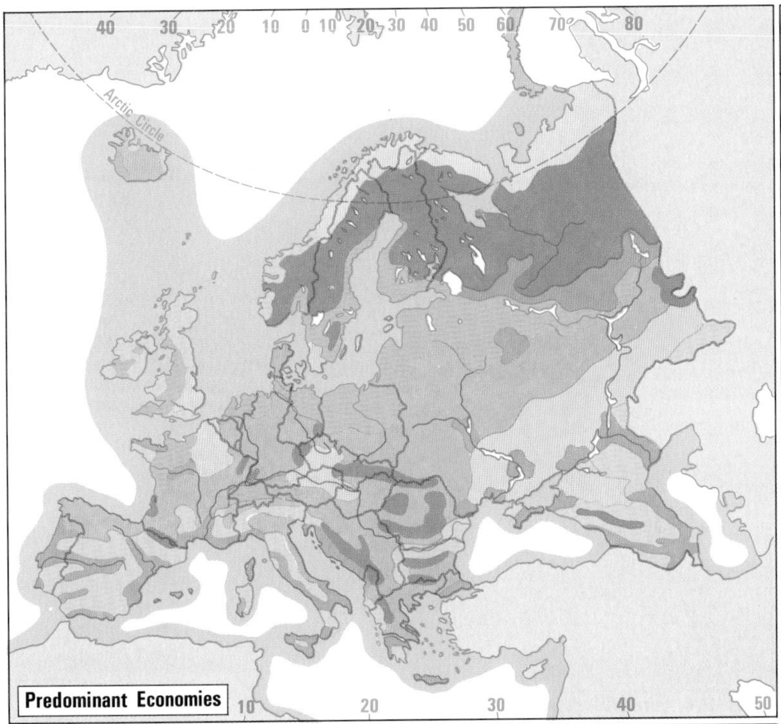

Commercial Economies

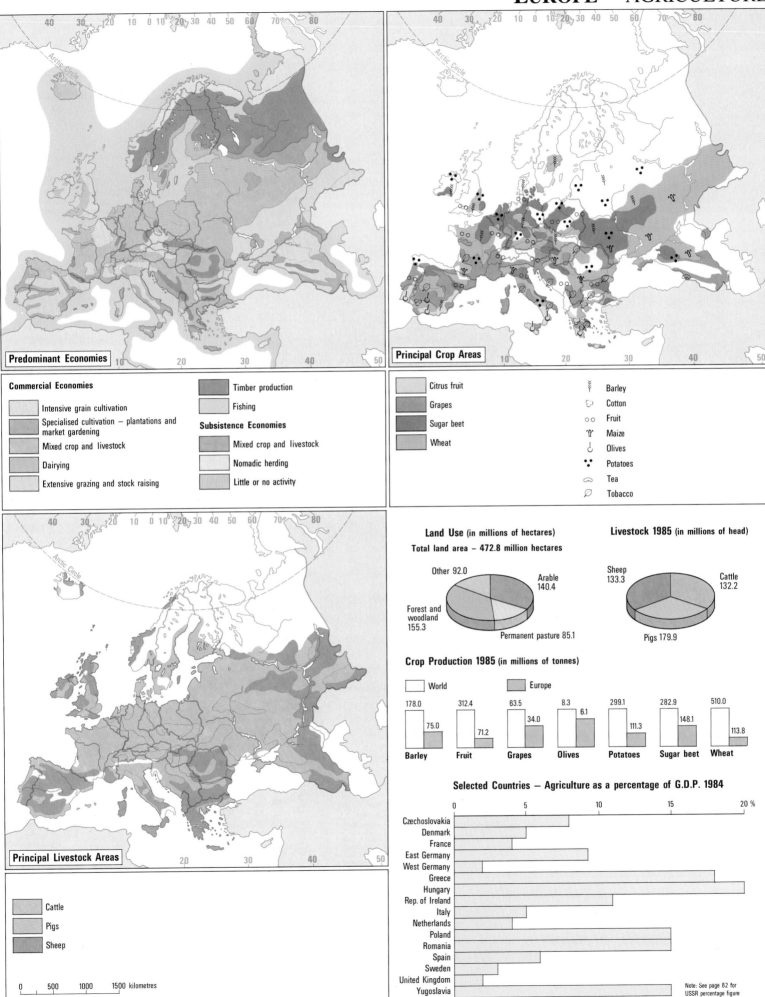

- Intensive grain cultivation
- Specialised cultivation – plantations and market gardening
- Mixed crop and livestock
- Dairying
- Extensive grazing and stock raising
- Timber production
- Fishing

Subsistence Economies

- Mixed crop and livestock
- Nomadic herding
- Little or no activity

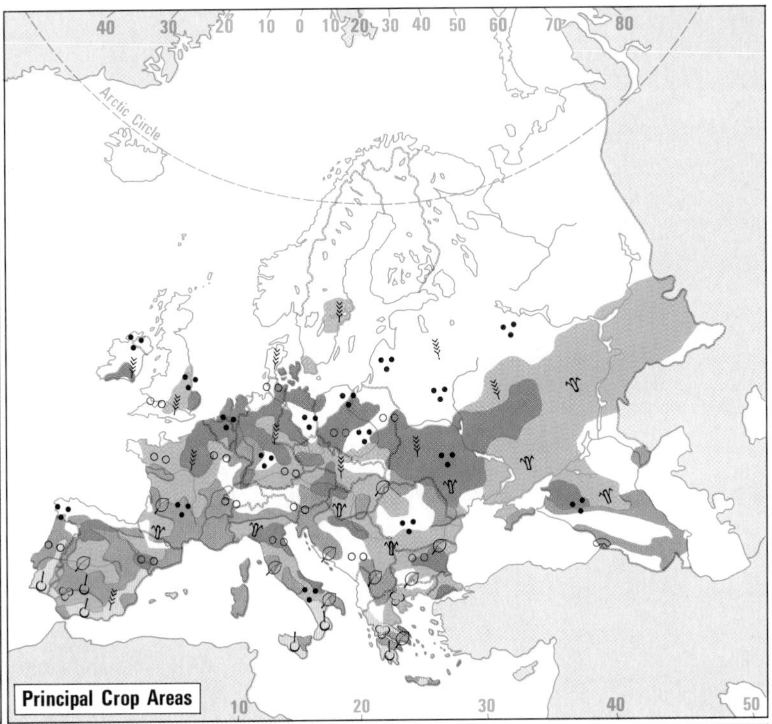

- Citrus fruit
- Grapes
- Sugar beet
- Wheat
- Barley
- Cotton
- Fruit
- Maize
- Olives
- Potatoes
- Tea
- Tobacco

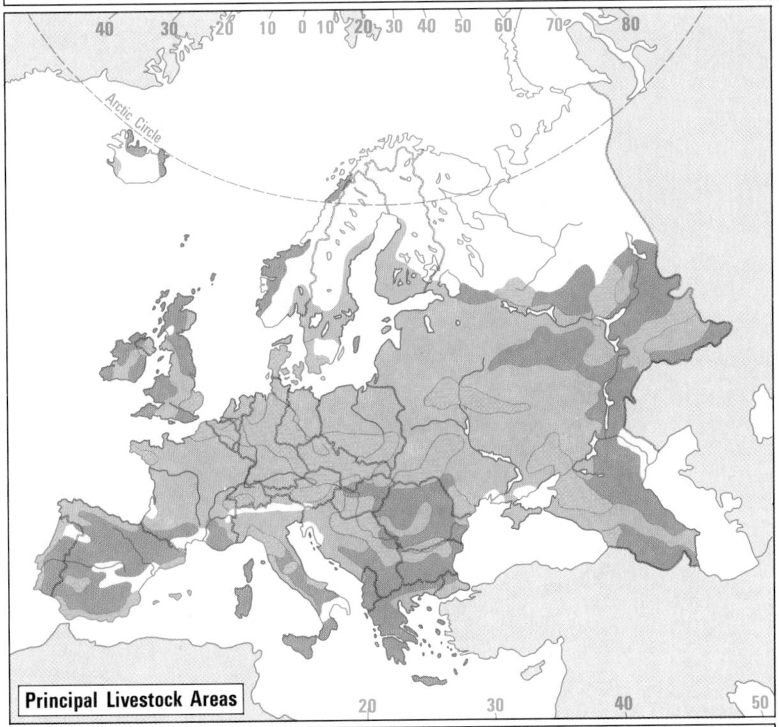

- Cattle
- Pigs
- Sheep

0 500 1000 1500 kilometres

Land Use (in millions of hectares)

Total land area – 472.8 million hectares

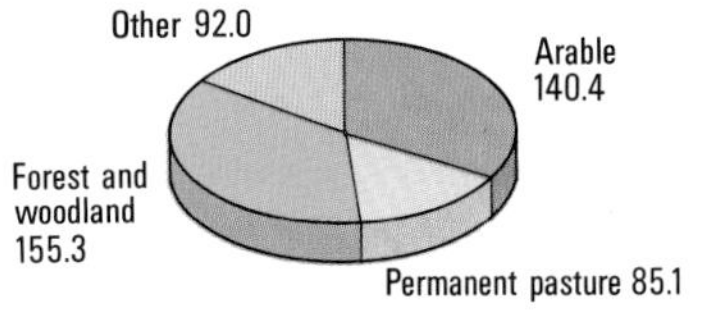

Livestock 1985 (in millions of head)

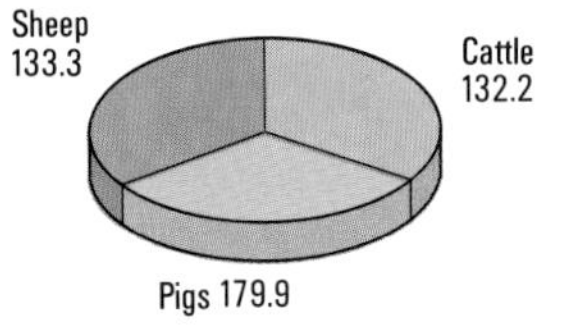

Crop Production 1985 (in millions of tonnes)

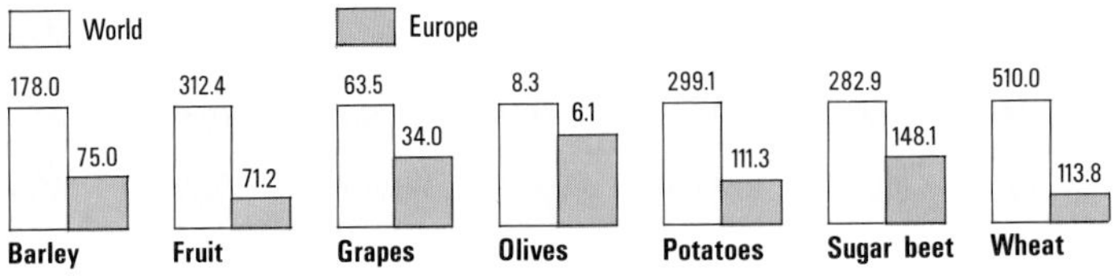

Selected Countries – Agriculture as a percentage of G.D.P. 1984

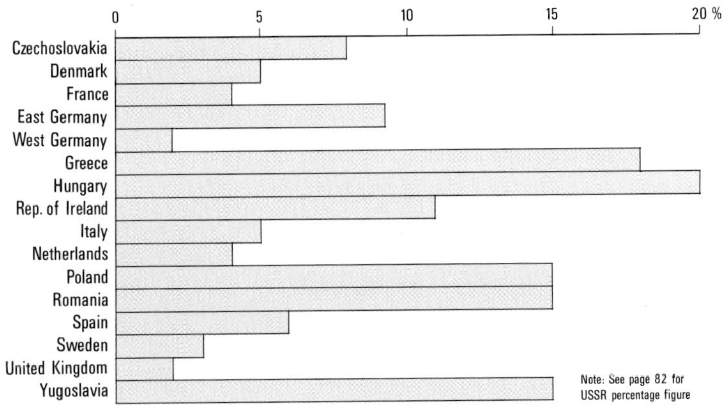

Note: See page 82 for USSR percentage figure

EUROPE – MINERALS AND INDUSTRY

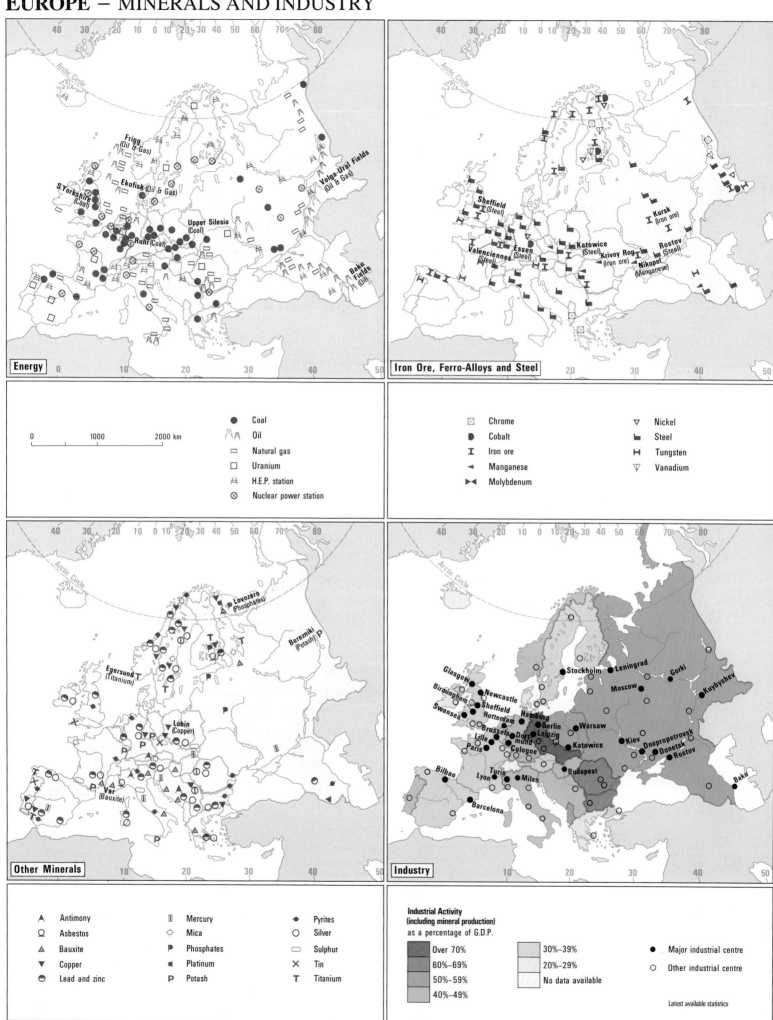

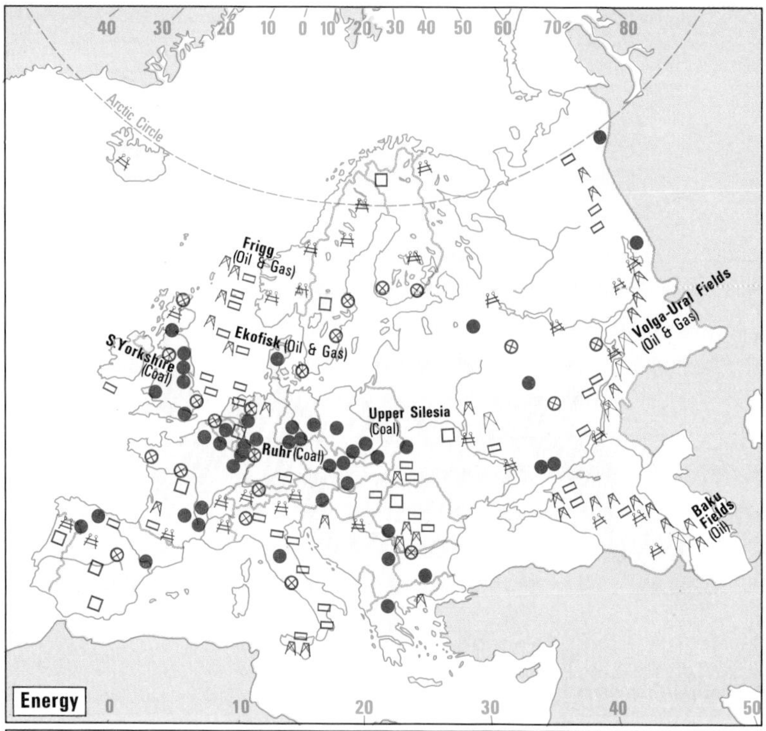

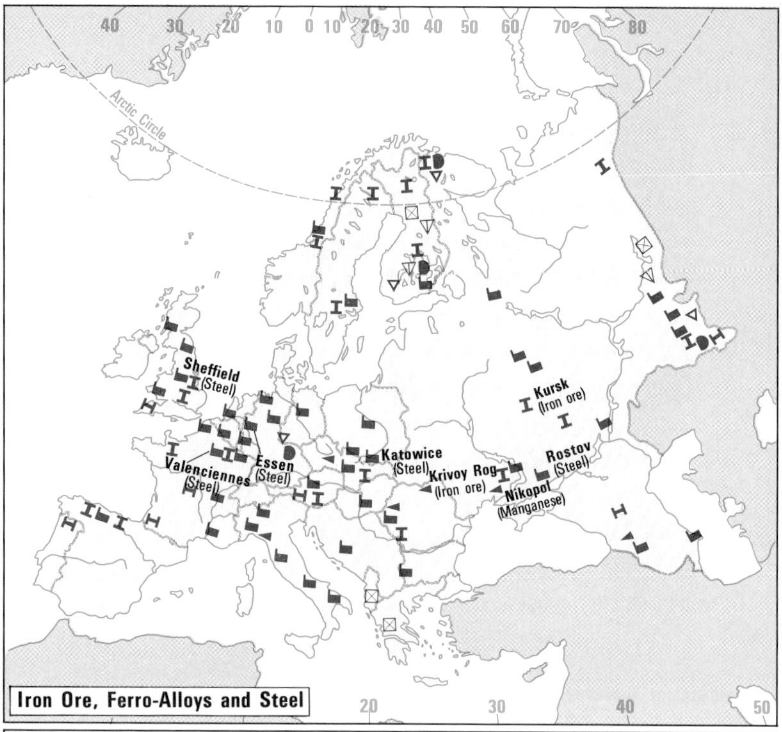

0 1000 2000 km

- Coal
- Oil
- Natural gas
- Uranium
- H.E.P. station
- Nuclear power station

- Chrome
- Cobalt
- Iron ore
- Manganese
- Molybdenum
- Nickel
- Steel
- Tungsten
- Vanadium

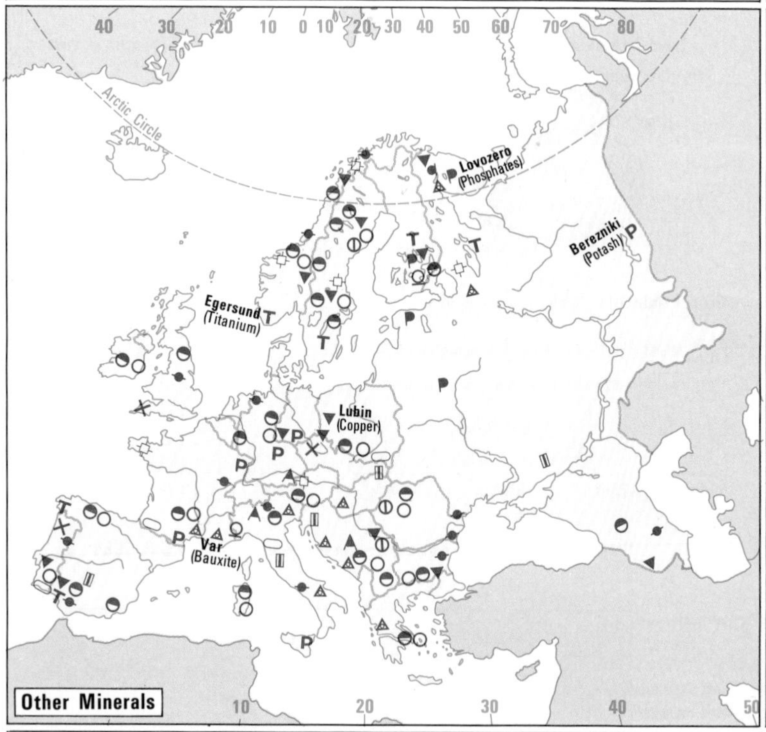

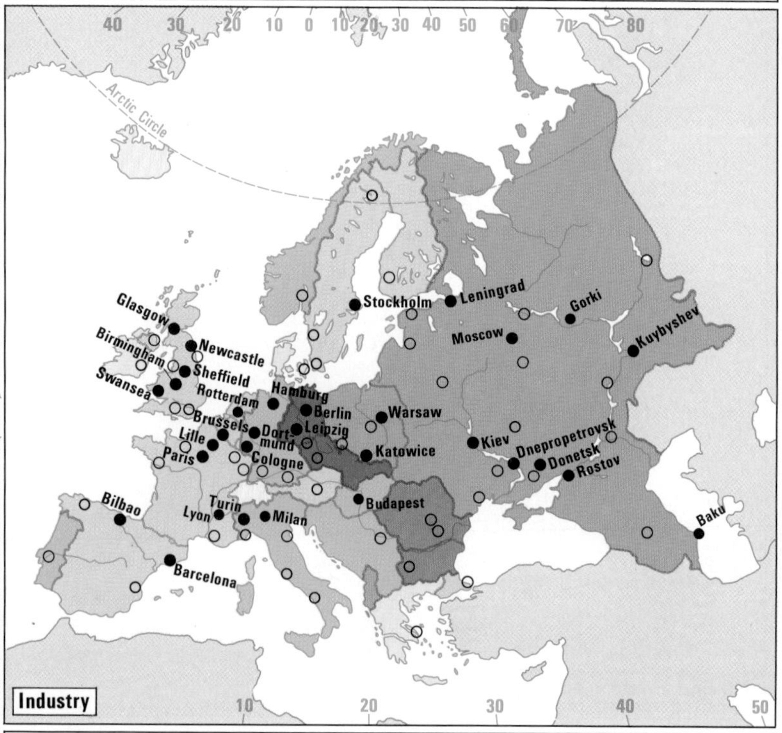

- Antimony
- Asbestos
- Bauxite
- Copper
- Lead and zinc
- Mercury
- Mica
- Phosphates
- Platinum
- Potash
- Pyrites
- Silver
- Sulphur
- Tin
- Titanium

Industrial Activity
(including mineral production)
as a percentage of G.D.P.

- Over 70%
- 60%–69%
- 50%–59%
- 40%–49%
- 30%–39%
- 20%–29%
- No data available

- Major industrial centre
- Other industrial centre

Latest available statistics

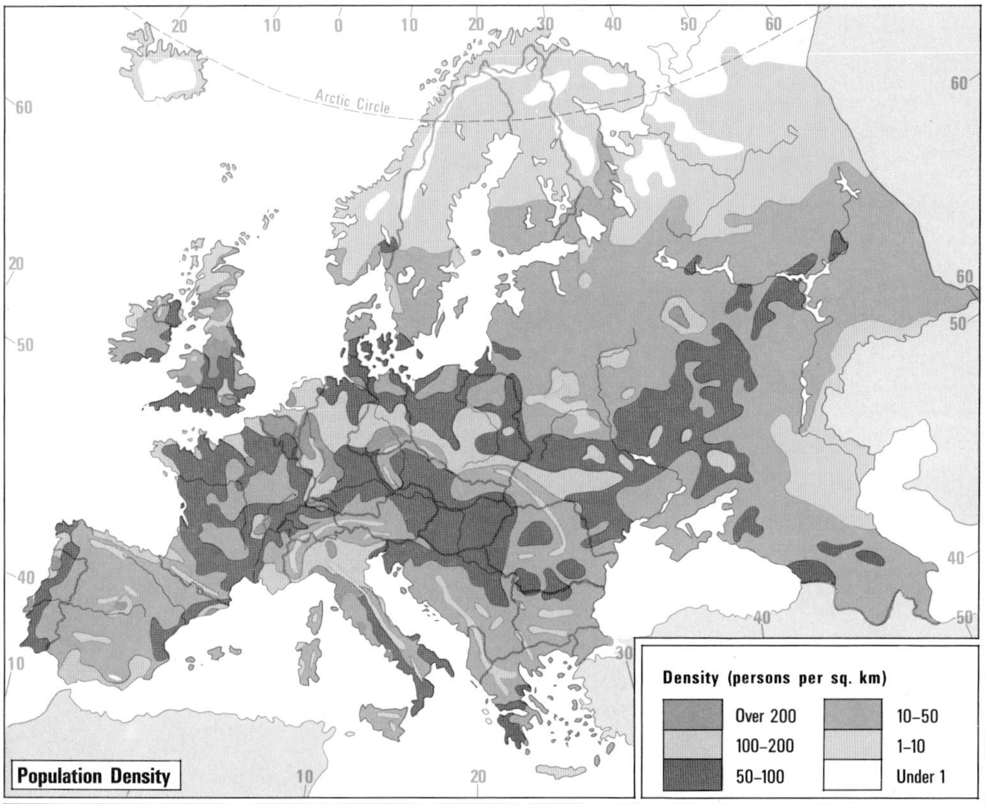

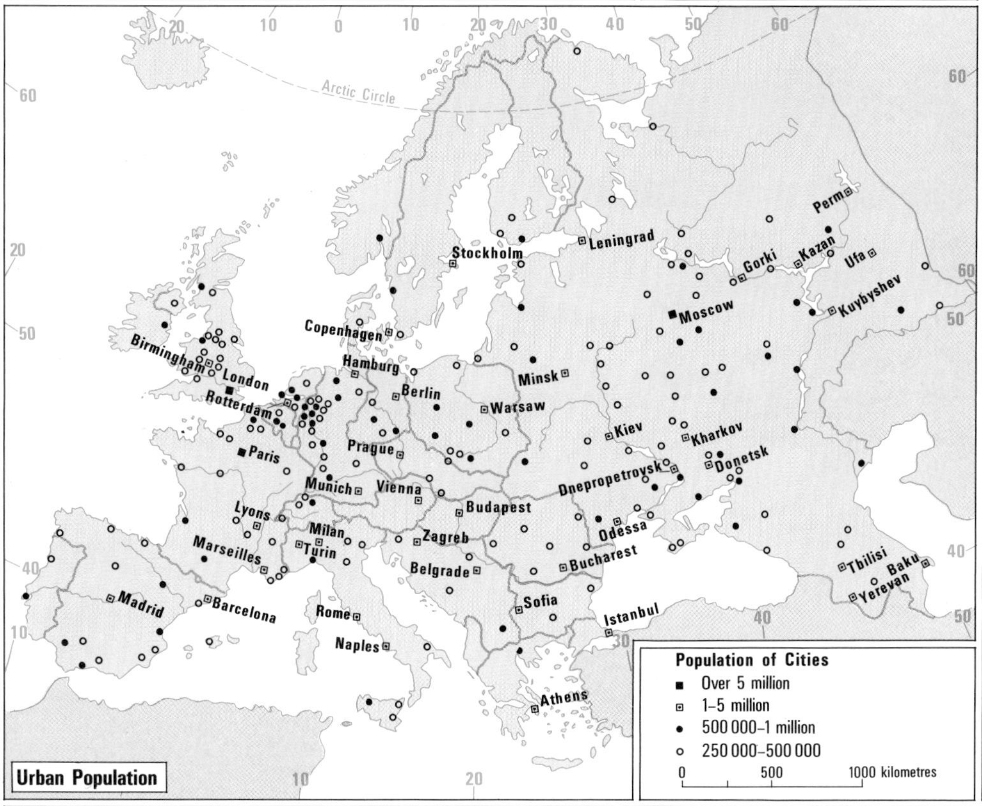

Life Expectancy by Selected Country

Country	Life Expectancy at birth Male	Female
W. Germany	72	78
Italy	74	79
United Kingdom	72	78
France	74	80
Spain	74	80
Poland	67	76
Yugoslavia	66	73
Romania	69	74
E. Germany	68	75
Czechoslovakia	66	74
Netherlands	73	80
Hungary	67	74
Portugal	71	77
Greece	72	78

Population by Selected Country

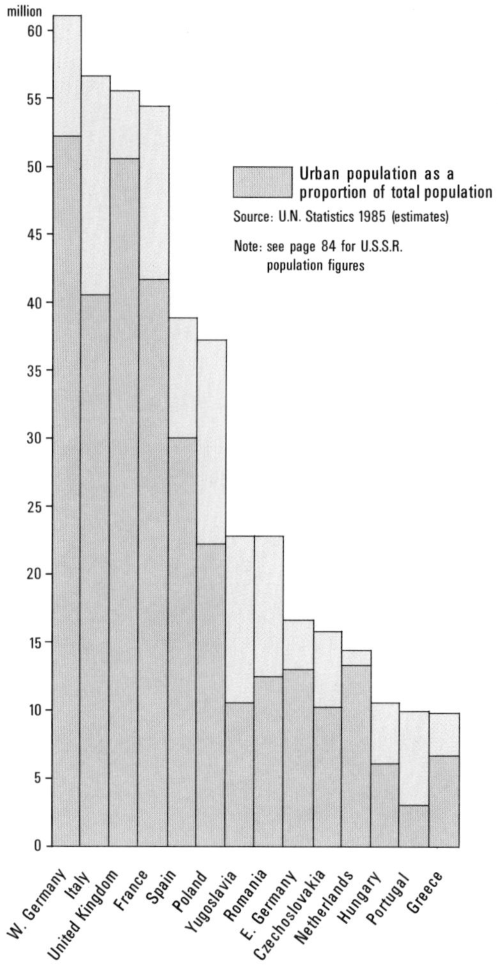

Percentage Population Breakdown by Age and Sex

Source: UN Demographic Yearbook

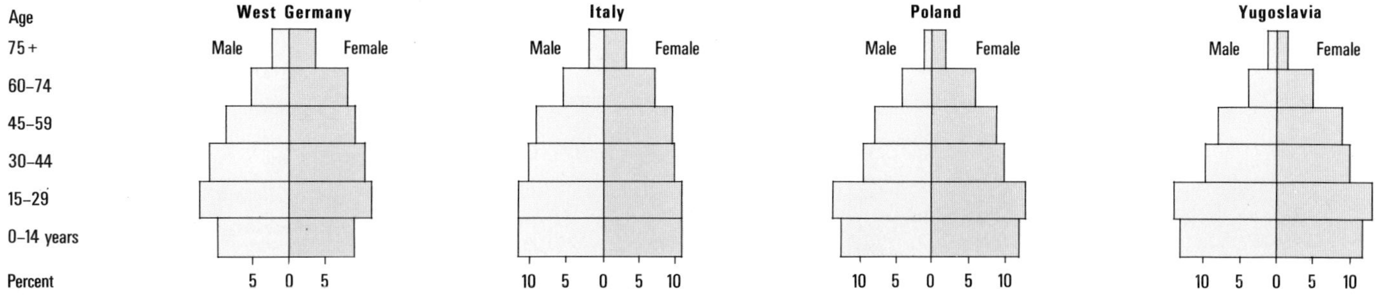

ARCTIC OCEAN
ATLANTIC OCEAN
NORWEGIAN SEA
Arctic Circle
Spitsbergen (Norway)
Franz Josef Land
BARENTS SEA
North Cape
Kola Peninsula
White Sea
Novaya Zemlya
KARA SEA
Severnaya Zemlya
C. Chelyuskin
LAPTEV SEA
Taymyr Pen.
New Siberian Is.
EAST SIBERIAN SEA
Wrangel
Bering
Gydr Peninsula
Norilsk
Vorkuta
Salekhard
Central Siberian Uplands
Verkhoyansk
Verkhoyansk Range
Cherskogo Range
Yakutsk
Okhotsk
Aldan
Stanovoy Range
West Siberian Plain
Surgut
Tobolsk
UNION OF SOVIET SOCIALIST REPUBLICS
Ural Mountains
Yablonovy Ra.
Skovorodino
Bratsk
Tomsk
Krasnoyarsk
Novosibirsk
Omsk
Kurgan
Sverdlovsk
Chelyabinsk
Magnitogorsk
Perm
Kazan
Gorki
Yaroslavl
Moscow
Kirov
Kotlas
Arkhangelsk
Petrozavodsk
Murmansk
Leningrad
Kalinin
Central Russian Uplands
Tula
Penza
Voronezh
Kuybyshev
Ufa
Orenburg
Orsk
Saratov
Volgograd
Kharkov
Donetsk
Rostov
Krasnodar
Astrakhan
Guryev
Kirghiz Steppe
Tselinograd
Pavlodar
Barnaul
Novokuznetsk
Sayan Mts.
Kyzyl
Irkutsk
Lake Baikal
Ulan-Ude
Chita
Blagoveshchensk
Manzhouli
Semipalatinsk
Karaganda
Balkhash
Lake Balkhash
Altai Mts.
Ulan Bator
MONGOLIA
Saynshand
Gobi Desert
Da Hinggan Ling
Aralsk
Aral Sea
Kzyl-Orda
Alma-Ata
Frunze
Chimkent
Tashkent
Samarkand
Dushanbe
Ashkhabad
Krasnovodsk
Urumqi
Tian Shan
Pik Pobedy 7439m
Tarim Basin
Lop Nur
Altun Shan
Qilian Shan
Kunlun Shan
Tibetan Plateau
Karakoram Ra.
Mt. K2 8611m
JAMMU AND KASHMIR
Hindu Kush
Srinagar
Xining
Lanzhou
Huang He
CHINA
Himalayas
Lhasa
Mt. Everest 8848m
NEPAL
Kathmandu
Thimphu
BHUTAN
Dibrugarh
Gongga Shan 7556m
Chengdu
Chongqing
Zigong
Guiyang
Kunming
Xi'an
Qin Ling
Zhengzhou
Taiyuan
Beijing
Tianjin
Hohhot
Jinan
Qingdao
Xuzhou
Nanjing
Hefei
Wuhan
Nanchang
Changsha
Hengyang
Guangzhou
Liuzhou
Nanning
Macau (Port.)
Hong Kong (U.K.)
Haikou
Hainan
Shenyang
Anshan
Tangshan
Lüda
Changchun
Harbin
Qiqihar
Glasgow
Dublin
REP. OF IRELAND
UNITED KINGDOM
London
Amsterdam
NETH.
Brussels
BELG.
LUX.
Paris
FRANCE
Bordeaux
Lyons
Marseilles
PORTUGAL
SPAIN
Zaragoza
Barcelona
Bern
SWITZ.
Alps
Milan
ITALY
Rome
Sardinia
Naples
Palermo
Sicily
Tunis
TUNISIA
Bergen
Oslo
NORWAY
SWEDEN
Stockholm
Göteborg
Helsinki
FINLAND
Oulu
Copenhagen
DENMARK
NORTH SEA
Baltic Sea
Hamburg
WEST GERMANY
EAST GERMANY
Bonn
Berlin
Munich
AUSTRIA
Vienna
CZECH.
Prague
POLAND
Warsaw
Kraków
North European Plain
Riga
Vilnius
Minsk
Kiev
Lvov
HUNG.
Budapest
Zagreb
YUGOSLAVIA
Belgrade
ROMANIA
Bucharest
Dnepropetrovsk
Odessa
BULG.
Sofia
ALB.
Tirane
GREECE
Thessaloniki
Athens
Crete
MEDITERRANEAN SEA
BLACK SEA
Sevastopol
Istanbul
Izmir
Ankara
TURKEY
Taurus Mts.
Adana
Mt. Elbrus 5642m
Caucasus Mts.
Groznyy
Tbilisi
Mt. Ararat 5165m
Yerevan
Baku
CASPIAN SEA
CYPRUS
Nicosia
Aleppo
SYRIA
LEB.
Beirut
Damascus
ISRAEL
Jerusalem
Amman
JORDAN
Syrian Desert
Mosul
Tabriz
Tehran
Elburz Mts.
Qom
IRAQ
Baghdad
IRAN
Esfahan
Zagros Mts.
Dasht-e Lut
Mashhad
Herat
Kabul
AFGHANISTAN
Qandahar
Kerman
Shiraz
Zahedan
Quetta
Basra
Abadan
KUWAIT
Kuwait
Persian Gulf
LIBYA
Benghazi
Alexandria
Cairo
Suez
EGYPT
Aswan
RED SEA
SAUDI ARABIA
Madinah
Riyadh
Dammam
BAHRAIN
QATAR
Doha
Abu Dhabi
UNITED ARAB EMIRATES
Dubai
Jiddah
Makkah
Port Sudan
SUDAN
Khartoum
Asmara
Rub' al Khali
YEMEN ARAB REPUBLIC
Sana'a
P.D.R. YEMEN
Aden
Taiz
Mukalla
OMAN
Muscat
G. of Oman
Makran
DJIBOUTI
Djibouti
ETHIOPIA
Ethiopian Highlands
Addis Ababa
Gulf of Aden
Berbera
Hargeisa
C. Guardafui
Socotra (P.D.R.Y.)
SOMALIA
Mogadishu
Kismayu
KENYA
Mombasa
Equator
Amirante Is.
Aldabra Is.
COMOROS
Mozambique Channel
MADAGASCAR
C. Bobaomby
Rawalpindi
Islamabad
Faisalabad
Lahore
Amritsar
Ludhiana
Multan
PAKISTAN
Sukkur
Karachi
Hyderabad
New Delhi
Delhi
Agra
Jaipur
Kanpur
Lucknow
Varanasi
Patna
INDIA
Indore
Narmada
Jabalpur
Ahmadabad
Nagpur
Calcutta
BANGLADESH
Dhaka
Chittagong
Bombay
Solapur
Hyderabad
Deccan
Western Ghats
Eastern Ghats
Vijayawada
Bangalore
Mangalore
Madras
Madurai
Trivandrum
C. Comorin
Jaffna
Colombo
SRI LANKA
ARABIAN SEA
Lakshadweep Is. (India)
MALDIVES
INDIAN OCEAN
Chagos Archipelago (U.K.)
BAY OF BENGAL
Andaman Is. (India)
ANDAMAN SEA
Nicobar Is. (India)
BURMA
Mandalay
Chiang Mai
Pegu
Rangoon
Moulmeim
LAOS
Vientiane
THAILAND
Bangkok
VIETNAM
Hanoi
Hai-phong
Da Nang
KAMPUCHEA
Phnom Penh
Ho Chi Minh
Nha Trang
Gulf of Thailand
Songkhla
Georgetown
Ipoh
Kuala Lumpur
MALAYSIA
SINGAPORE
Singapore
Medan
Str. of Malacca
Sumatra
Padang
Kepulauan Mentawai
Jambi
Palembang
Bangka
Tanjungkarang Telukbetung
JAVA SEA
Jakarta
Semarang
Surabaya
Java
INDONESIA
Kalimantan
Borneo
SOUTH CHINA SEA
Projection: Lambert's Azimuthal Equal Area
1 : 40 000 000
0
500
1000
1500 km

Asia-Political
1:85 000 000
0 1000 2000 km
Europe
Africa
UNION OF SOVIET SOCIALIST REPUBLICS
MONGOLIA
CHINA
INDIA
Moscow
Ulan Bator
Beijing
Tokyo
JAPAN
NORTH KOREA
Pyongyang
Seoul
SOUTH KOREA
T'ai-pei
TAIWAN
Ankara
TURKEY
SYRIA
CYPRUS
LEB.
Damascus
ISRAEL
JORDAN
IRAQ
Baghdad
Tehran
IRAN
Kabul
AFGHANISTAN
SAUDI ARABIA
Riyadh
KUWAIT
BAHRAIN
QATAR
U.A.E.
OMAN
YEMEN A.R.
P.D.R. YEMEN
Islamabad
PAKISTAN
New Delhi
NEPAL
Kathmandu
BHUTAN
BANGLADESH
Dhaka
BURMA
Rangoon
Hanoi
Vientiane
LAOS
VIETNAM
THAILAND
Bangkok
KAMPUCHEA
Phnom Penh
Manila
PHILIPPINES
SRI LANKA
Colombo
MALDIVES
BRUNEI
MALAYSIA
Kuala Lumpur
SINGAPORE
INDONESIA
Jakarta
Australia
INDIAN OCEAN
PACIFIC OCEAN
Arctic Circle
Tropic of Cancer
Equator
Asia - Geographical Statistics
Largest Countries
U.S.S.R. 22 402 000km²*
China 9 597 000km²
* includes European U.S.S.R.
Longest Rivers
Chang Jiang, China 5 470km
Ob-Irtysh, Asia 5 410km
Highest Volcanoes (extinct)
Mt. Elbrus, U.S.S.R. 5642m
Mt. Fujiyama, Japan 3776m
Most Populated Cities (1982)
Shanghai, China 11 940 000
Tokyo, Japan 11 676 000
Largest Inland Seas
Caspian Sea, U.S.S.R./Iran 372 000km²
Aral Sea, U.S.S.R. 66 500km²
Deepest Ocean Trenches
Japan Trench, Pac. Oc. 10 595m
Kuril Trench, Pac. Oc. 10 542m
Highest Mountains
Mt. Everest, Nepal/China 8848m
Mt. K2, Pakistan/China 8611m
Most Populated Countries (1984)
China 1 049.7 million
India 745.0 million
Largest Islands
Borneo, S.E. Asia 752 000km²
Sumatra, Indonesia 422 000km²
Mt. Elbrus 5642m
Mt. K2 8611m
Mt. Everest 8848m
Mt. Fujiyama 3776m
10 542m Kuril Trench
10 595m Japan Trench
Shanghai
Chang Jiang
Ob
Irtysh
Aral Sea
Caspian Sea
Borneo
Sumatra
metres
6000
5000
4000
3000
2000
1000
500
200
0
Land below sea level
200
2000
4000
6000
8000
Aleutian Islands
BERING SEA
Kamchatka Peninsula
Petropavlovsk-Kamchatskiy
Sakhalin
Kuril Islands
PACIFIC OCEAN
Vladivostok
Hokkaido
Sapporo
Hakodate
Aomori
Sendai
Honshu
Tokyo
Yokohama
Mt. Fujiyama 3776m
Nagoya
Kyoto
Kobe
Osaka
Shikoku
Kitakyushu
Kyushu
Kagoshima
SEA OF JAPAN
Pusan
EAST CHINA SEA
Ryukyu Is.
T'ai-pei
TAIWAN
Kaohsiung
Luzon
Quezon City
Manila
PHILIPPINES
PHILIPPINE SEA
Mindoro
Samar
Panay
Cebu
Butuan
Mindanao
Davao
Zamboanga
CELEBES SEA
Halmahera
Sarmi
Jayapura
Irian Jaya
Pk. Jaya 5030m
Seram
Moluccas
Maluku
Samarinda
Sulawesi
Palopo
BANDA SEA
ARAFURA SEA
Pandang
FLORES SEA
Flores
Timor
TIMOR SEA
Sumba
Darwin
AUSTRALIA

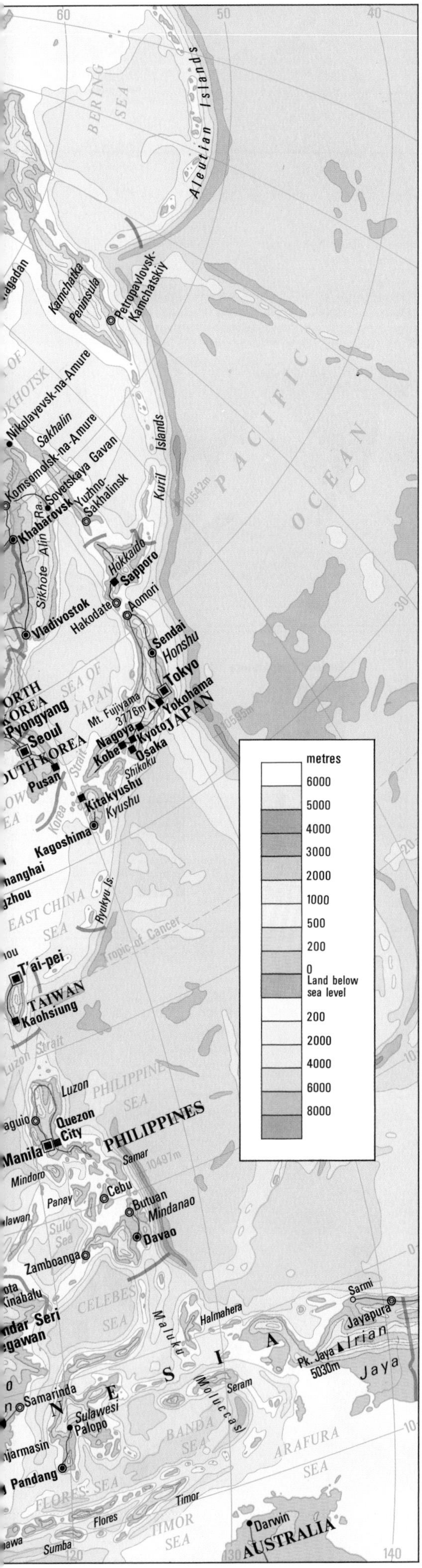

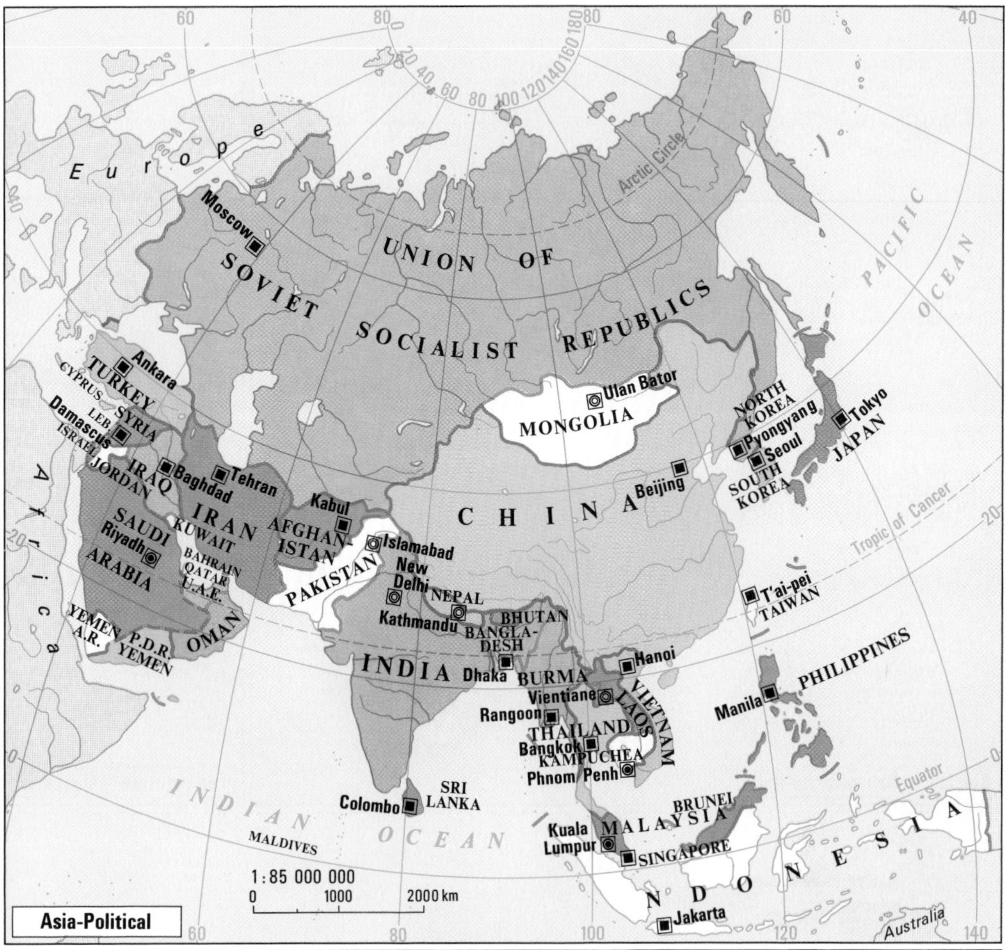

ASIA – CLIMATE

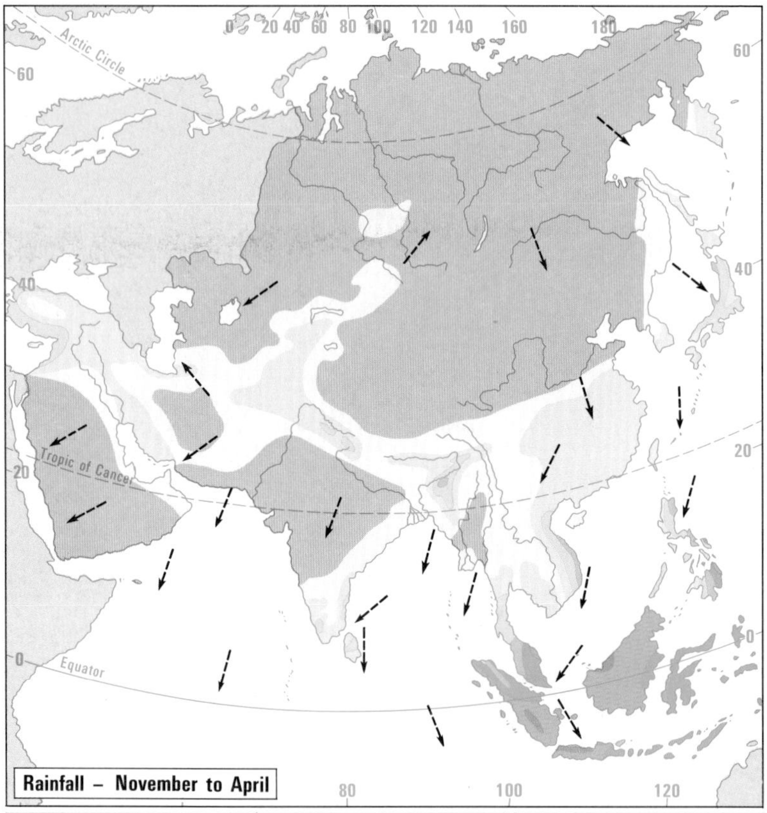

Rainfall – November to April

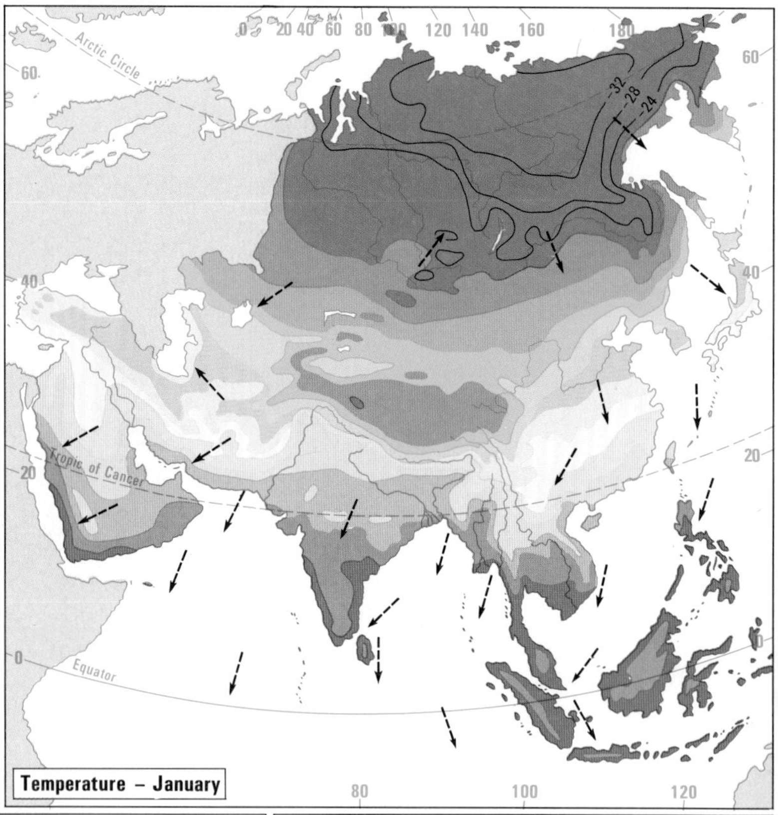

Temperature – January

Rainfall in millimetres

- Over 2000
- 1500–2000
- 1000–1500
- 750–1000
- 500–750
- 250–500
- 125–250
- 0–125

0 1000 2000 3000 kilometres

Prevailing winds

- November to April
- May to October

Surface temperature °C

32
28
24
20
16
12
8
4

0
−4
−8
−12
−16
below −20

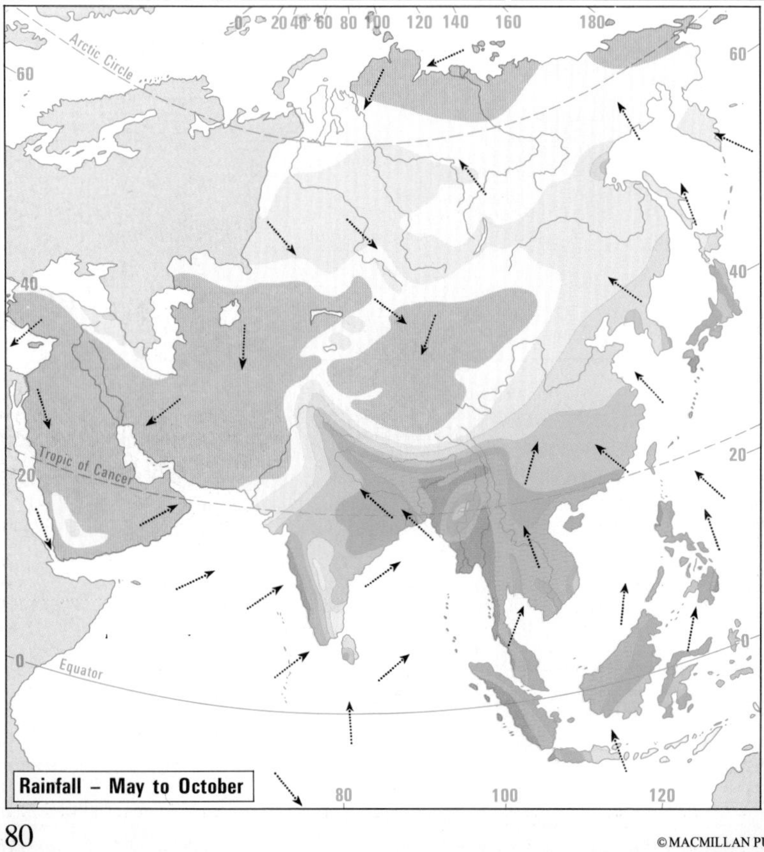

Rainfall – May to October

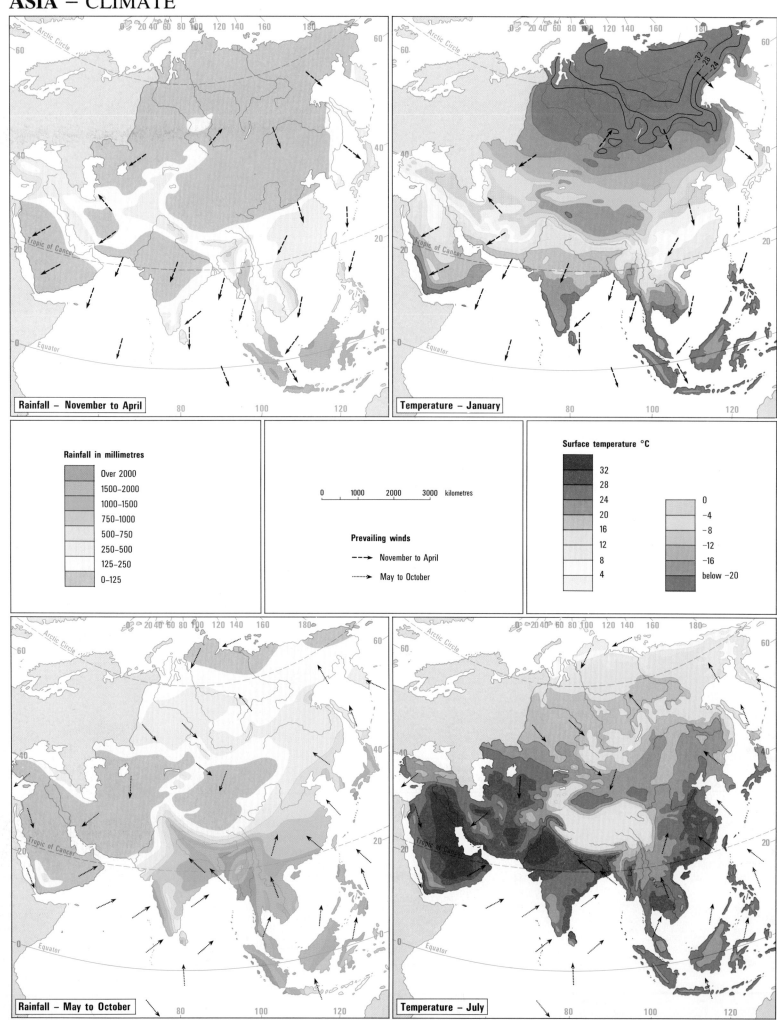

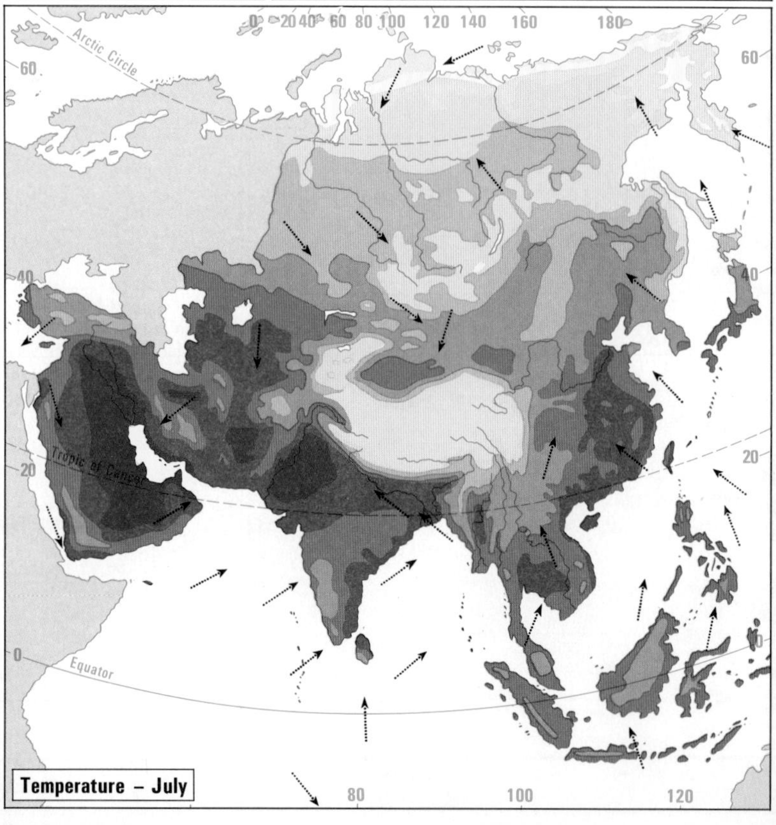

Temperature – July

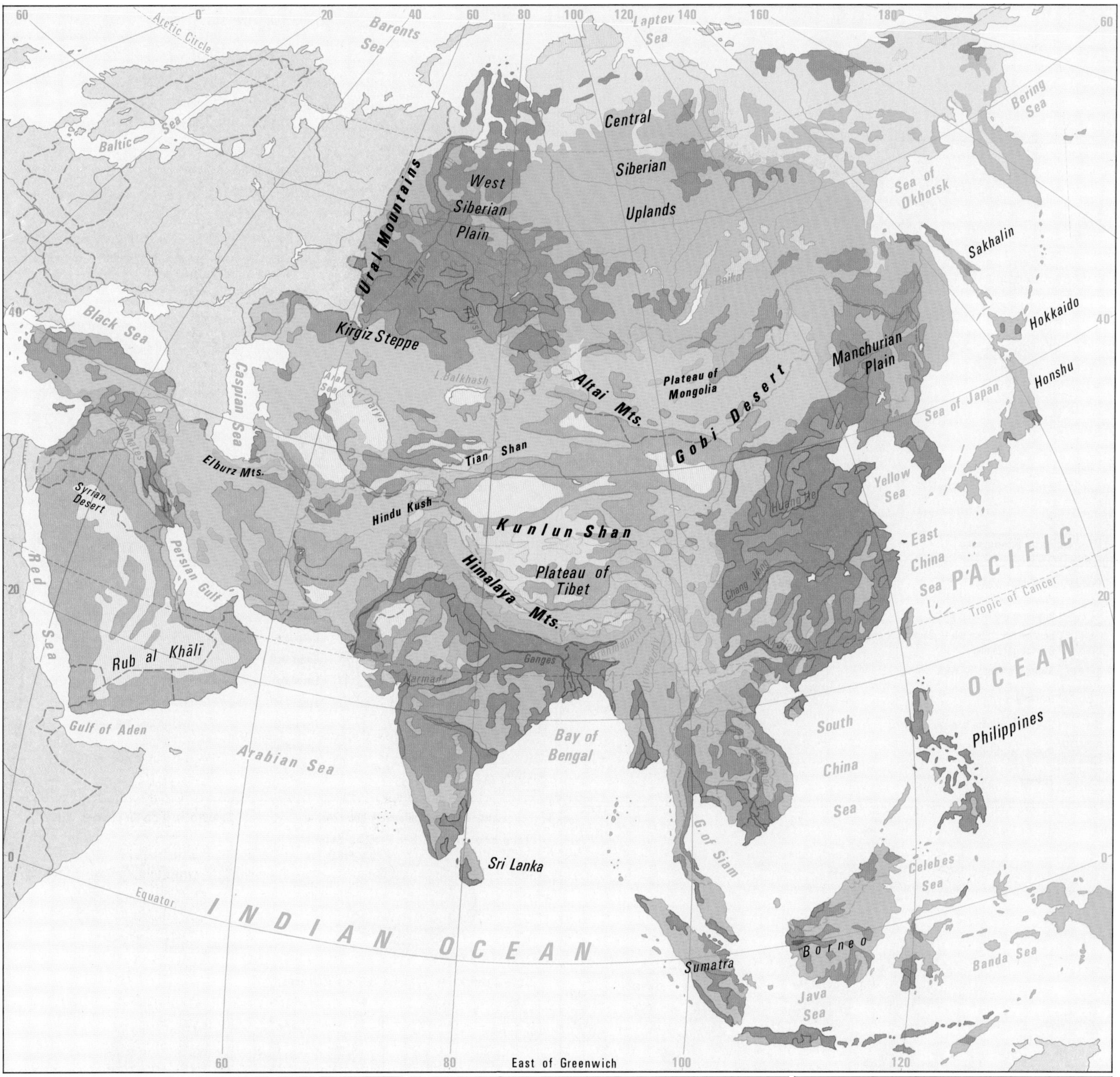
Arctic Circle
Barents Sea
Laptev Sea
Bering Sea
Baltic Sea
Central Siberian Uplands
West Siberian Plain
Ural Mountains
Sea of Okhotsk
Sakhalin
L. Baikal
Hokkaido
Black Sea
Kirgiz Steppe
Manchurian Plain
Honshu
Caspian Sea
L. Balkhash
Aral Sea
Altai Mts.
Plateau of Mongolia
Gobi Desert
Sea of Japan
Tian Shan
Elburz Mts.
Syrian Desert
Hindu Kush
Huang He
Yellow Sea
Kunlun Shan
East China Sea
PACIFIC
Persian Gulf
Plateau of Tibet
Himalaya Mts.
Chang Jiang
Tropic of Cancer
Red Sea
Rub al Khālī
Ganges
Narmada
OCEAN
Gulf of Aden
Arabian Sea
Bay of Bengal
South China Sea
Philippines
Sri Lanka
G. of Siam
Celebes Sea
Equator
INDIAN OCEAN
Borneo
Sumatra
Banda Sea
Java Sea
East of Greenwich

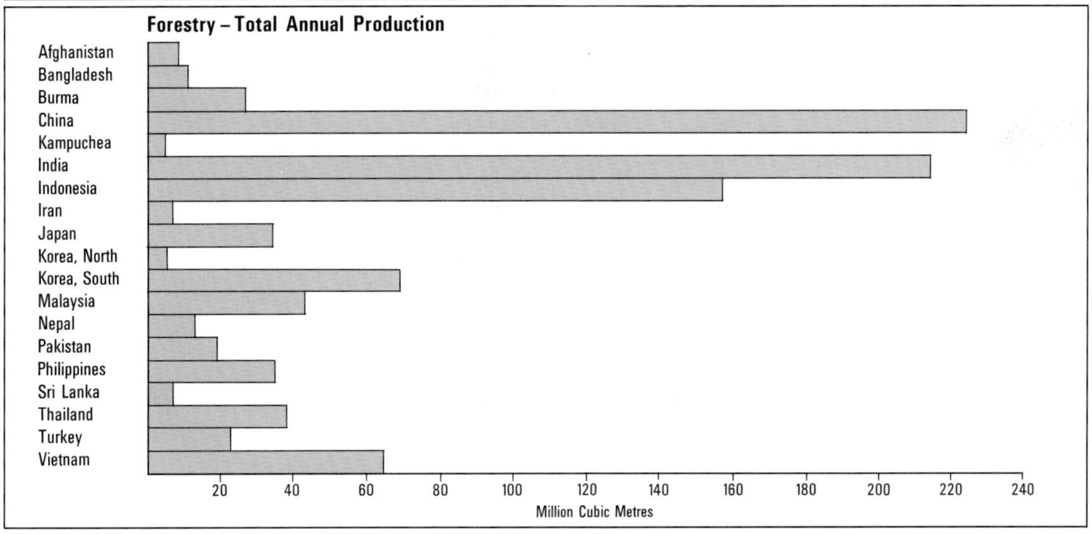
Forestry – Total Annual Production
Afghanistan
Bangladesh
Burma
China
Kampuchea
India
Indonesia
Iran
Japan
Korea, North
Korea, South
Malaysia
Nepal
Pakistan
Philippines
Sri Lanka
Thailand
Turkey
Vietnam
20 40 60 80 100 120 140 160 180 200 220 240
Million Cubic Metres

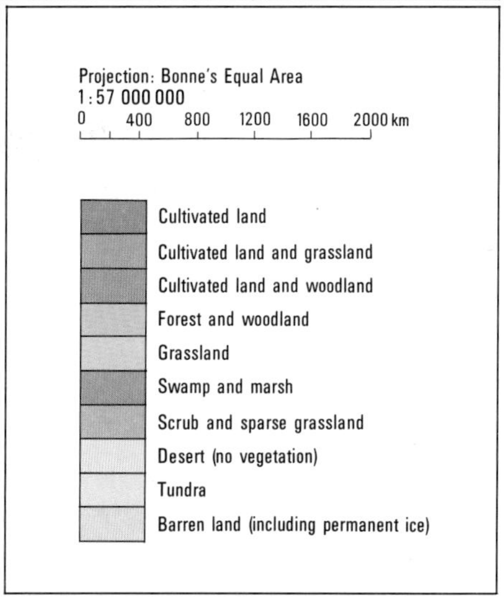
Projection: Bonne's Equal Area
1:57 000 000
0 400 800 1200 1600 2000 km
Cultivated land
Cultivated land and grassland
Cultivated land and woodland
Forest and woodland
Grassland
Swamp and marsh
Scrub and sparse grassland
Desert (no vegetation)
Tundra
Barren land (including permanent ice)

ASIA – AGRICULTURE

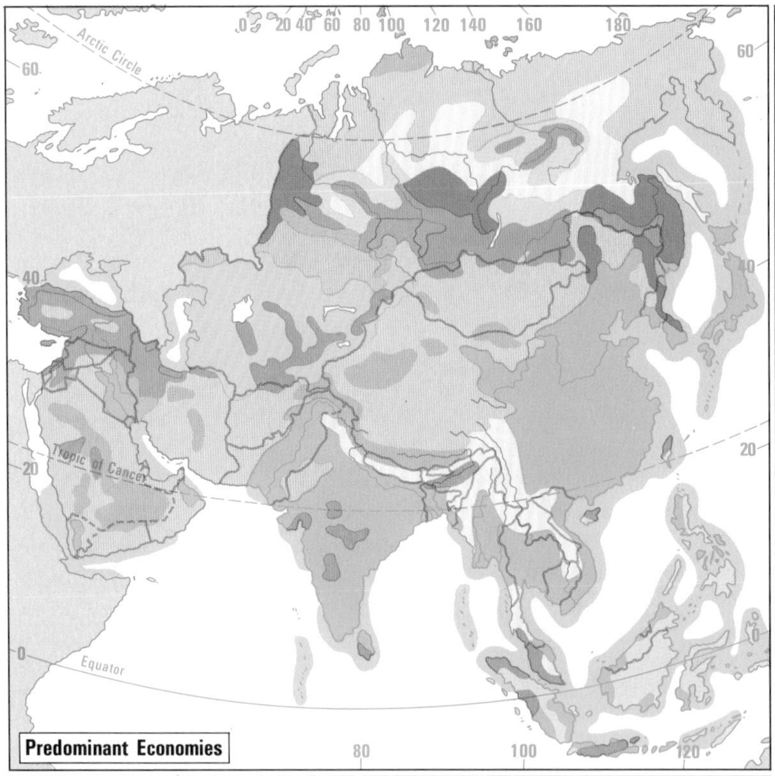

Predominant Economies

Commercial Economies

- Intensive grain cultivation
- Specialised cultivation – plantations and market gardening
- Mixed crop and livestock
- Dairying
- Extensive grazing and stock raising
- Timber production
- Fishing

Subsistence Economies

- Gathering, hunting, fishing and primitive cultivation
- Shifting and marginal cultivation
- Intensive crop cultivation
- Mixed crop and livestock
- Nomadic herding
- Little or no activity

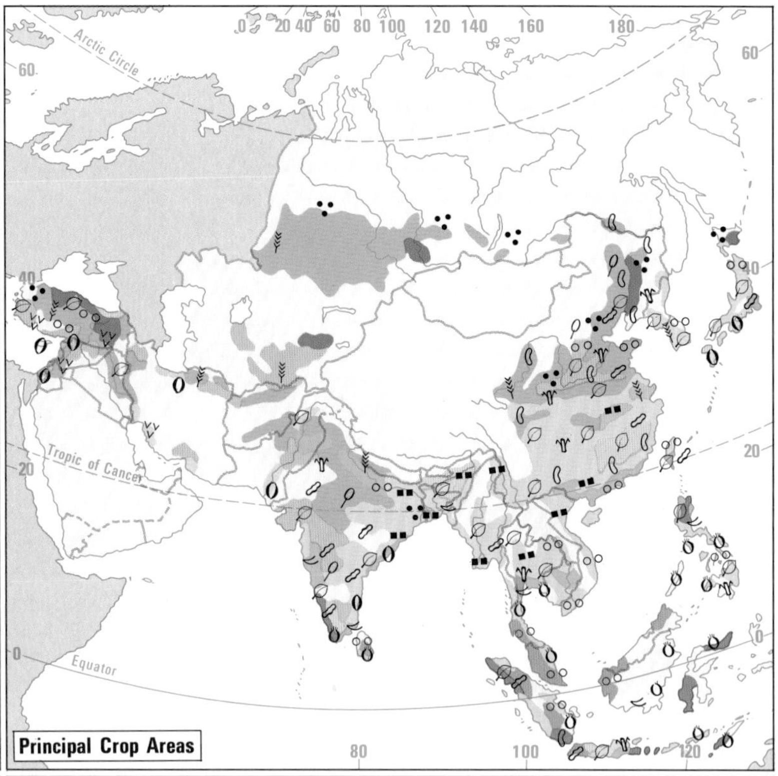

Principal Crop Areas

- Coffee
- Cotton
- Rice
- Rubber
- Sugar beet
- Sugar cane
- Tea
- Wheat
- Barley
- Bananas
- Citrus fruit
- Coconuts
- Fruit
- Grapes
- Groundnuts
- Jute
- Maize
- Millet
- Potatoes
- Soya beans
- Tobacco

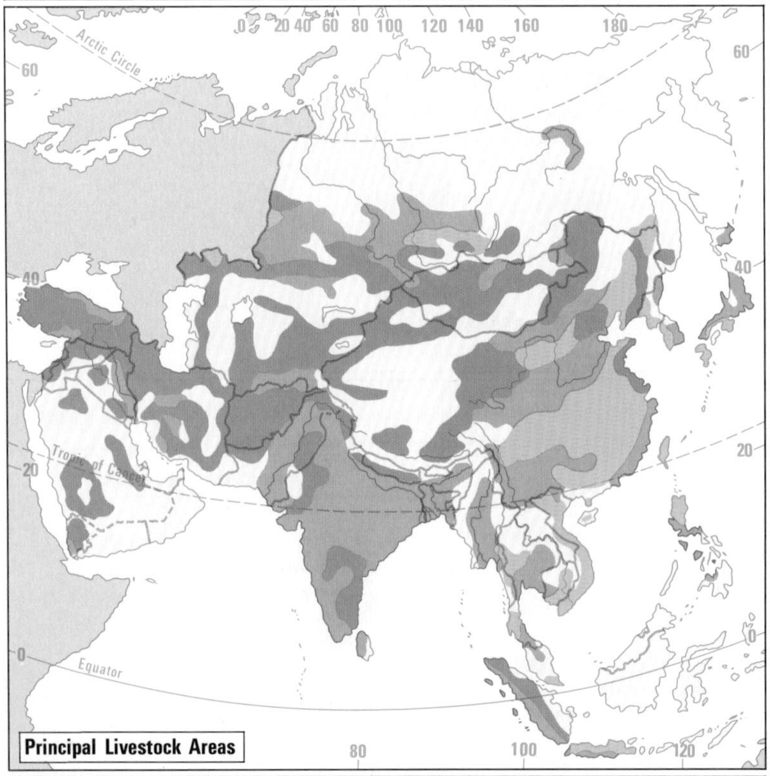

Principal Livestock Areas

0 1000 2000 3000 kilometres

- Cattle
- Pigs
- Sheep

Land Use (in millions of hectares)

Total land area – 2 679.1 million hectares

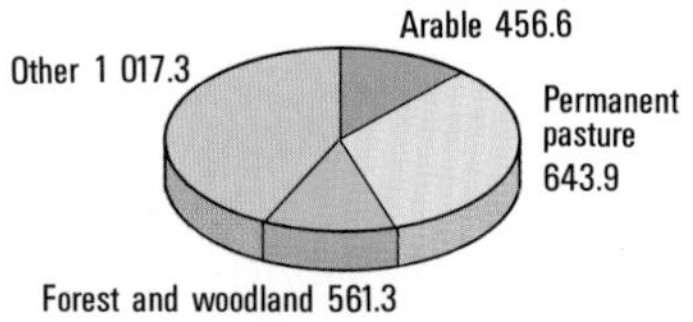

Livestock 1985 (in millions of head)

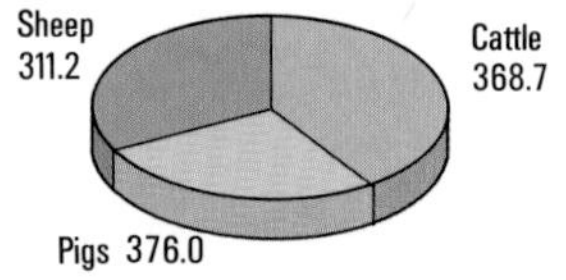

Crop Production 1985 (in millions of tonnes)

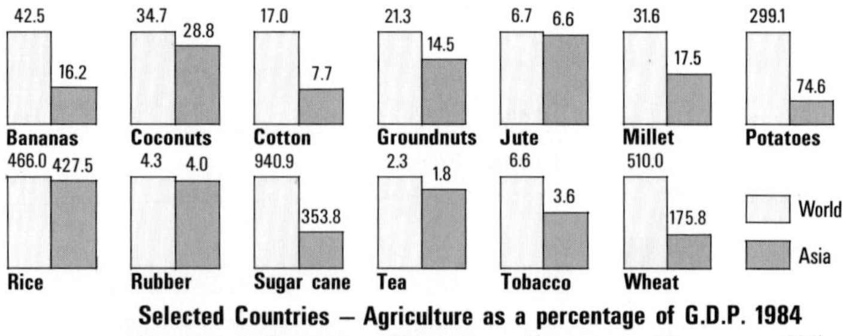

Selected Countries – Agriculture as a percentage of G.D.P. 1984

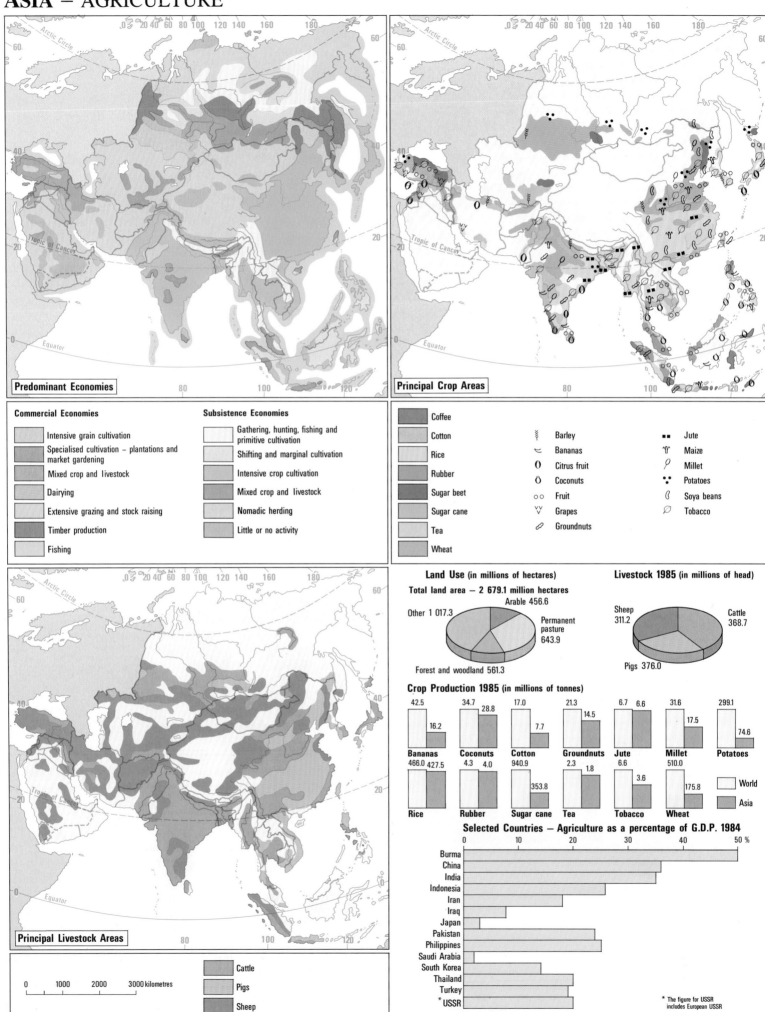

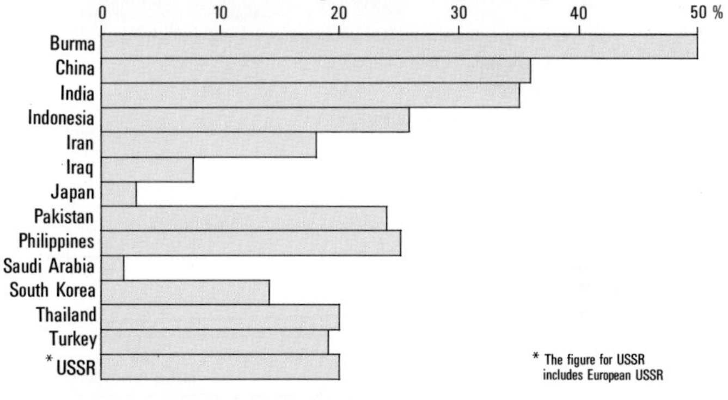

* The figure for USSR includes European USSR

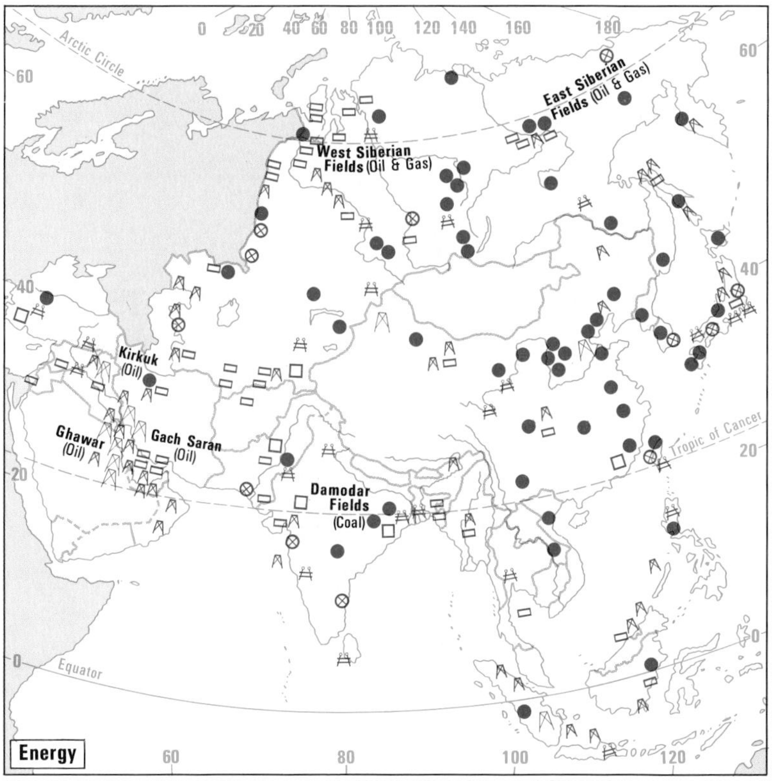

- Coal
- Natural gas
- Oil
- Uranium
- H.E.P. Station
- Nuclear power station

0 1000 2000 3000 4000 km

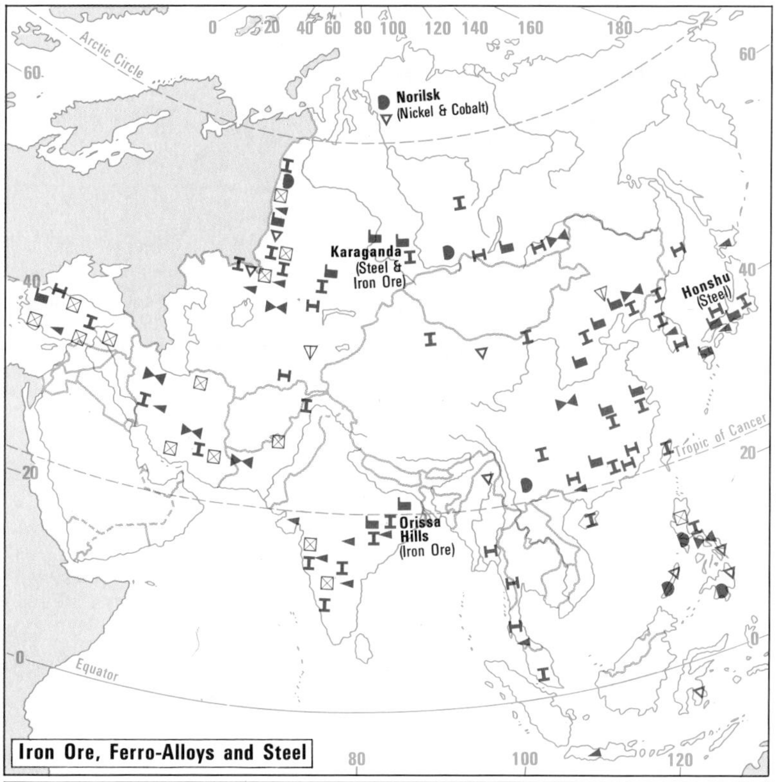

- Chrome
- Cobalt
- Iron ore
- Manganese
- Molybdenum
- Nickel
- Steel
- Tungsten
- Vanadium

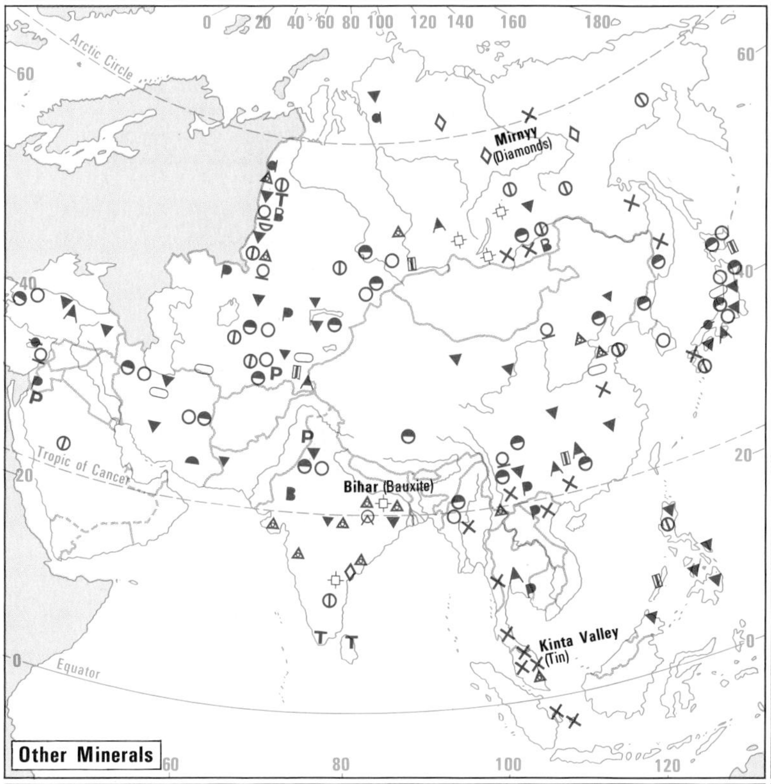

- Antimony
- Asbestos
- Bauxite
- Beryllium
- Copper
- Diamonds
- Gold
- Lead and zinc
- Mercury
- Mica
- Nitrates
- Phosphates
- Platinum
- Potash
- Pyrites
- Silver
- Sulphur
- Tin
- Titanium

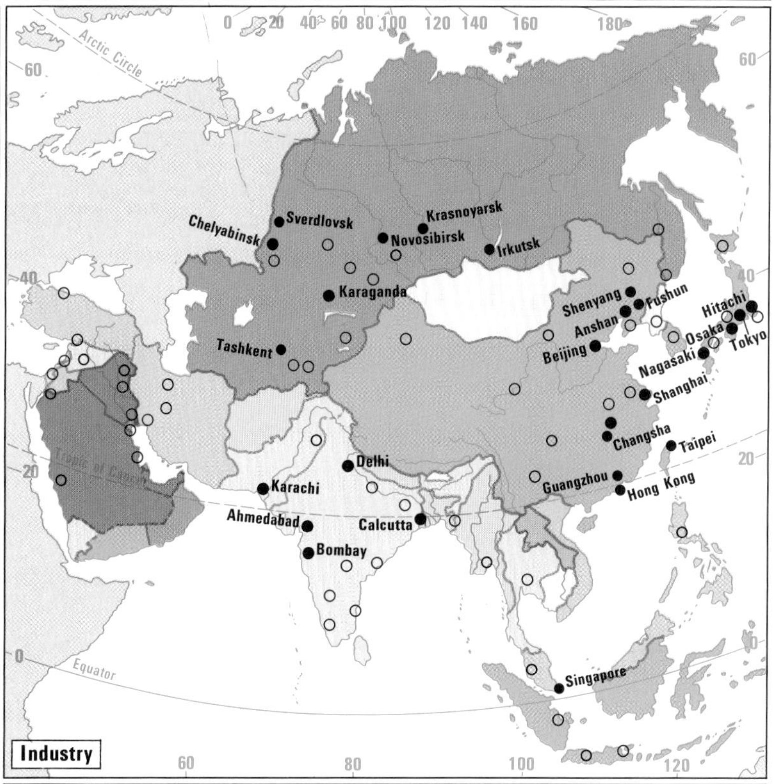

Industrial Activity (including mineral production) as a percentage of G.D.P.

- 60% – 69%
- 50% – 59%
- 40% – 49%
- 30% – 39%
- 20% – 29%
- 10% – 19%
- 0% – 9%
- No data available

- Major industrial centre
- Other industrial centre

Latest available statistics

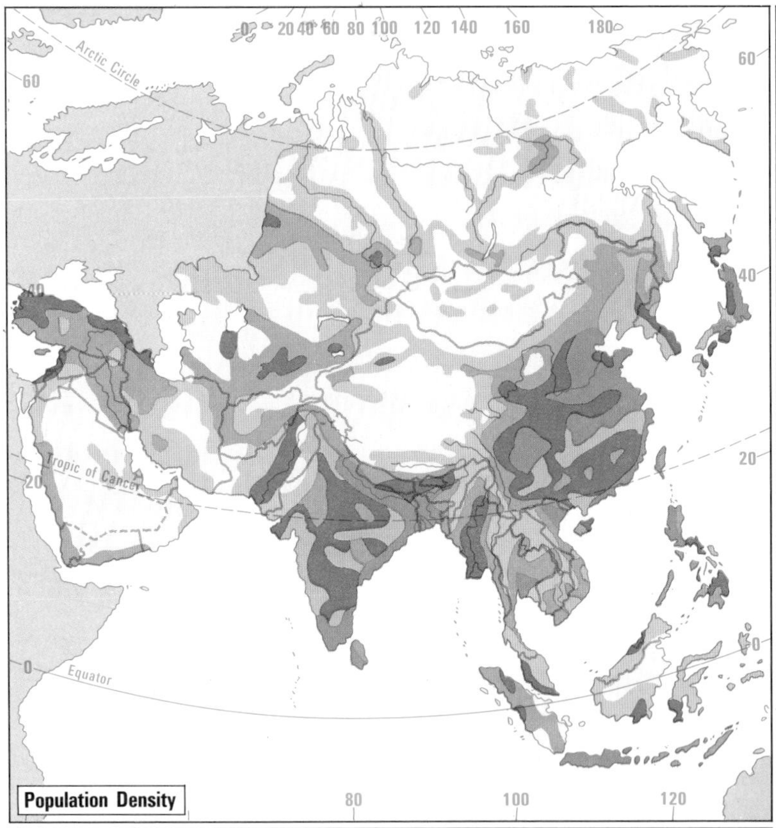

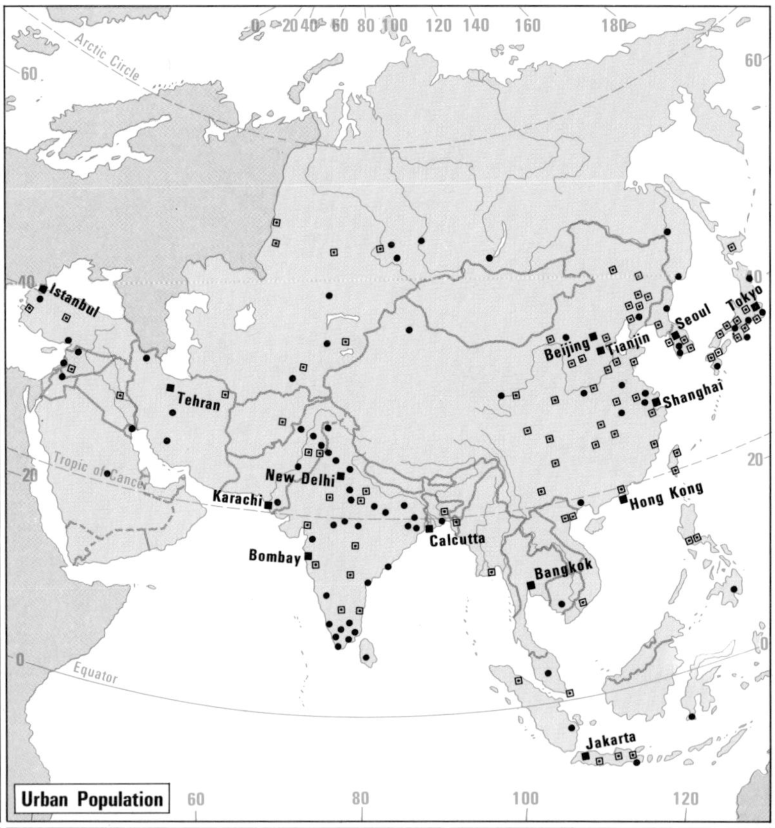

Density (persons per sq km)

- Over 200
- 100–200
- 50–100
- 10–50
- 1–10
- Under 1

0 1000 2000 3000 kilometres

Population of Cities

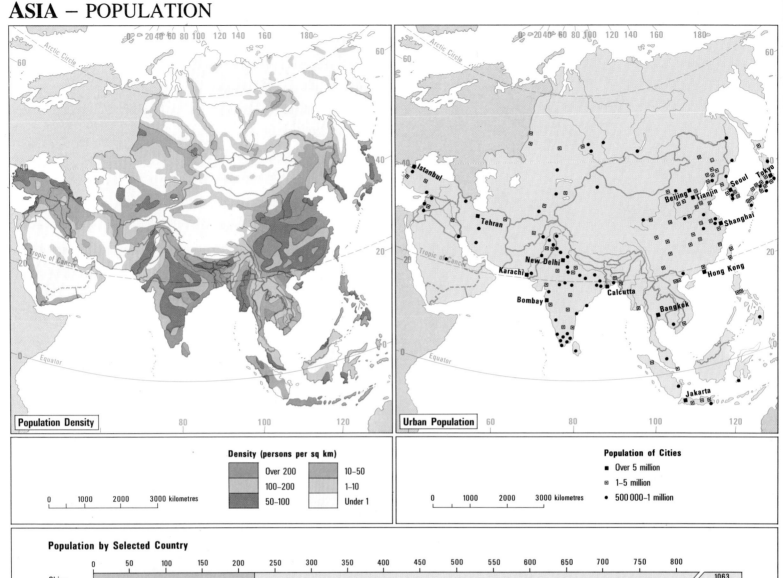

- ■ Over 5 million
- ⊡ 1–5 million
- • 500 000–1 million

0 1000 2000 3000 kilometres

Population by Selected Country

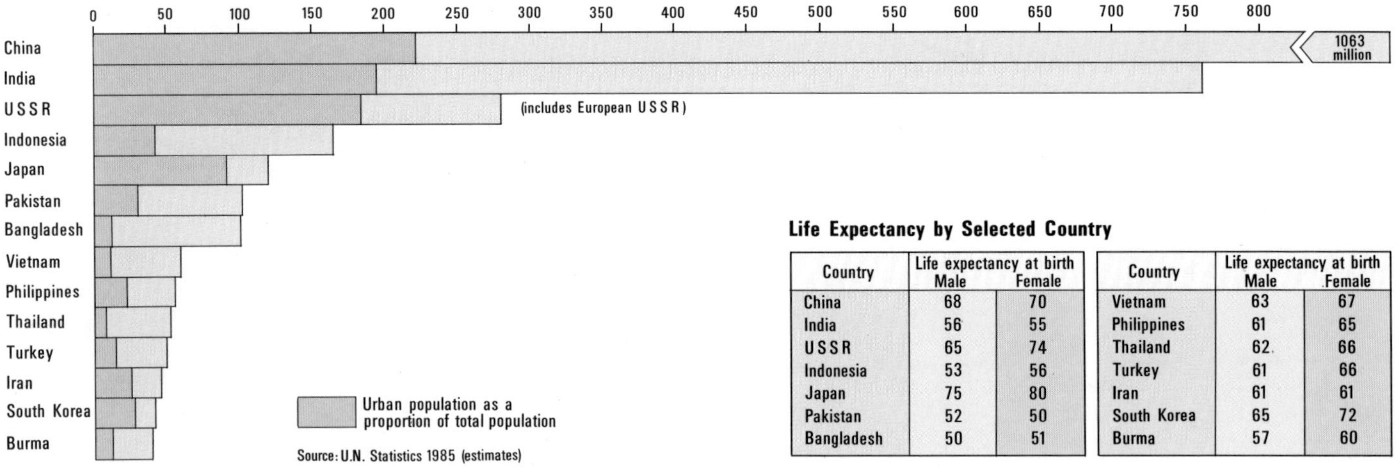

Source: U.N. Statistics 1985 (estimates)

Life Expectancy by Selected Country

Country	Life expectancy at birth Male	Female
China	68	70
India	56	55
USSR	65	74
Indonesia	53	56
Japan	75	80
Pakistan	52	50
Bangladesh	50	51

Country	Life expectancy at birth Male	Female
Vietnam	63	67
Philippines	61	65
Thailand	62	66
Turkey	61	66
Iran	61	61
South Korea	65	72
Burma	57	60

Percentage Population Breakdown by Age and Sex

Source: UN Demographic Yearbook

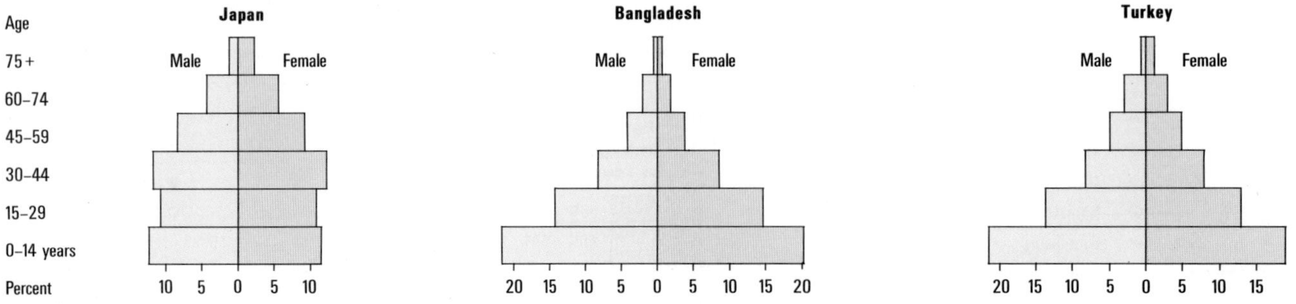

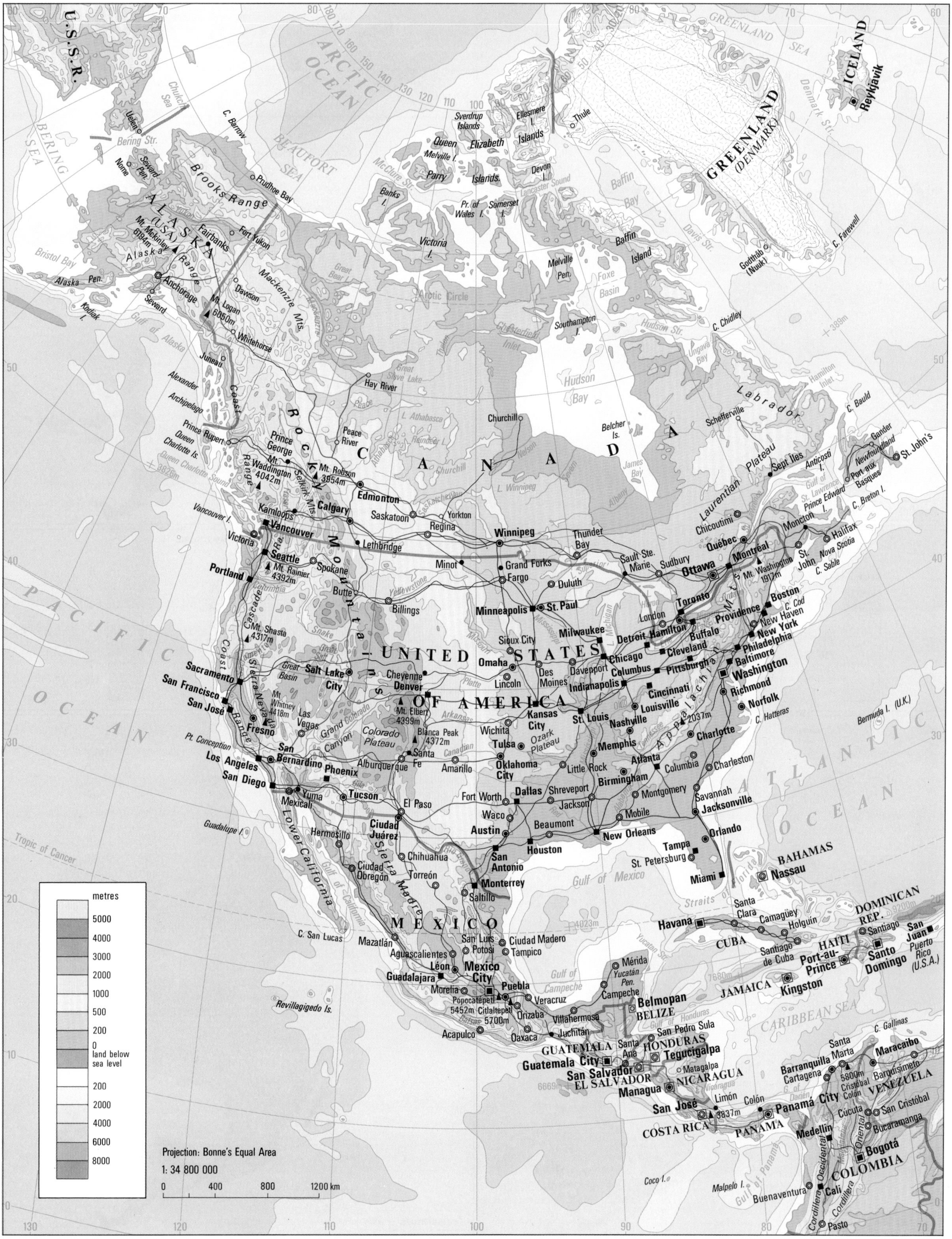
U.S.S.R.
ARCTIC OCEAN
BERING SEA
Chukchi Sea
Bering Str.
Uelen
Nome
Seward Pen.
C. Barrow
BEAUFORT SEA
Prudhoe Bay
Brooks Range
ALASKA (USA)
Fairbanks
Fort Yukon
Mt. McKinley 6194m
Alaska Range
Bristol Bay
Alaska Pen.
Anchorage
Seward
Kodiak I.
Gulf of Alaska
Mt. Logan 6050m
Dawson
Mackenzie Mts.
Whitehorse
Juneau
Alexander Archipelago
Coast Mts.
Sverdrup Islands
Ellesmere I.
Queen Elizabeth Islands
Melville I.
Parry Islands
Devon I.
McClure Str.
Banks I.
Lancaster Sound
Pr. of Wales I.
Somerset I.
Victoria I.
Great Bear L.
Arctic Circle
Thule
GREENLAND (DENMARK)
GREENLAND SEA
Denmark Str.
ICELAND
Reykjavík
Baffin Bay
Baffin Island
Davis Str.
C. Farewell
Godthåb (Nuuk)
Melville Pen.
Foxe Basin
Southampton I.
Hudson Str.
C. Chidley
Ungava Bay
Chesterfield Inlet
Great Slave Lake
Hay River
Hudson Bay
Labrador
Hamilton Inlet
C. Bauld
Churchill
Belcher Is.
Schefferville
James Bay
Rocky Mountains
Prince Rupert
Queen Charlotte Is.
Queen Charlotte Sound
Peace River
Prince George
Mt. Waddington 4042m
Mt. Robson 3954m
Selkirk Mts.
L. Athabasca
Reindeer L.
Churchill
CANADA
Edmonton
Calgary
Saskatoon
Yorkton
Regina
Kamloops
Vancouver I.
Vancouver
Victoria
Lethbridge
L. Winnipeg
Winnipeg
Thunder Bay
Laurentian Plateau
Sept Îles
Anticosti I.
Gulf of St. Lawrence
Newfoundland
Gander
St. John's
Port aux Basques
C. Breton I.
Prince Edward I.
Chicoutimi
Québec
Moncton
Halifax
Nova Scotia
C. Sable
St. John
Montréal
Mt. Washington 1917m
Ottawa
Sudbury
Sault Ste. Marie
Seattle
Mt. Rainier 4392m
Spokane
Portland
Minot
Grand Forks
Fargo
Duluth
Butte
Billings
Yellowstone
Minneapolis
St. Paul
Toronto
London
Hamilton
Buffalo
Boston
C. Cod
Providence
New Haven
New York
PACIFIC OCEAN
Cascade Range
Mt. Shasta 4317m
Coast Ranges
Snake
Great Basin
Salt Lake City
Sioux City
Milwaukee
Detroit
Chicago
Cleveland
Philadelphia
Baltimore
Washington
Pittsburgh
UNITED STATES OF AMERICA
Sacramento
San Francisco
San José
Sierra Nevada
Mt. Whitney 4418m
Fresno
Omaha
Cheyenne
Denver
Lincoln
Des Moines
Davenport
Columbus
Indianapolis
Cincinnati
Richmond
Norfolk
C. Hatteras
Las Vegas
Grand Canyon
Mt. Elbert 4399m
Colorado Plateau
Blanca Peak 4372m
Wichita
Kansas City
St. Louis
Louisville
Nashville
Appalachian Mts.
2037m
Bermuda I. (U.K.)
Pt. Conception
San Bernardino
Los Angeles
San Diego
Santa Fe
Tulsa
Ozark Plateau
Memphis
Charlotte
Albuquerque
Amarillo
Oklahoma City
Little Rock
Atlanta
Columbia
Charleston
Phoenix
Birmingham
Tucson
Yuma
Mexicali
Fort Worth
Dallas
Shreveport
Montgomery
Savannah
Jacksonville
ATLANTIC OCEAN
El Paso
Waco
Jackson
Mobile
Guadalupe I.
Ciudad Juárez
Austin
Beaumont
New Orleans
Orlando
Hermosillo
Tampa
St. Petersburg
Tropic of Cancer
Lower California
Chihuahua
San Antonio
Houston
Miami
BAHAMAS
Nassau
Sierra Madre
Ciudad Obregón
Torreón
Gulf of California
Gulf of Mexico
Monterrey
Saltillo
Straits of Florida
MEXICO
Havana
Santa Clara
Camagüey
Holguín
DOMINICAN REP.
C. San Lucas
4023m
San Luis Potosí
Ciudad Madero
Tampico
CUBA
Santiago de Cuba
HAITI
Santiago
San Juan
Puerto Rico (U.S.A.)
Mazatlán
Aguascalientes
Yucatán Pen.
Mérida
Port-au-Prince
Santo Domingo
León
Guadalajara
Mexico City
Gulf of Campeche
Campeche
7680m
JAMAICA
Kingston
Morelia
Puebla
Veracruz
Belmopan
BELIZE
Popocatépetl 5452m
Citlaltépetl 5700m
Orizaba
Revillagigedo Is.
Villahermosa
Gulf of Honduras
CARIBBEAN SEA
San Pedro Sula
Acapulco
Oaxaca
Juchitán
C. Gallinas
GUATEMALA
Santa Ana
HONDURAS
Tegucigalpa
Santa Marta
Maracaibo
Guatemala City
Matagalpa
Barranquilla
Cartagena
5800m Cristóbal Colón
Barquisimeto
San Salvador
EL SALVADOR
NICARAGUA
VENEZUELA
6669m
Managua
Limón
Colón
Panamá City
Cúcuta
San Cristóbal
San José
3837m
Medellín
Bucaramanga
COSTA RICA
PANAMA
Bogotá
COLOMBIA
Coco I.
Malpelo I.
Gulf of Panamá
Buenaventura
Cali
Cordillera Occidental
Cordillera Oriental
Pasto
metres
5000
4000
3000
2000
1000
500
200
0
land below sea level
200
2000
4000
6000
8000
Projection: Bonne's Equal Area
1: 34 800 000
0
400
800
1200 km
130
120
110
100
90
80
70
60
50
40
30
20
10
0

NORTH AMERICA – CLIMATE

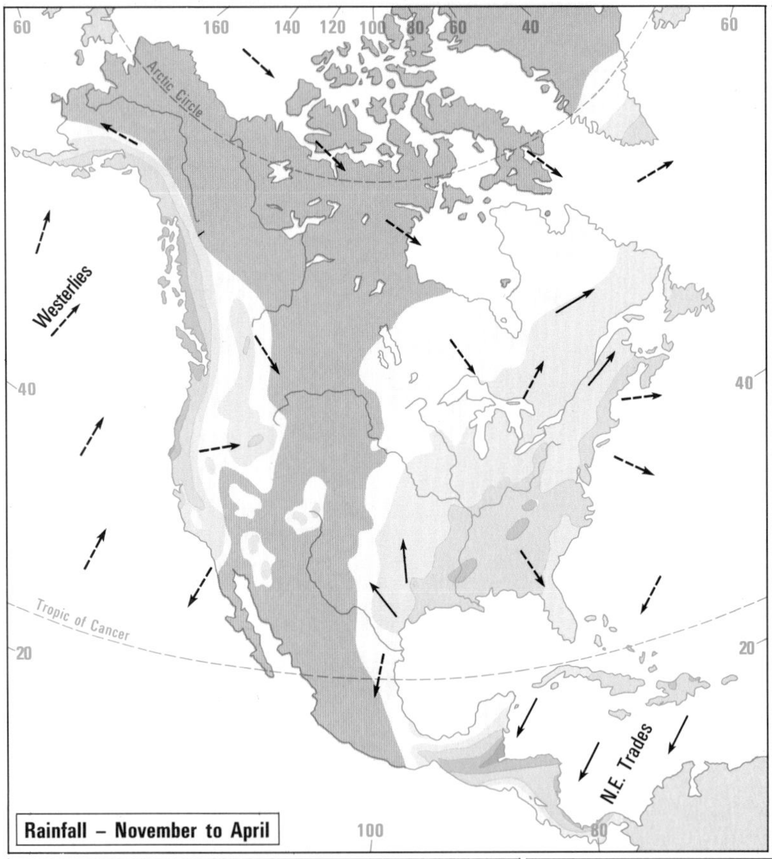

Rainfall – November to April

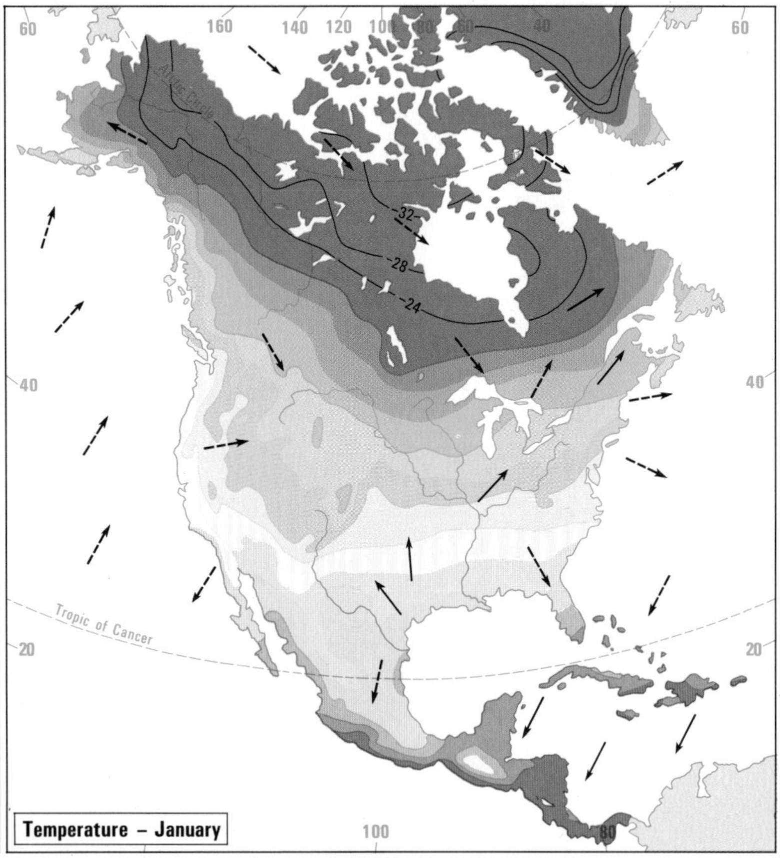

Temperature – January

Rainfall in millimetres

- 1000–1500
- 750–1000
- 500–750
- 250–500
- 125–250
- 0–125

0 1000 2000 3000 kilometres

Prevailing winds

- Blowing all year
- November to April
- May to October

Surface temperature °C

- 32
- 28
- 24
- 20
- 16
- 12
- 8
- 4
- 0
- −4
- −8
- −12
- −16
- below −20

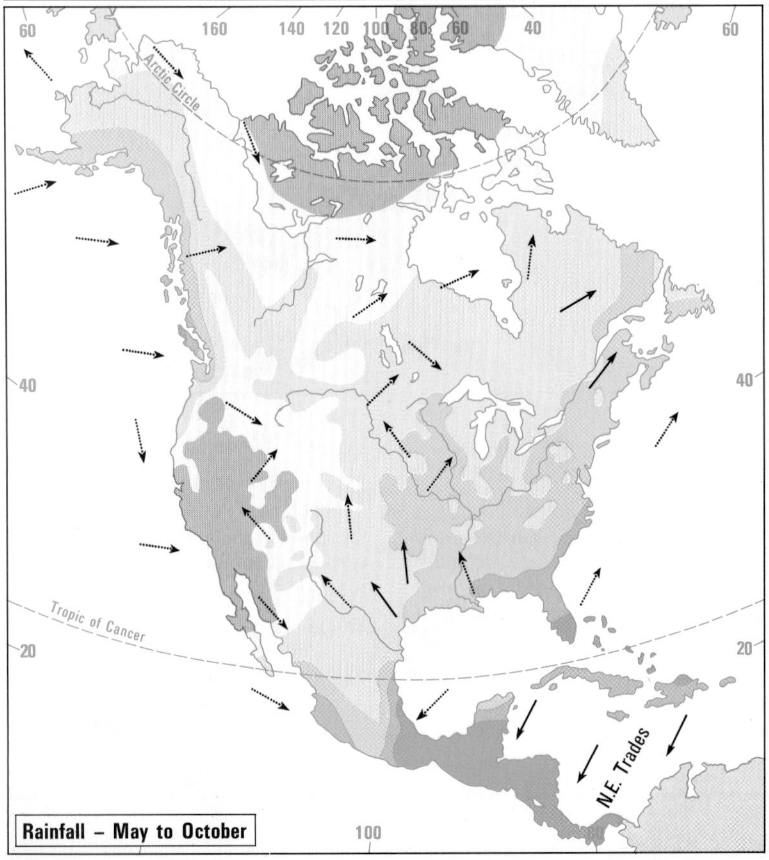

Rainfall – May to October

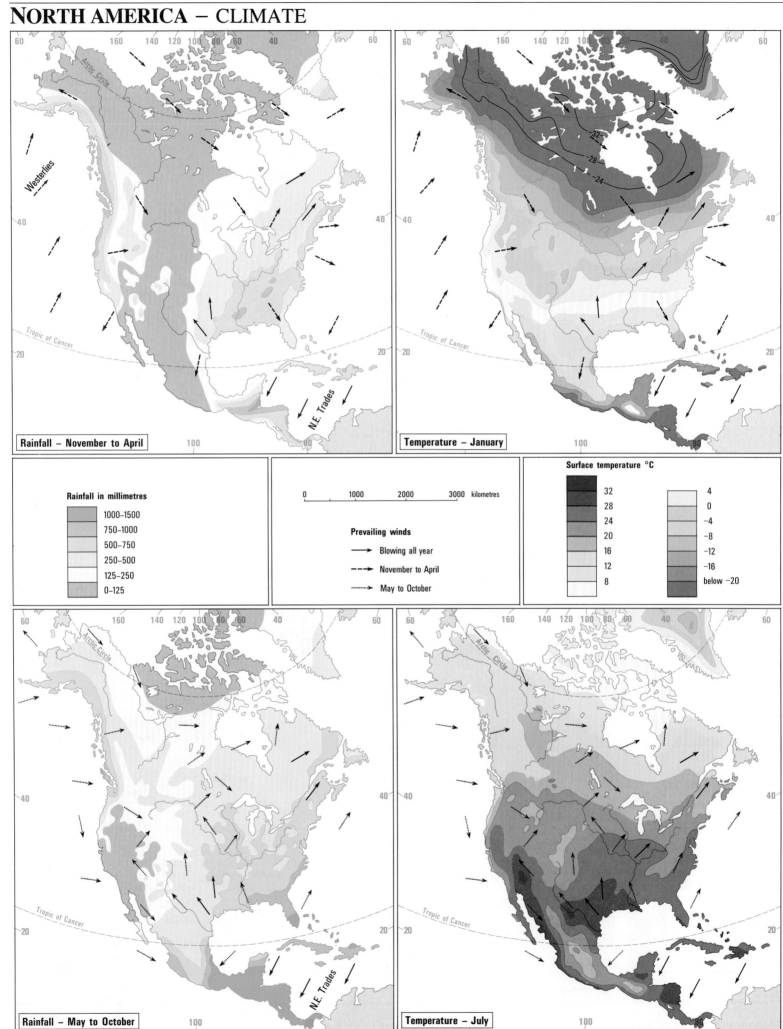

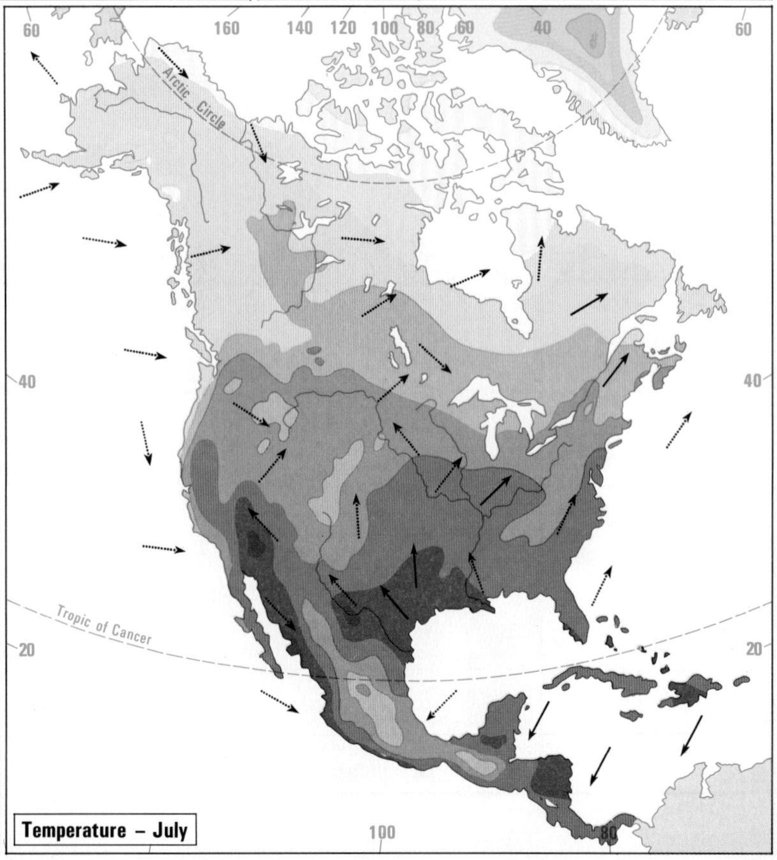

Temperature – July

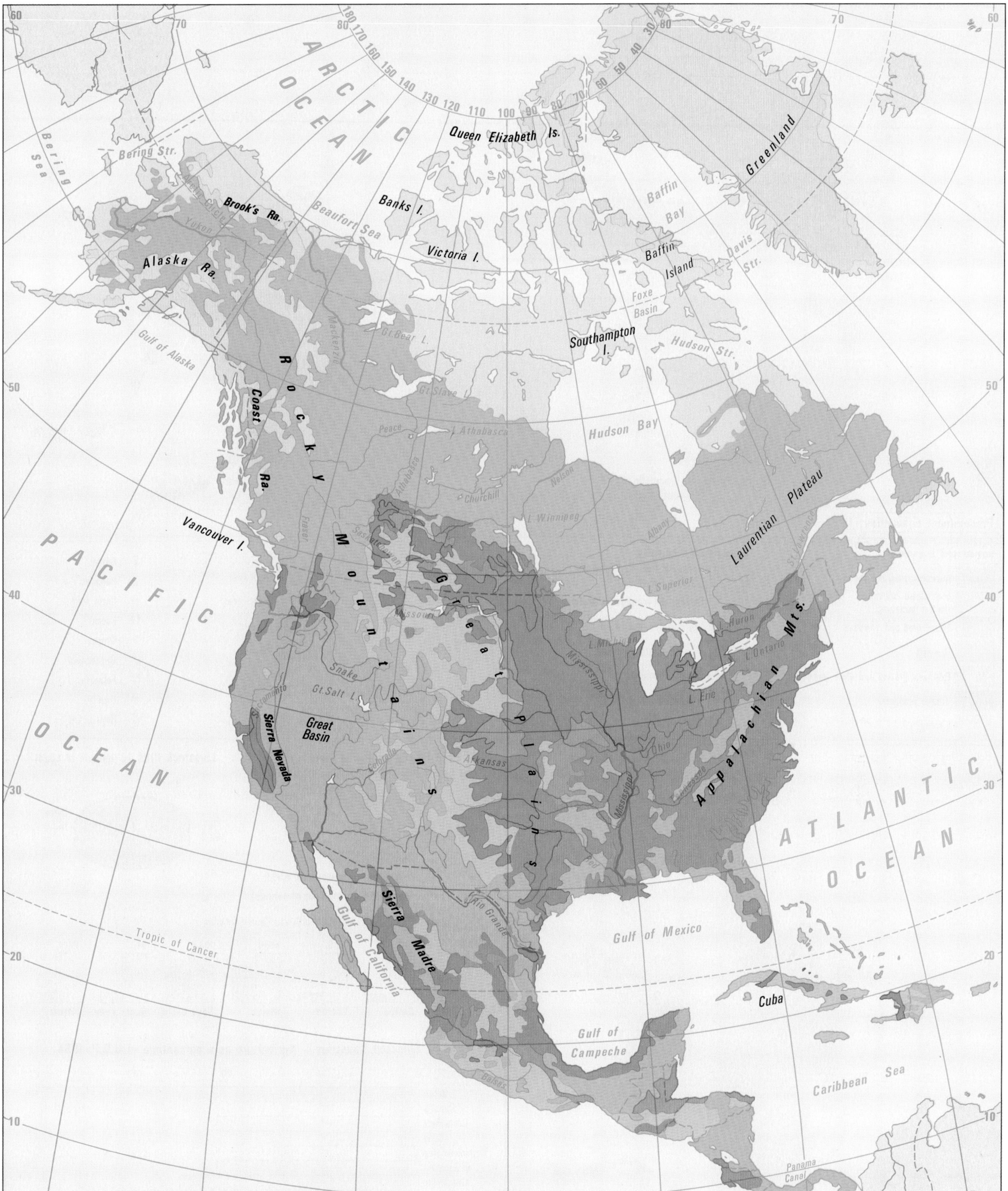

Projection: Bonne's Equal Area
1 : 34 800 000
0 400 800 1200 km

- Cultivated land
- Cultivated land and grassland
- Cultivated land and woodland
- Forest and woodland
- Grassland
- Swamp and marsh
- Scrub and sparse grassland
- Tundra
- Barren land (including permanent ice)

NORTH AMERICA – AGRICULTURE

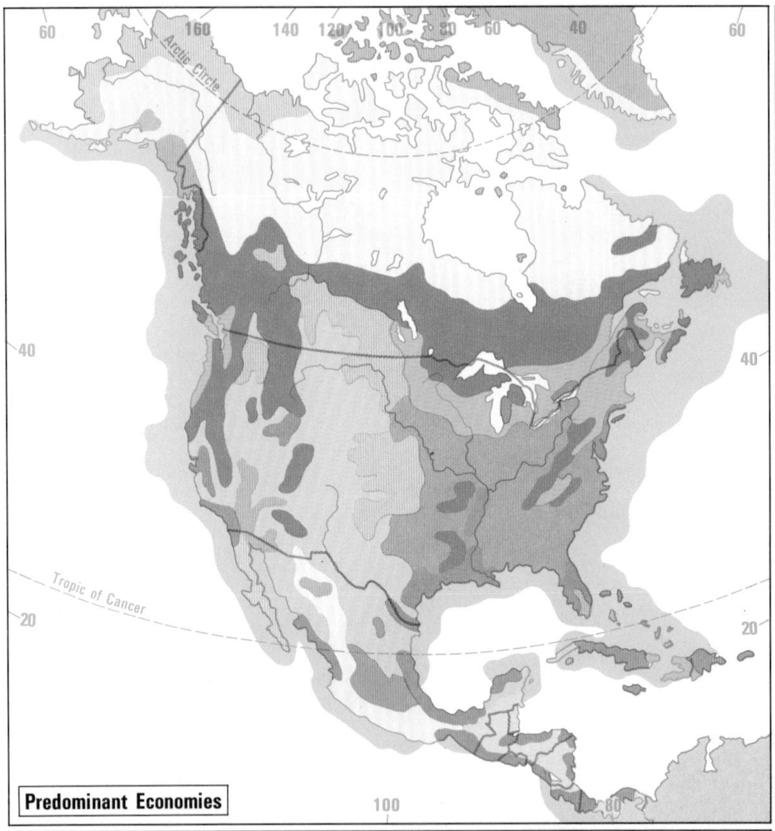

Predominant Economies

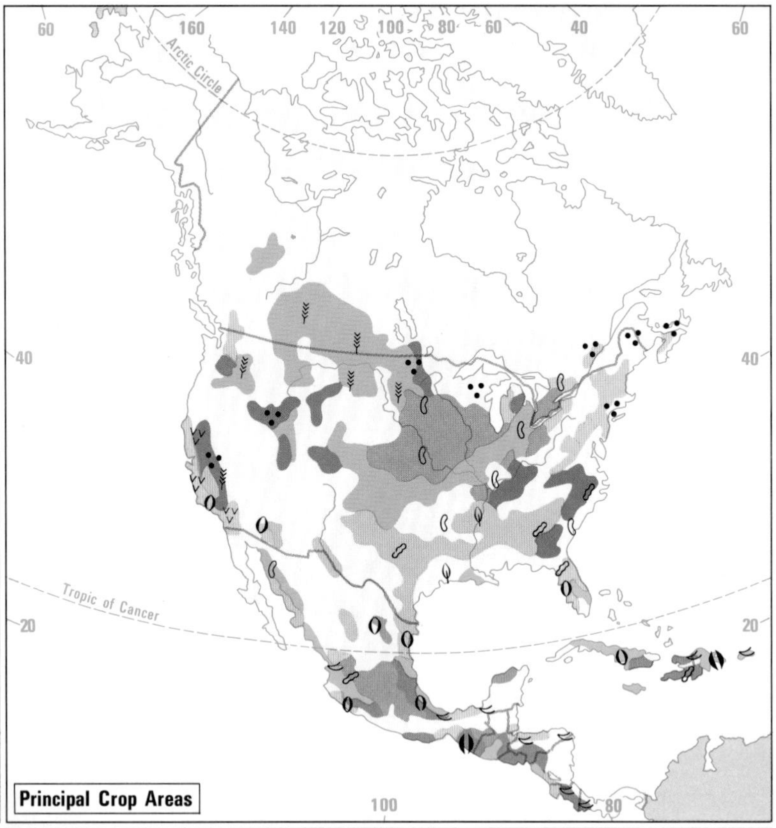

Principal Crop Areas

Commercial Economies

- Intensive grain cultivation
- Specialised cultivation – plantations and market gardening
- Mixed crop and livestock
- Dairying
- Extensive grazing and stock raising
- Timber production
- Fishing

Subsistence Economies

- Gathering, hunting, fishing and primitive cultivation
- Shifting and marginal cultivation
- Intensive crop cultivation
- Mixed crop and livestock
- Nomadic herding
- Little or no activity

- Coffee
- Cotton
- Fruit
- Maize
- Sugar beet
- Sugar cane
- Tobacco
- Wheat

- Barley
- Bananas
- Citrus fruit
- Cocoa
- Grapes
- Groundnuts
- Potatoes
- Rice
- Soya beans

0 1000 2000 kilometres

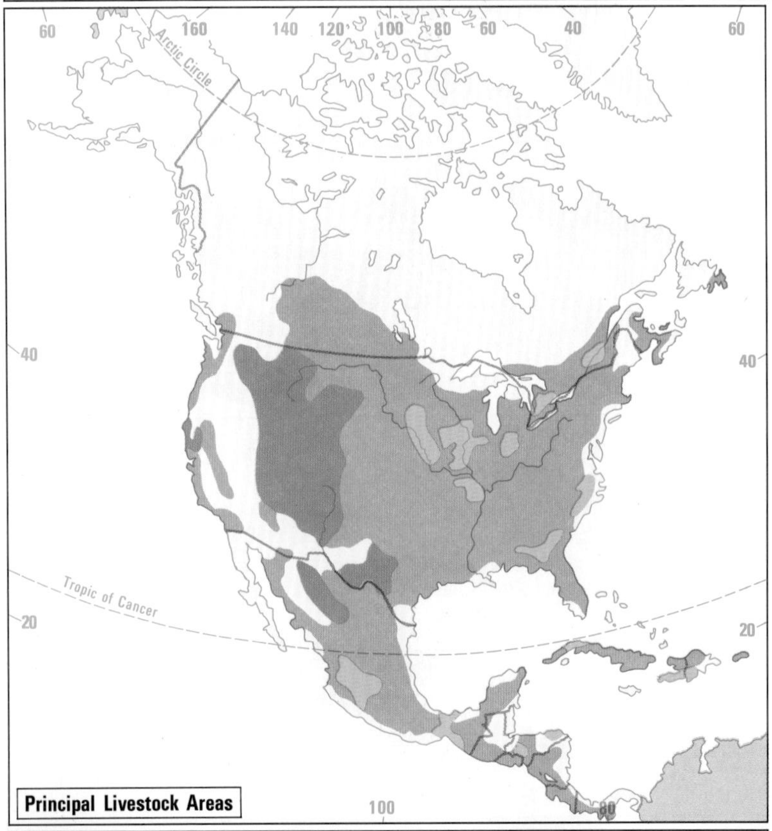

Principal Livestock Areas

- Cattle
- Pigs
- Sheep

Land Use (in millions of hectares)
Total land area – 2 139.2 million hectares

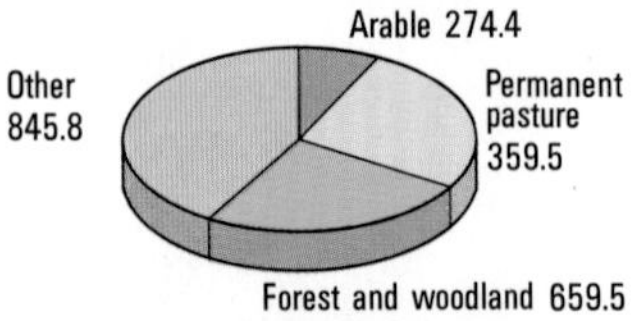

Livestock 1985 (in millions of head)

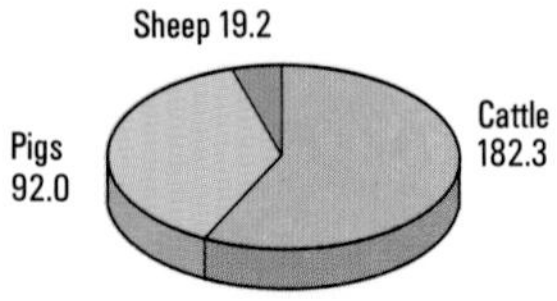

Crop Production 1985 (in millions of tonnes)

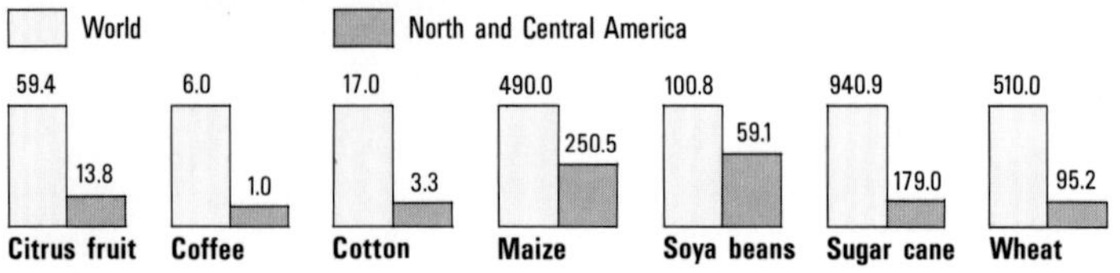

Selected Countries – Agriculture as a percentage of G.D.P. 1984

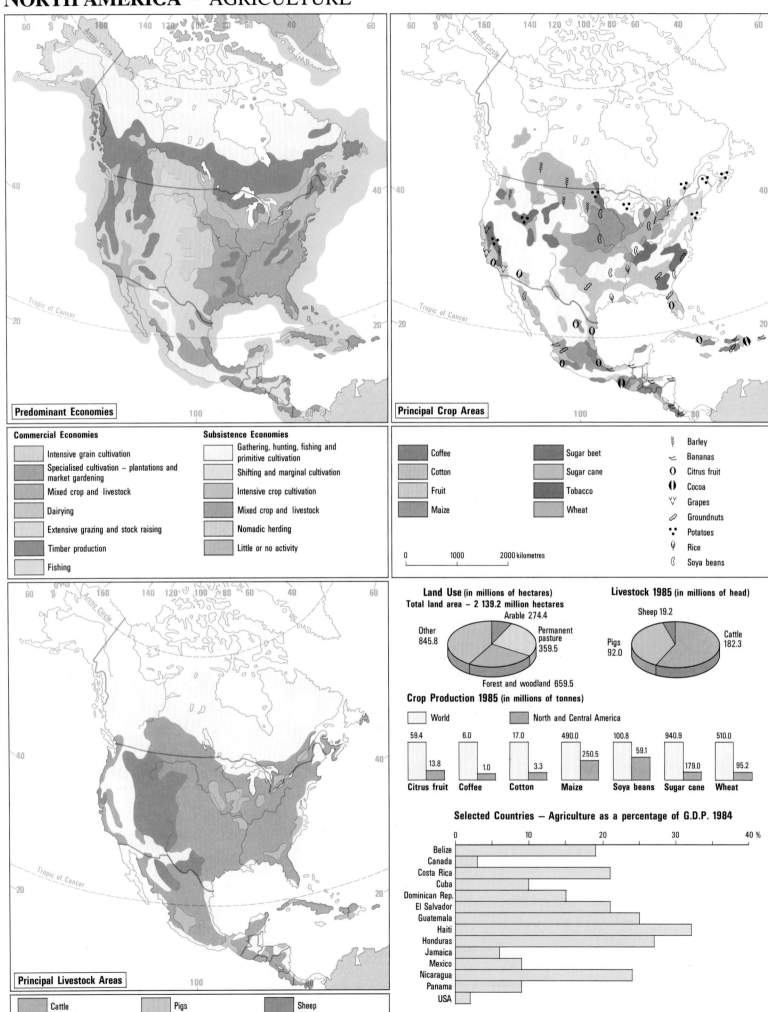

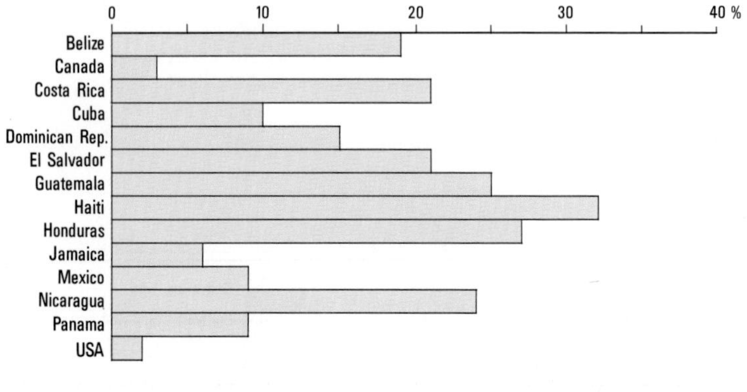

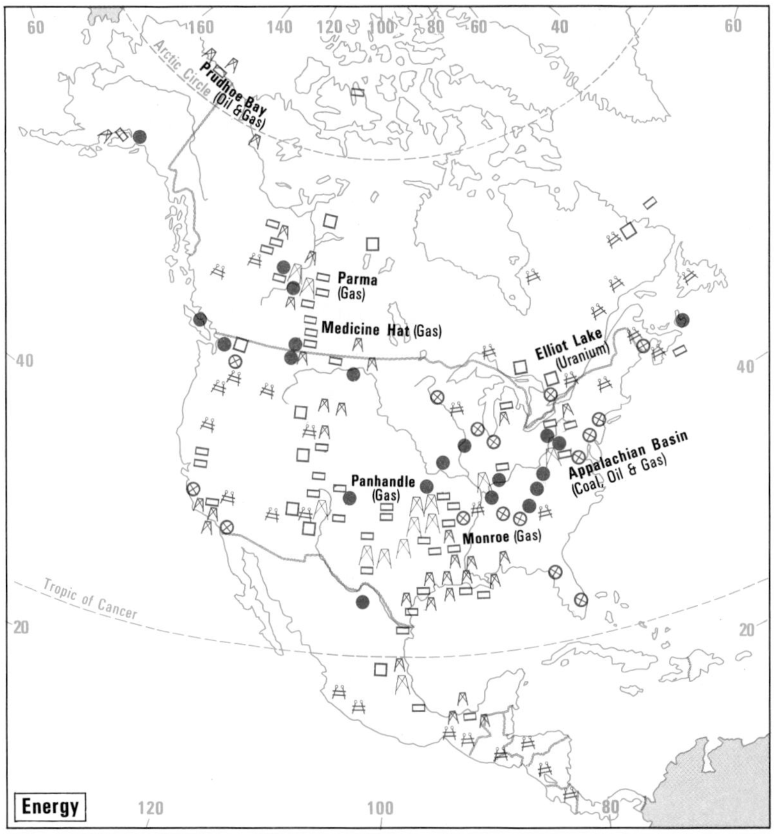

● Coal | ▭ Natural gas | Oil | ☐ Uranium | H.E.P. station | ⊗ Nuclear power station

0 1000 2000 3000km

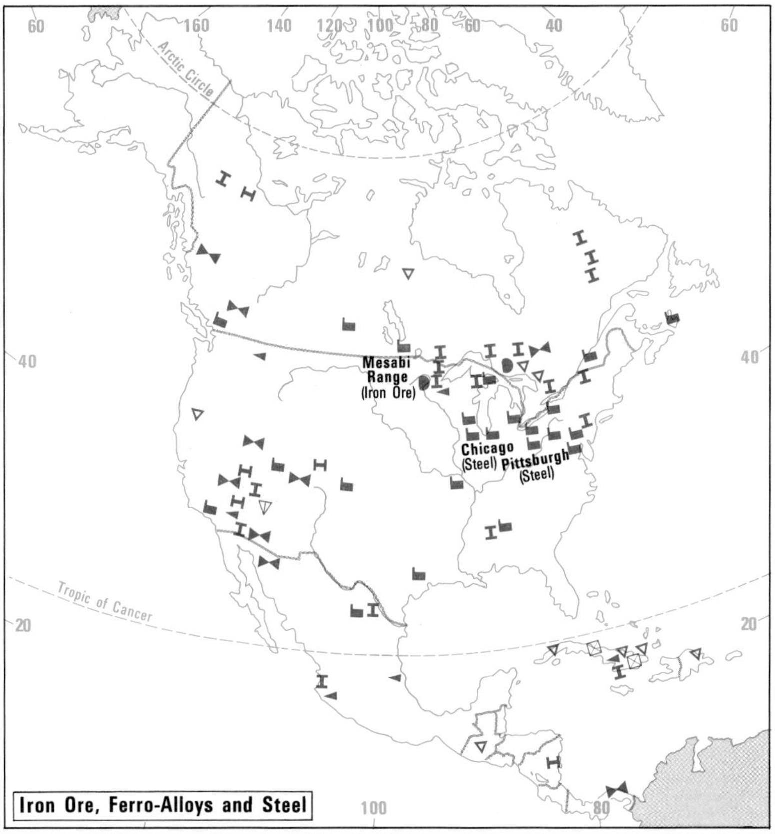

Chrome | Cobalt | Iron ore | Manganese | Molybdenum | Nickel | Steel | Tungsten | Vanadium

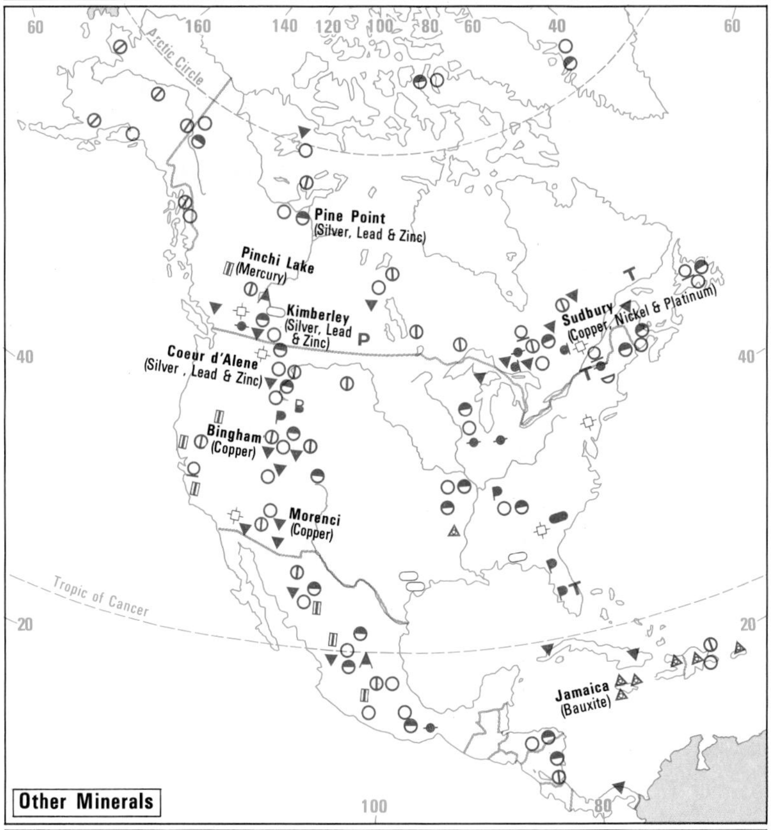

Antimony | Asbestos | Bauxite | Beryllium | Copper | Gold | Lead and zinc | Lithium | Mercury | Mica | Phosphates | Platinum | Potash | Pyrites | Silver | Sulphur | Titanium

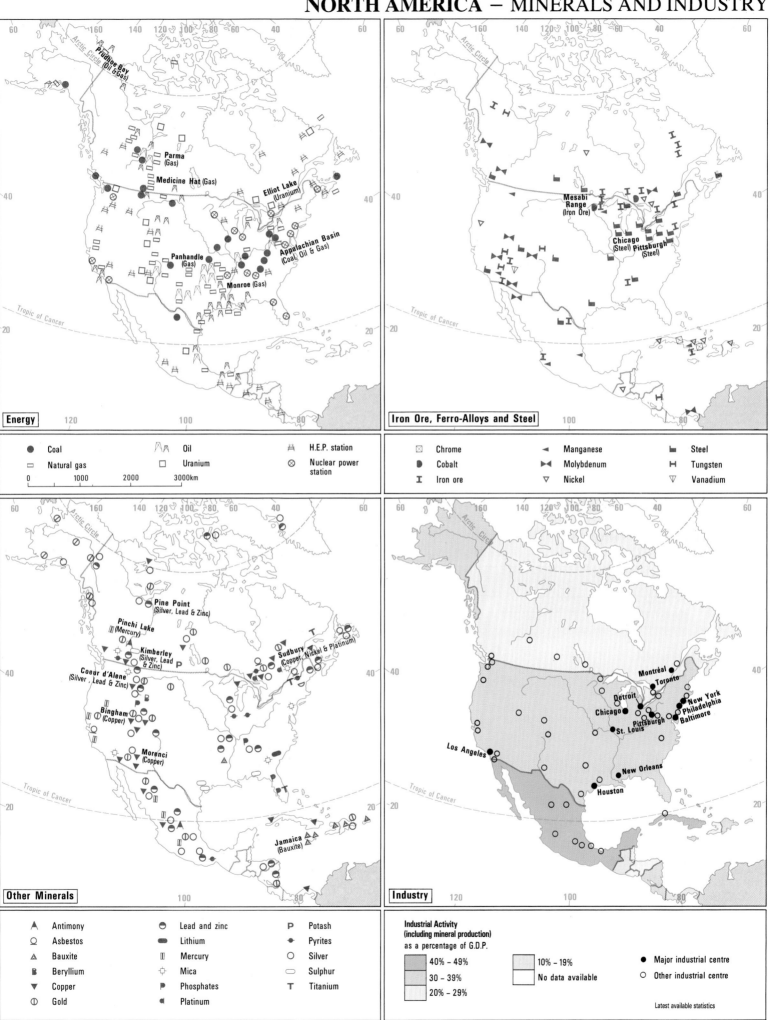

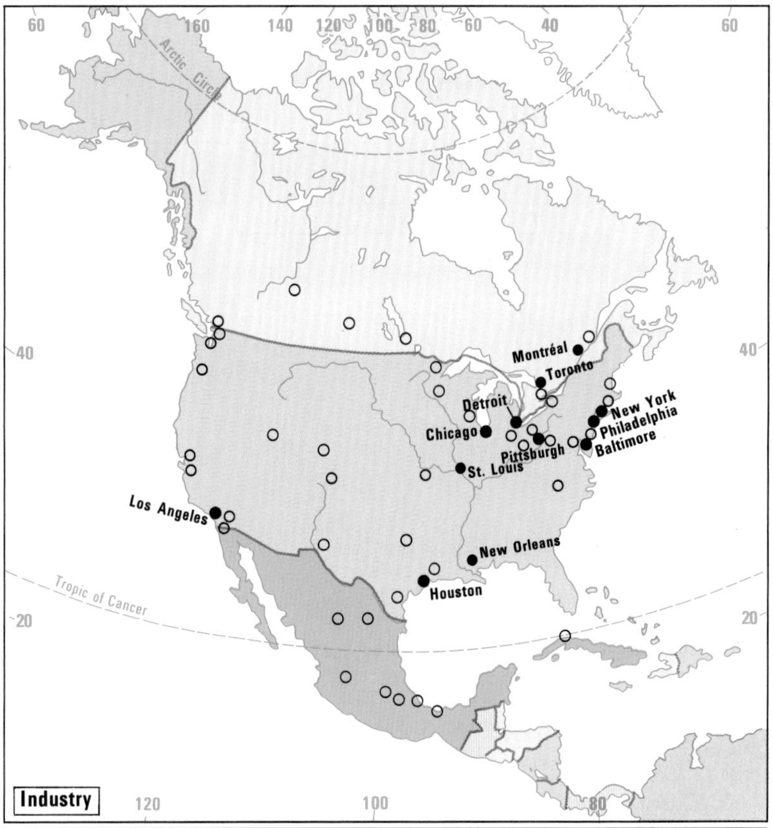

Industrial Activity (including mineral production) as a percentage of G.D.P.

40% – 49% | 30 – 39% | 20% – 29% | 10% – 19% | No data available

● Major industrial centre | ○ Other industrial centre

Latest available statistics

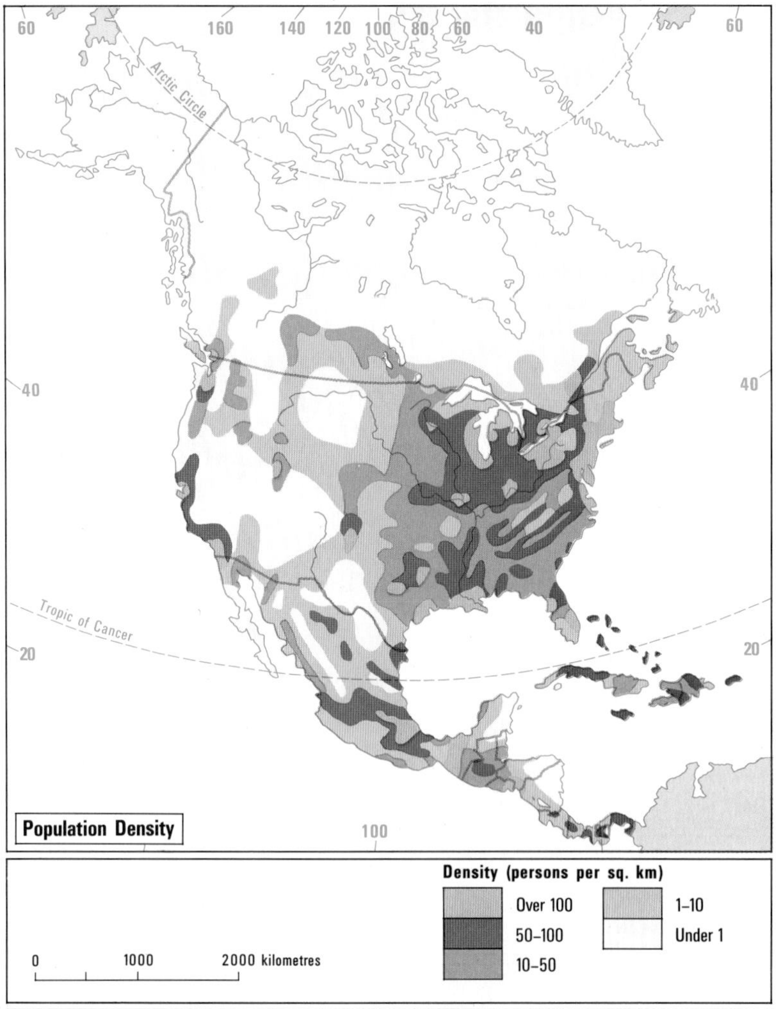

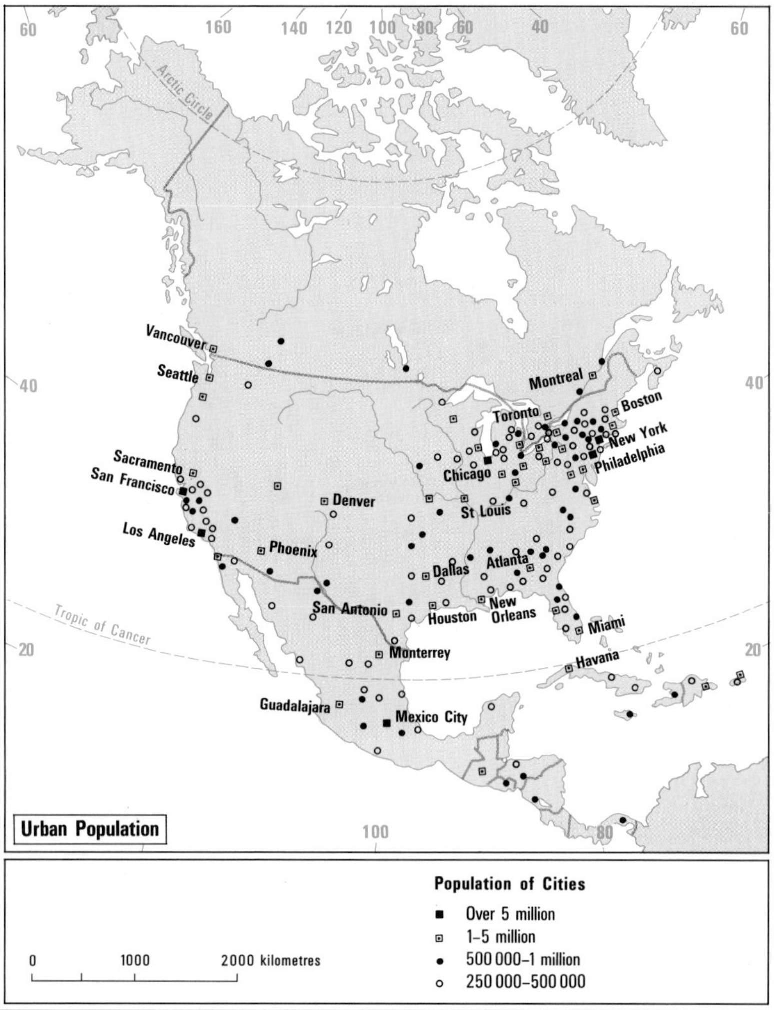

Population by Selected Country

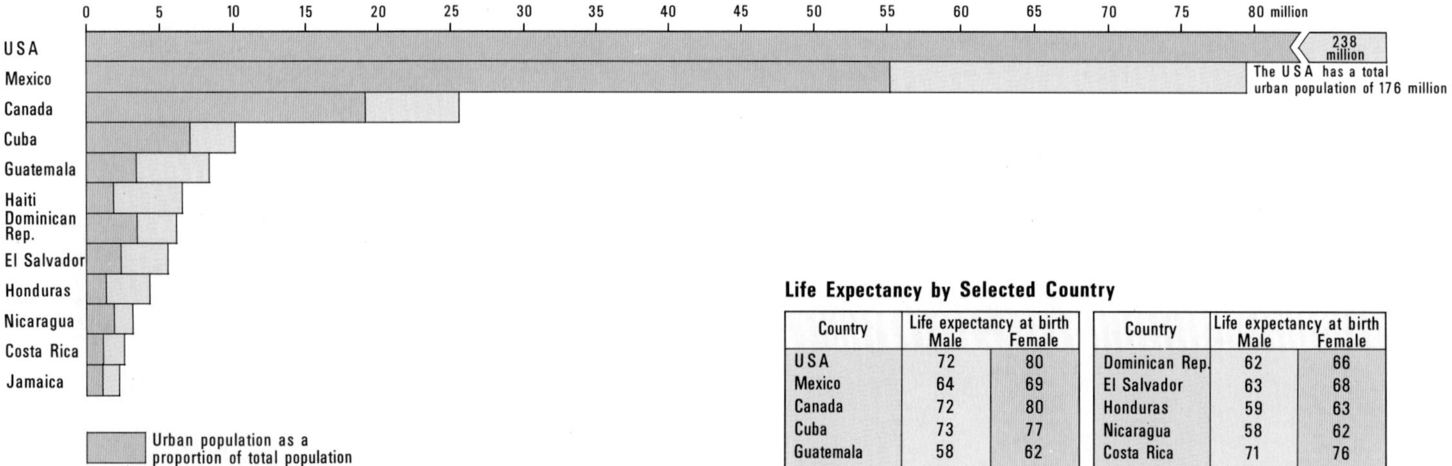

Source: U.N. Statistics 1985 (estimates)

Life Expectancy by Selected Country

Country	Life expectancy at birth Male	Female
USA	72	80
Mexico	64	69
Canada	72	80
Cuba	73	77
Guatemala	58	62
Haiti	53	57

Country	Life expectancy at birth Male	Female
Dominican Rep.	62	66
El Salvador	63	68
Honduras	59	63
Nicaragua	58	62
Costa Rica	71	76
Jamaica	71	76

Percentage Population Breakdown by Age and Sex

Source: UN Demographic Yearbook

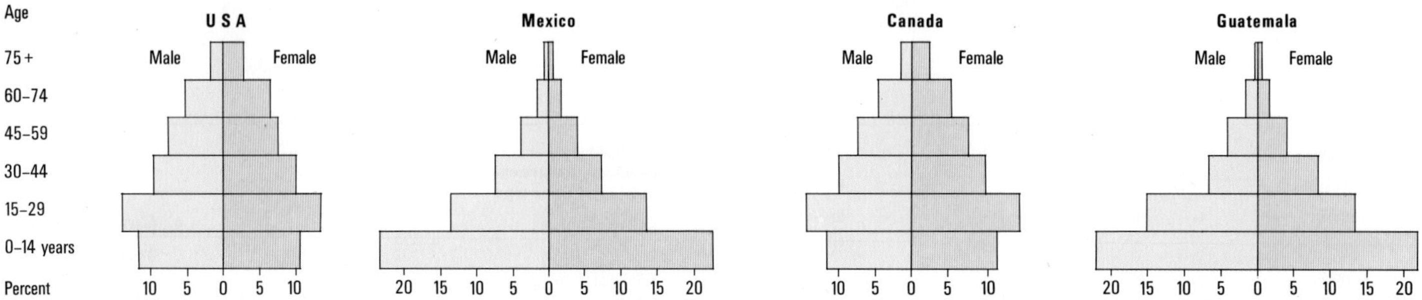

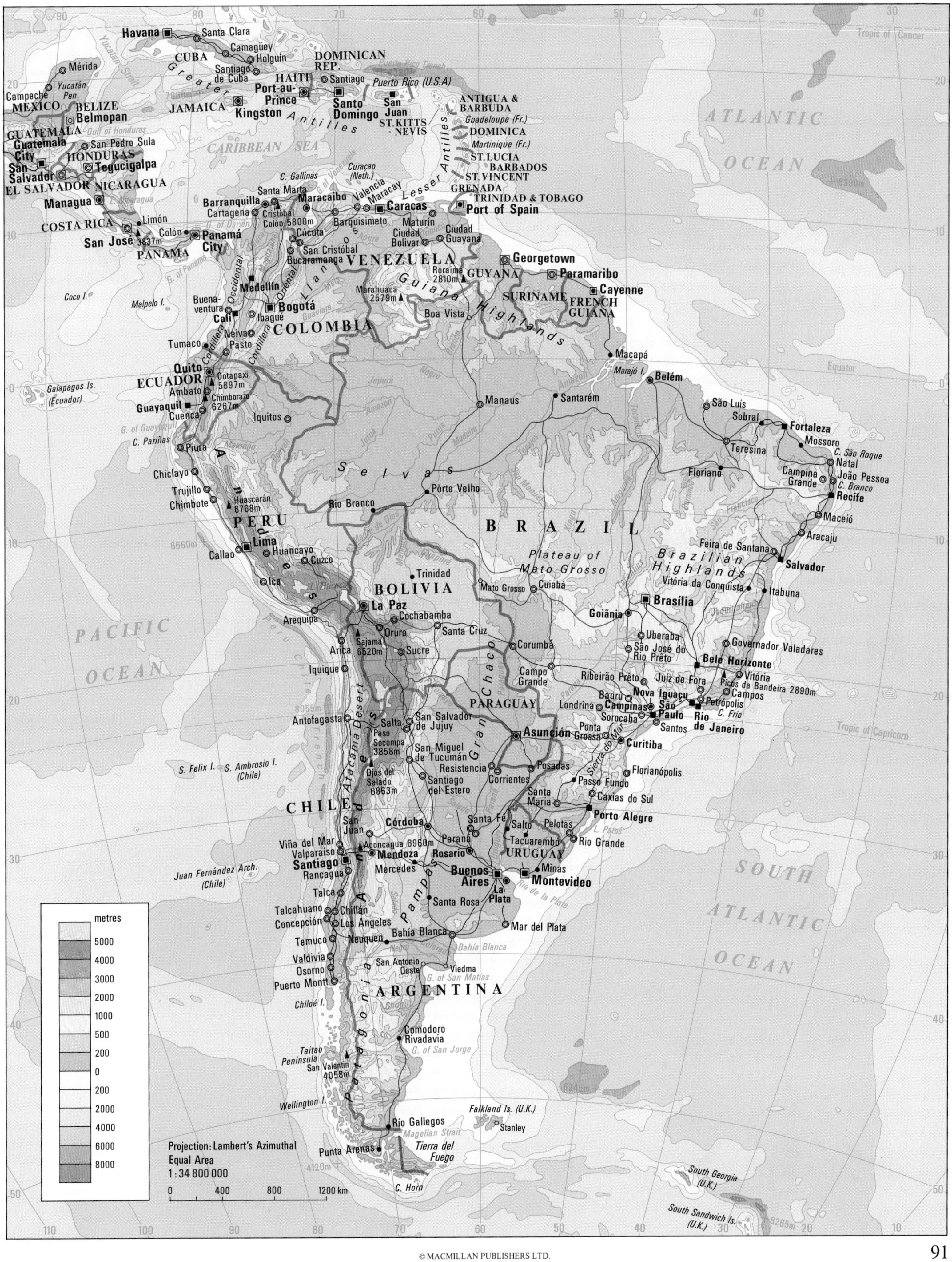
Havana
Santa Clara
Camagüey
Holguín
CUBA
Santiago de Cuba
Greater Antilles
DOMINICAN REP.
HAITI
Santiago
Port-au-Prince
Santo Domingo
Puerto Rico (U.S.A)
San Juan
ST.KITTS - NEVIS
ANTIGUA & BARBUDA
Guadeloupe (Fr.)
DOMINICA
Martinique (Fr.)
ST.LUCIA
BARBADOS
ST.VINCENT
GRENADA
TRINIDAD & TOBAGO
Port of Spain
Lesser Antilles
JAMAICA
Kingston
Mérida
Yucatán Pen.
Campeche
MEXICO
BELIZE
Belmopan
GUATEMALA
Guatemala City
San Pedro Sula
HONDURAS
Tegucigalpa
San Salvador
EL SALVADOR
NICARAGUA
Managua
COSTA RICA
Limón
San José
PANAMA
Colón
Panamá City
CARIBBEAN SEA
Gulf of Honduras
Curaçao (Neth.)
C. Gallinas
Santa Marta
Barranquilla
Cartagena
Maracaibo
Valencia
Maracay
Caracas
Barquisimeto
Maturín
Ciudad Bolívar
Ciudad Guayana
Cristóbal Colón 5800m
Cúcuta
San Cristóbal
Bucaramanga
VENEZUELA
Roraima 2810m
GUYANA
Georgetown
Paramaribo
SURINAME
Cayenne
FRENCH GUIANA
Marahuaca 2579m
Guiana Highlands
Boa Vista
Coco I.
Malpelo I.
Medellín
Buenaventura
Cali
Bogotá
Ibagué
Neiva
COLOMBIA
Tumaco
Pasto
Quito
ECUADOR
Cotopaxi 5897m
Ambato
Chimborazo 6267m
Guayaquil
Cuenca
Galapagos Is. (Ecuador)
G. of Guayaquil
C. Pariñas
Piura
Iquitos
Chiclayo
Trujillo
Chimbote
Huascarán 6768m
PERU
Lima
Callao
Huancayo
Cuzco
Ica
Arequipa
Macapá
Marajó I.
Belém
Manaus
Santarém
São Luís
Sobral
Fortaleza
Mossoró
Teresina
C. São Roque
Natal
João Pessoa
Campina Grande
C. Branco
Recife
Floriano
Maceió
Aracaju
Selvas
Pôrto Velho
Rio Branco
BRAZIL
Feira de Santana
Salvador
Brazilian Highlands
Vitória da Conquista
Itabuna
Plateau of Mato Grosso
Mato Grosso
Cuiabá
Trinidad
BOLIVIA
La Paz
Cochabamba
Santa Cruz
Oruro
Sucre
Sajama 6520m
Arica
Iquique
Brasília
Goiânia
Uberaba
São José do Rio Prêto
Governador Valadares
Belo Horizonte
Vitória
Pico da Bandeira 2890m
Campos
Corumbá
Campo Grande
Ribeirão Prêto
Juiz de Fora
Bauru
Nova Iguaçu
Petrópolis
Londrina
Campinas
São Paulo
Rio de Janeiro
C. Frio
Sorocaba
Santos
Ponta Grossa
Curitiba
Tropic of Capricorn
PARAGUAY
Asunción
Gran Chaco
Antofagasta
Salta
San Salvador de Jujuy
Paso Socompa 3858m
San Miguel de Tucumán
Atacama Desert
Ojos del Salado 6863m
Santiago del Estero
Resistencia
Corrientes
Posadas
Passo Fundo
Florianópolis
Caxias do Sul
Santa Maria
Porto Alegre
S. Felix I.
S. Ambrosio I. (Chile)
CHILE
San Juan
Córdoba
Santa Fé
Salto
Pelotas
L. Patos
Tacuarembó
Rio Grande
URUGUAY
Viña del Mar
Valparaíso
Aconcagua 6960m
Santiago
Mendoza
Paraná
Rosario
Rancagua
Mercedes
Buenos Aires
La Plata
Minas
Montevideo
Rio de la Plata
Juan Fernández Arch. (Chile)
Talca
Santa Rosa
Pampas
Talcahuano
Chillán
Concepción
Los Angeles
Bahía Blanca
Mar del Plata
Temuco
Neuquén
Valdivia
Osorno
San Antonio Oeste
Viedma
G. of San Matías
Puerto Montt
ARGENTINA
Chiloé I.
Patagonia
Comodoro Rivadavia
G. of San Jorge
Taitao Peninsula
San Valentín 4058m
Wellington I.
Río Gallegos
Magellan Strait
Falkland Is. (U.K.)
Stanley
Punta Arenas
Tierra del Fuego
C. Horn
South Georgia (U.K.)
South Sandwich Is. (U.K.)
Andes
PACIFIC OCEAN
ATLANTIC OCEAN
SOUTH ATLANTIC OCEAN
Tropic of Cancer
Equator
metres
5000
4000
3000
2000
1000
500
200
0
200
2000
4000
6000
8000
Projection: Lambert's Azimuthal Equal Area
1:34 800 000
0 400 800 1200 km

SOUTH AMERICA – CLIMATE

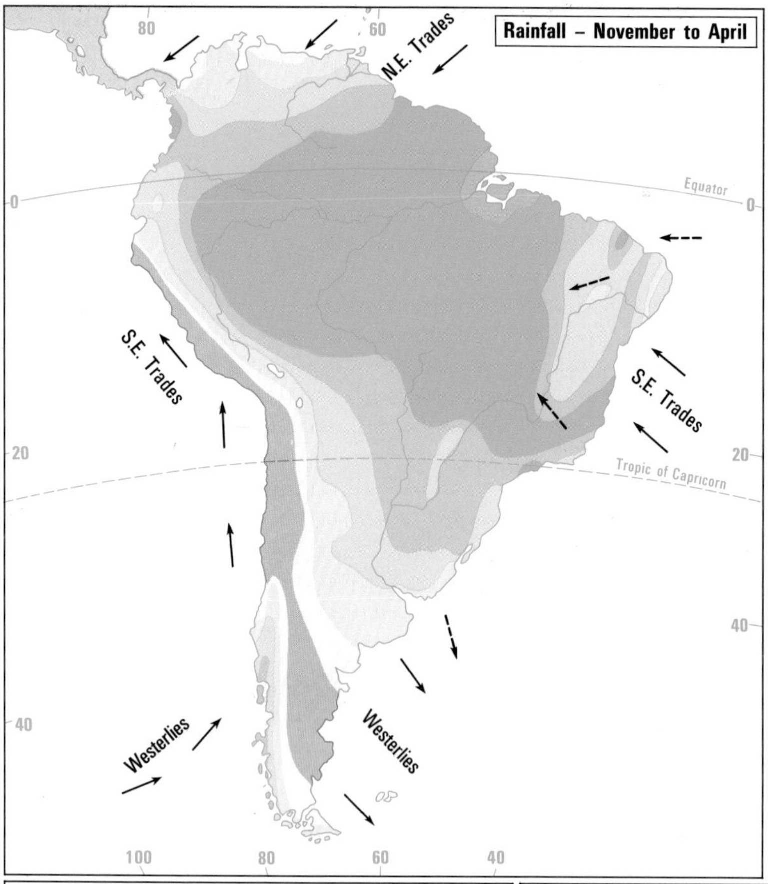

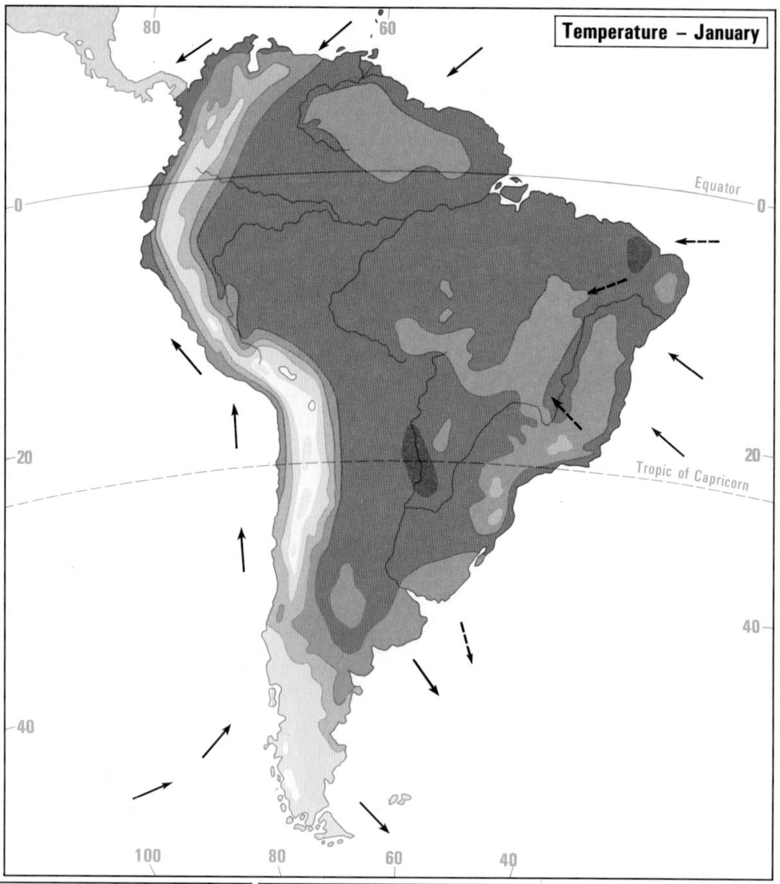

Rainfall in millimetres

1500–2000
1000–1500
750–1000
500–750
250–500
125–250
0–125

0 500 1500 kilometres

Prevailing winds

→ Blowing all year
- - → November to April
······→ May to October

Surface temperature °C

28
24
20
16
12
8
4
0

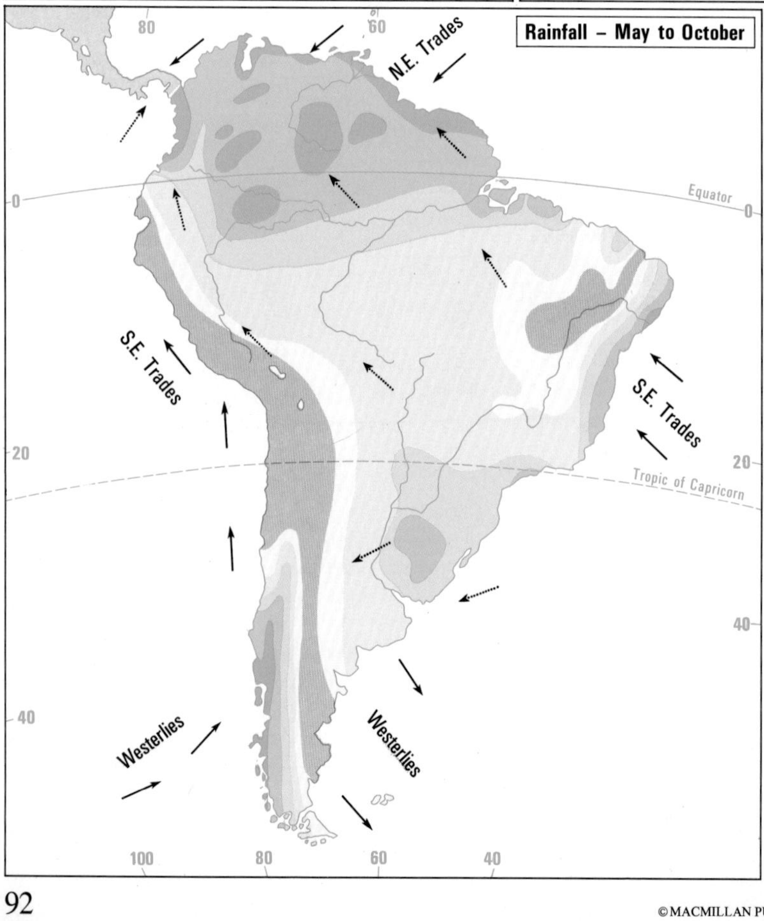

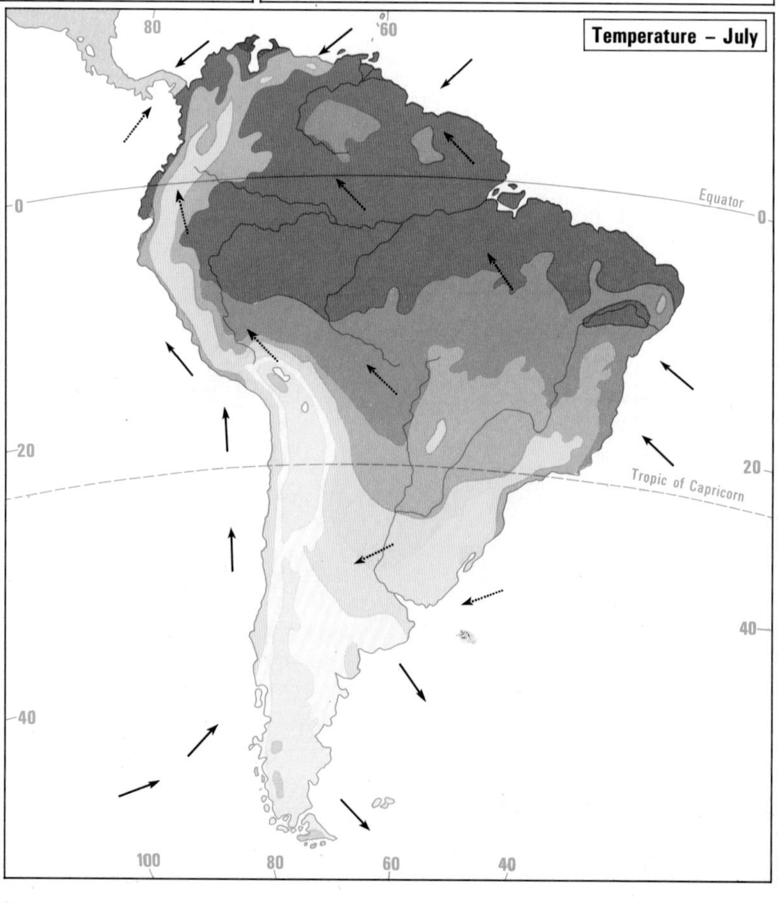

SOUTH AMERICA – NATURAL AND HUMAN ENVIRONMENT

Projection: Lambert's Azimuthal Equal Area.
1 : 34 800 000
0 400 800 1200 km

- Cultivated land
- Cultivated land and grassland
- Forest and woodland
- Grassland
- Swamp and marsh
- Scrub and sparse grassland
- Barren land (including permanent ice)

SOUTH AMERICA – AGRICULTURE

Predominant Economies

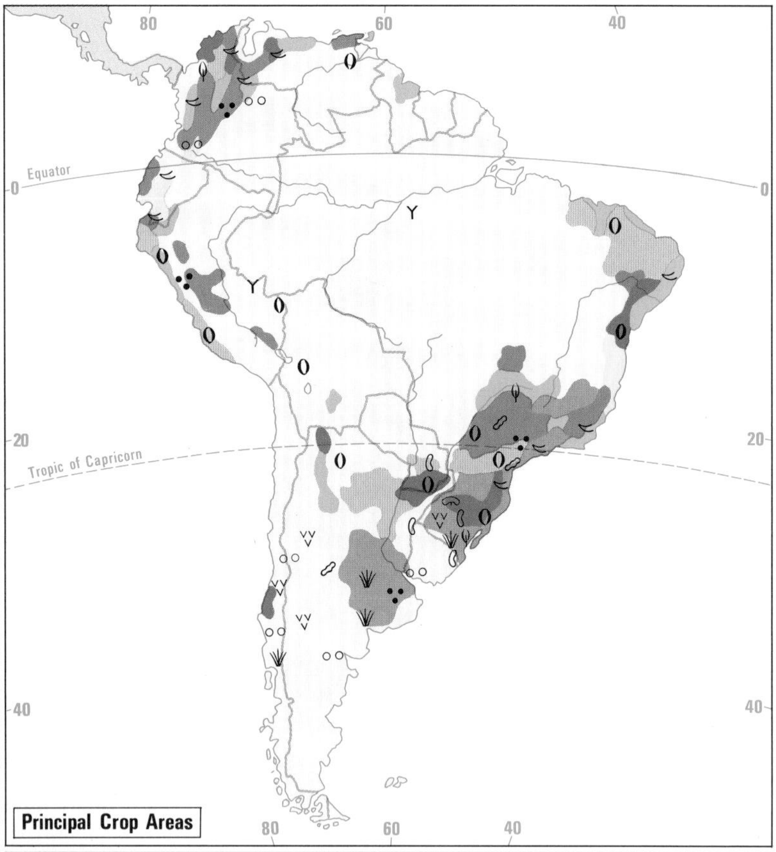

Principal Crop Areas

Commercial Economies

- Intensive grain cultivation
- Specialised cultivation – plantations and market gardening
- Mixed crop and livestock
- Extensive grazing and stock raising
- Timber production
- Fishing

Subsistence Economies

- Gathering, hunting, fishing and primitive cultivation
- Shifting and marginal cultivation
- Intensive crop cultivation
- Little or no activity

- Cocoa
- Coffee
- Cotton
- Maize
- Sugar cane
- Tobacco
- Bananas
- Citrus fruit
- Fruit
- Grapes
- Groundnuts
- Potatoes
- Rice
- Rubber
- Soya beans
- Tea
- Wheat

0 500 1000 kilometres

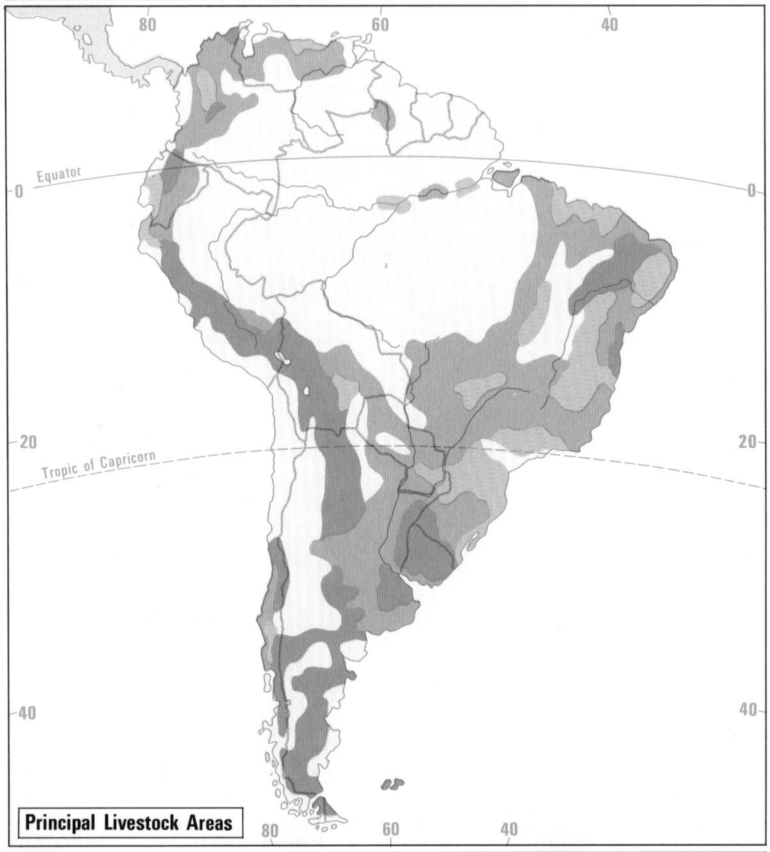

Principal Livestock Areas

- Cattle
- Pigs
- Sheep

Land Use (in millions of hectares)
Total land area – 1 753.5 million hectares

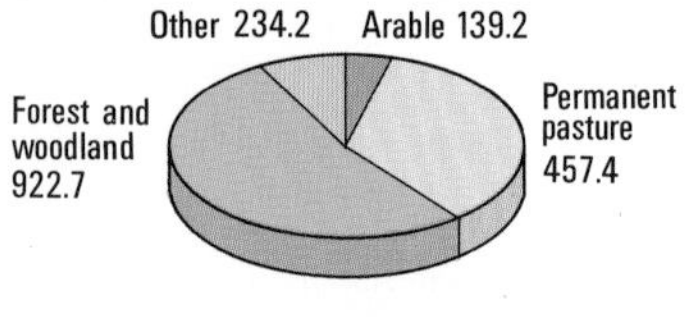

Livestock 1985 (in millions of head)

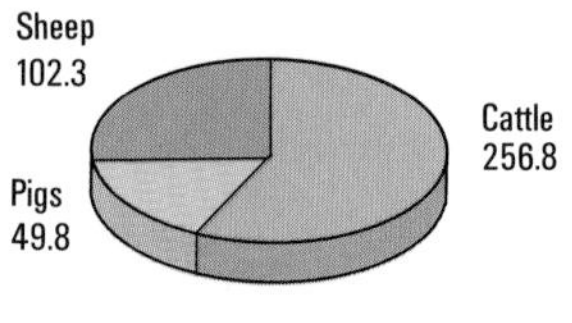

Crop Production 1985 (in millions of tonnes)

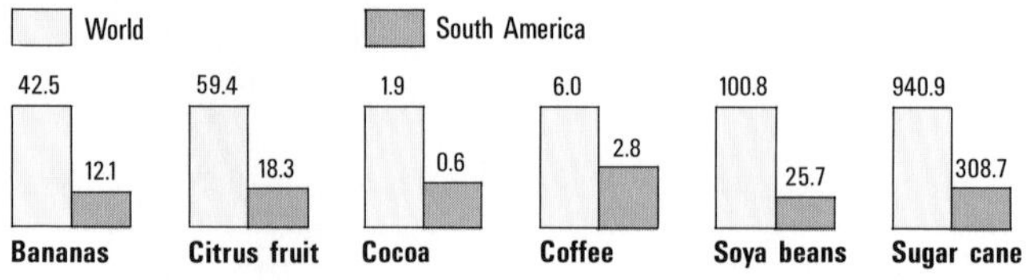

Selected Countries – Agriculture as a percentage of G.D.P. 1984

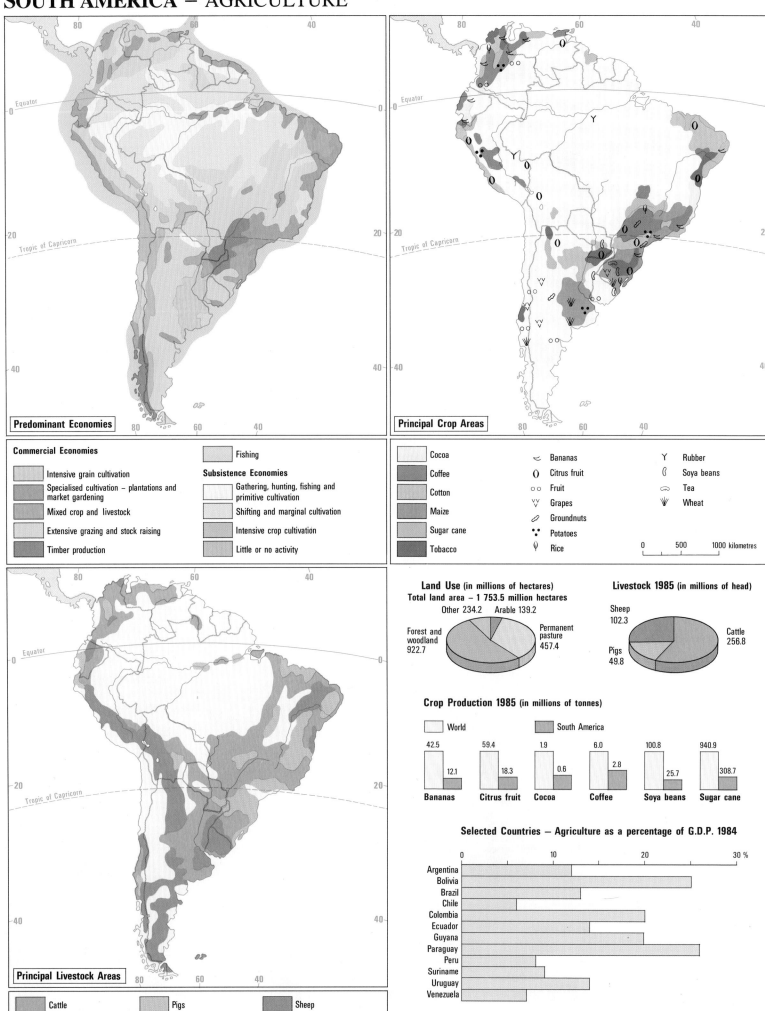

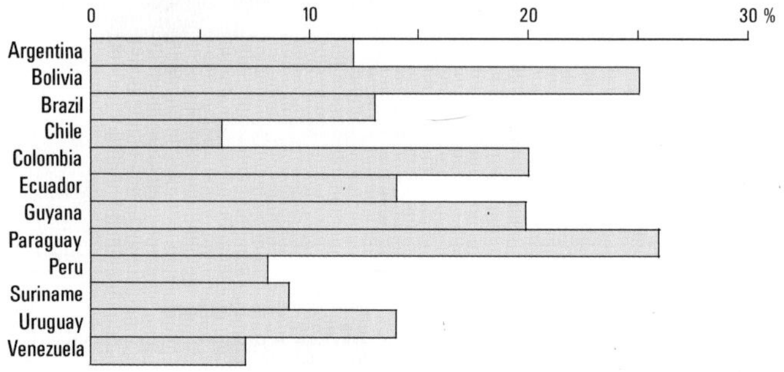

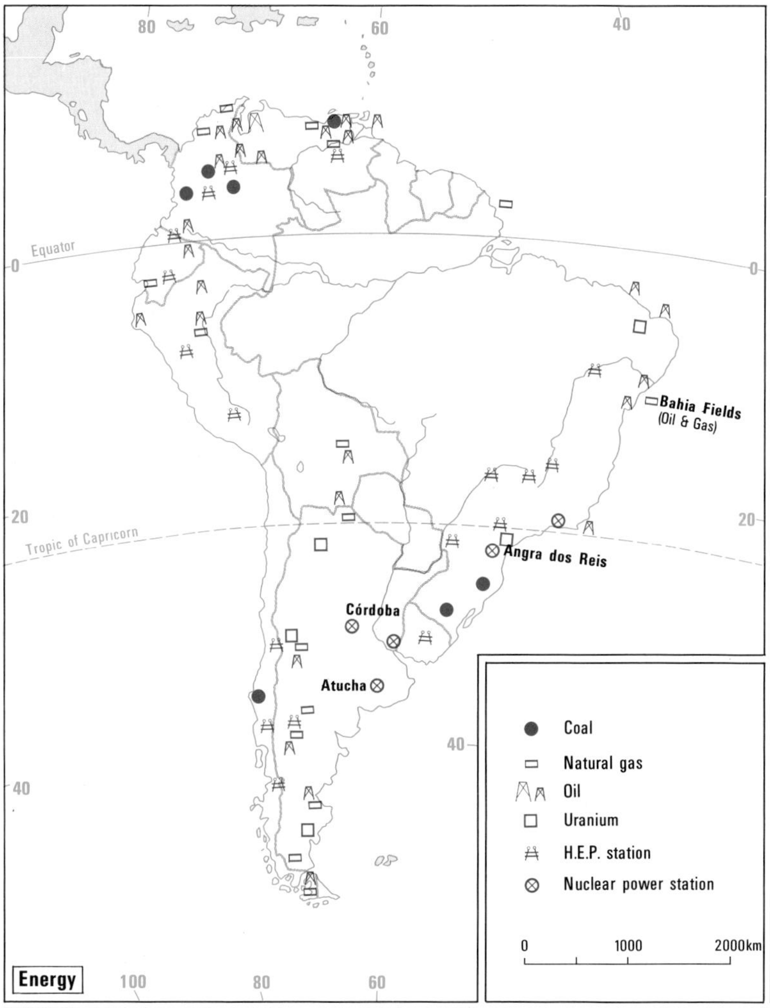

Energy
Equator
Tropic of Capricorn
Bahia Fields
(Oil & Gas)
Angra dos Reis
Córdoba
Atucha
Coal
Natural gas
Oil
Uranium
H.E.P. station
Nuclear power station
0
1000
2000km

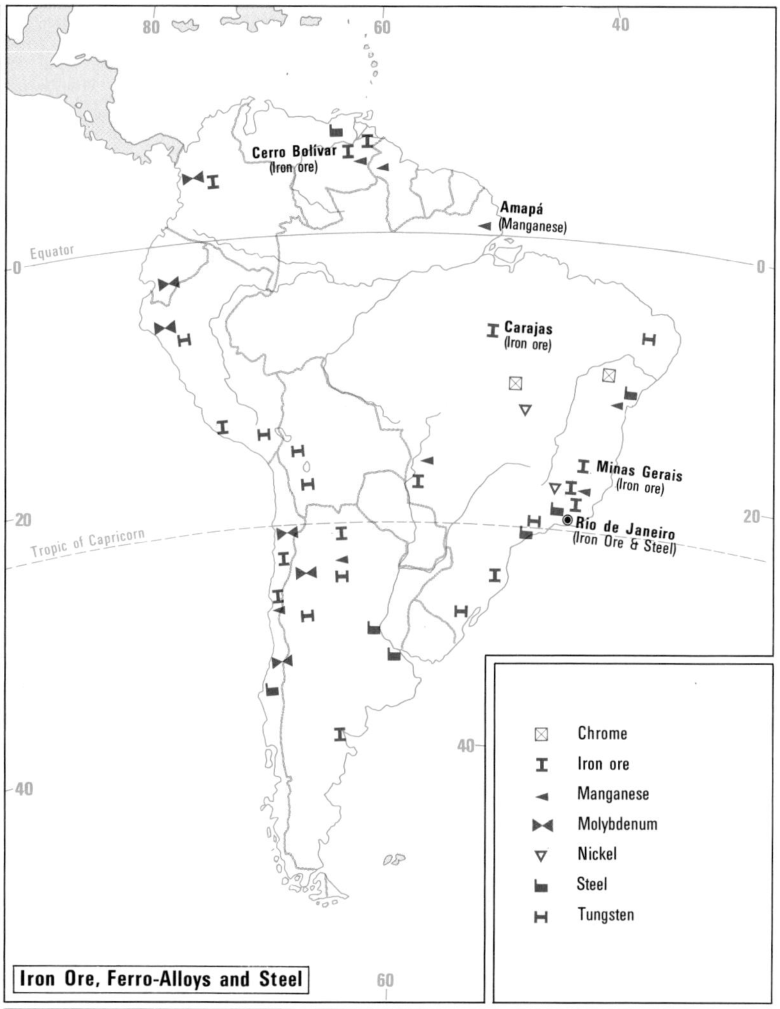

Iron Ore, Ferro-Alloys and Steel
Equator
Tropic of Capricorn
Cerro Bolivar
(Iron ore)
Amapá
(Manganese)
Carajas
(Iron ore)
Minas Gerais
(Iron ore)
Rio de Janeiro
(Iron Ore & Steel)
Chrome
Iron ore
Manganese
Molybdenum
Nickel
Steel
Tungsten

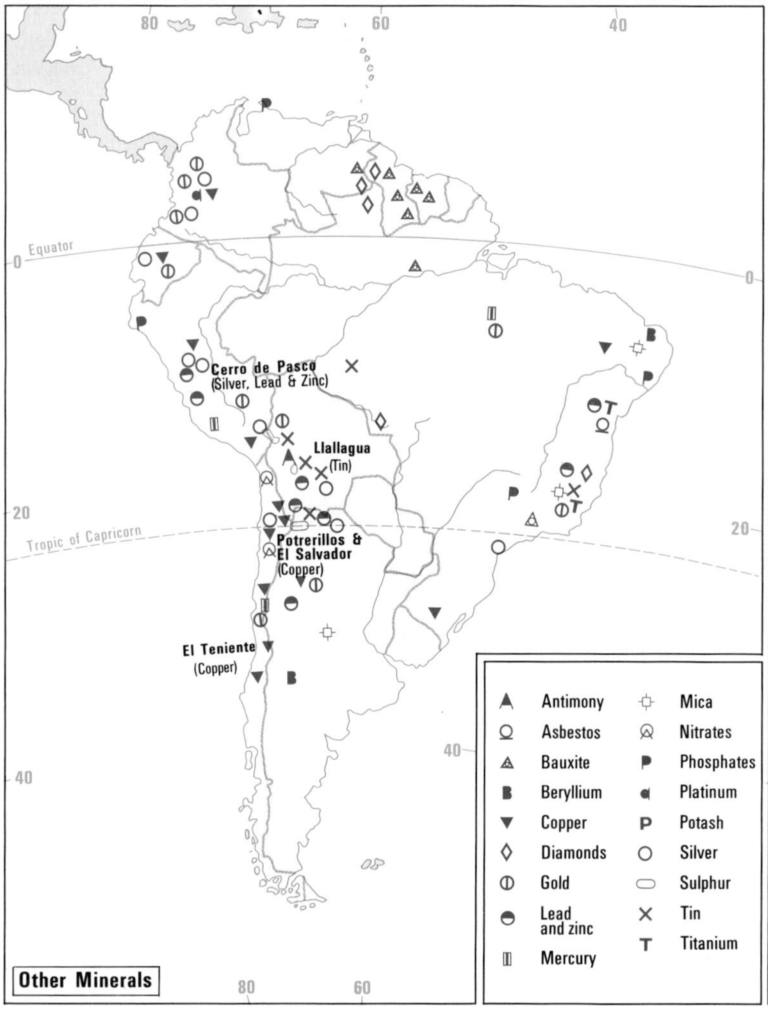

Other Minerals
Equator
Tropic of Capricorn
Cerro de Pasco
(Silver, Lead & Zinc)
Llallagua
(Tin)
Potrerillos &
El Salvador
(Copper)
El Teniente
(Copper)
Antimony
Asbestos
Bauxite
Beryllium
Copper
Diamonds
Gold
Lead and zinc
Mercury
Mica
Nitrates
Phosphates
Platinum
Potash
Silver
Sulphur
Tin
Titanium

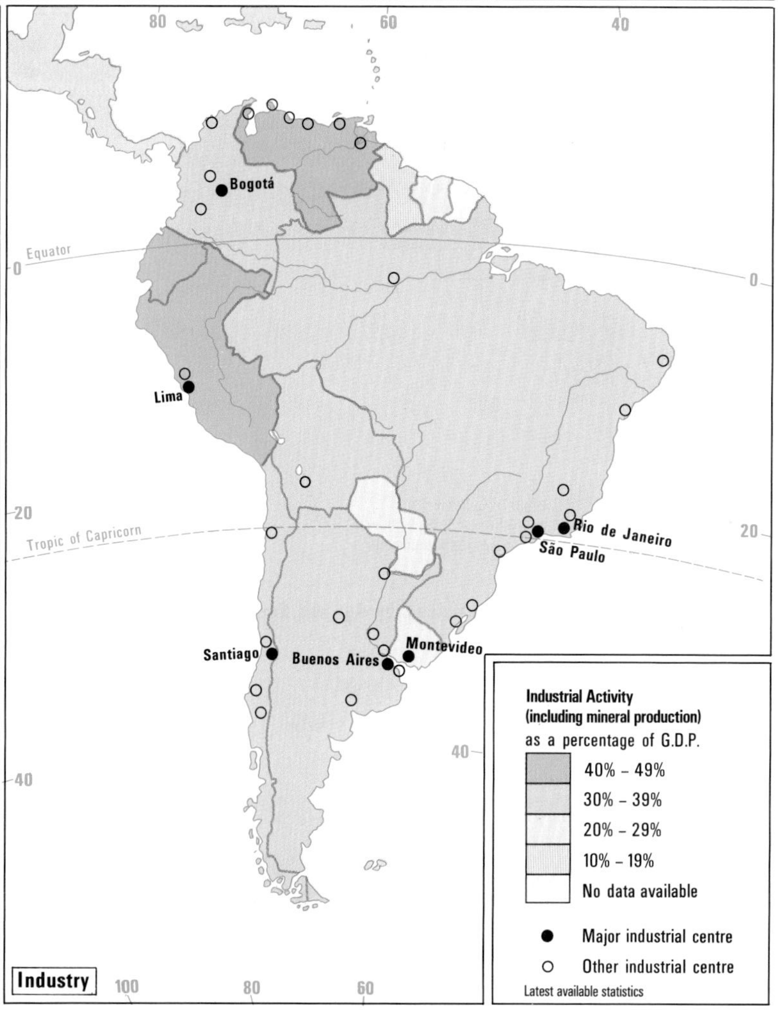

Industry
Equator
Tropic of Capricorn
Bogotá
Lima
Rio de Janeiro
São Paulo
Santiago
Buenos Aires
Montevideo
Industrial Activity
(including mineral production)
as a percentage of G.D.P.
40% – 49%
30% – 39%
20% – 29%
10% – 19%
No data available
Major industrial centre
Other industrial centre
Latest available statistics

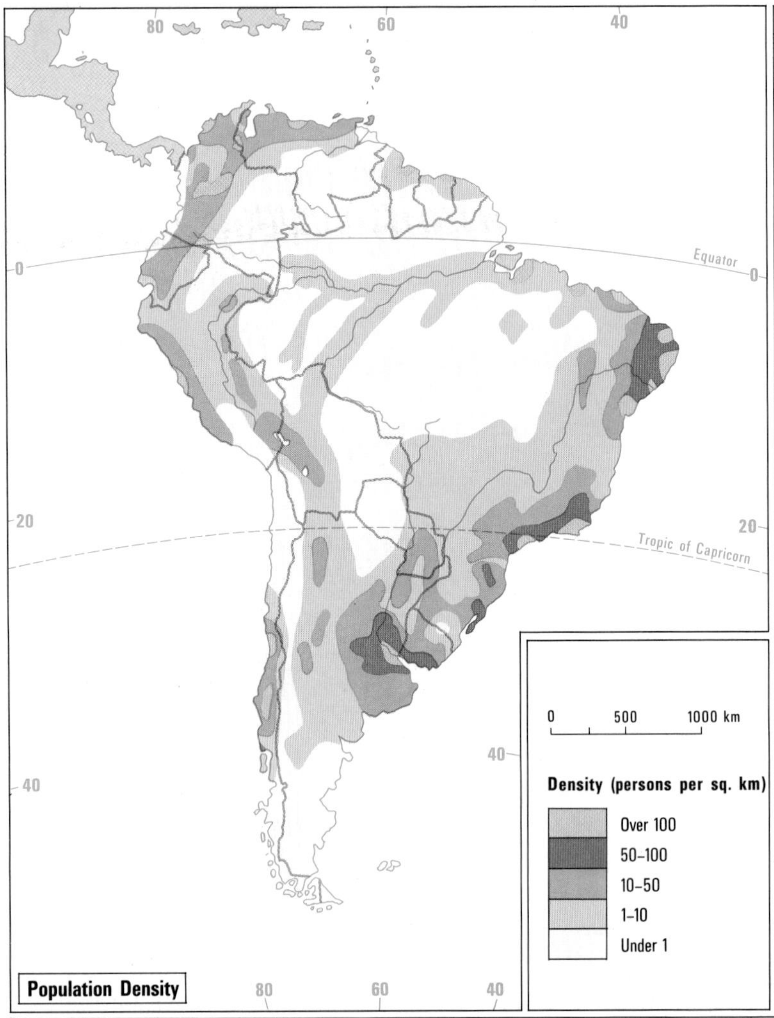

Population Density

Urban Population

Population by Selected Country

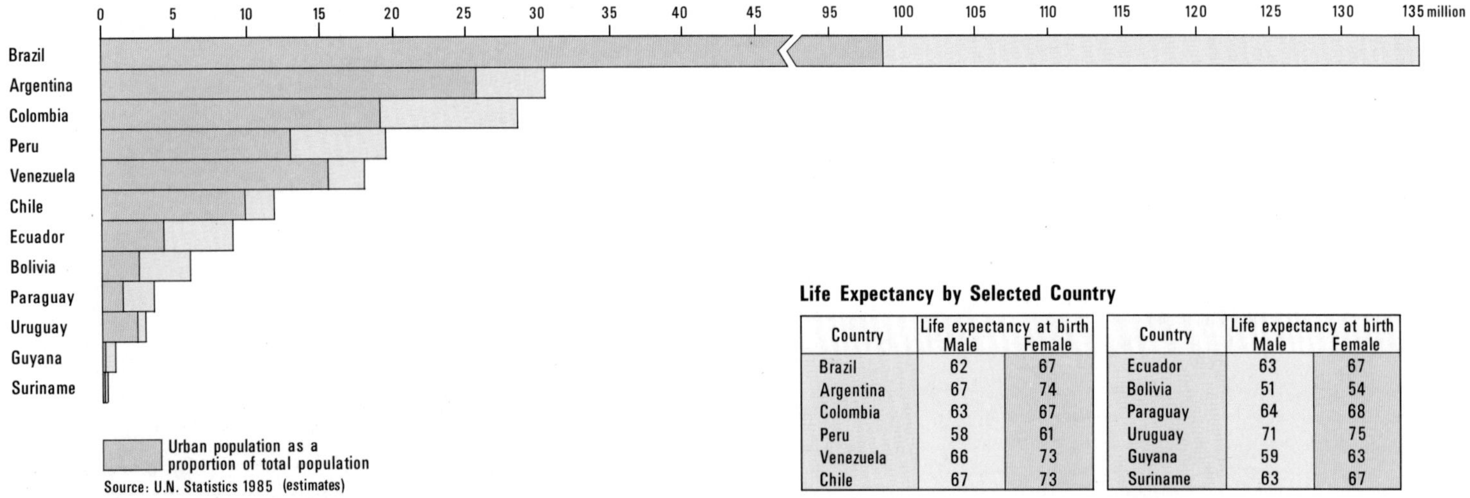

Source: U.N. Statistics 1985 (estimates)

Life Expectancy by Selected Country

Country	Life expectancy at birth Male	Female	Country	Life expectancy at birth Male	Female
Brazil	62	67	Ecuador	63	67
Argentina	67	74	Bolivia	51	54
Colombia	63	67	Paraguay	64	68
Peru	58	61	Uruguay	71	75
Venezuela	66	73	Guyana	59	63
Chile	67	73	Suriname	63	67

Percentage Population Breakdown by Age and Sex

Source: UN Demographic Yearbook

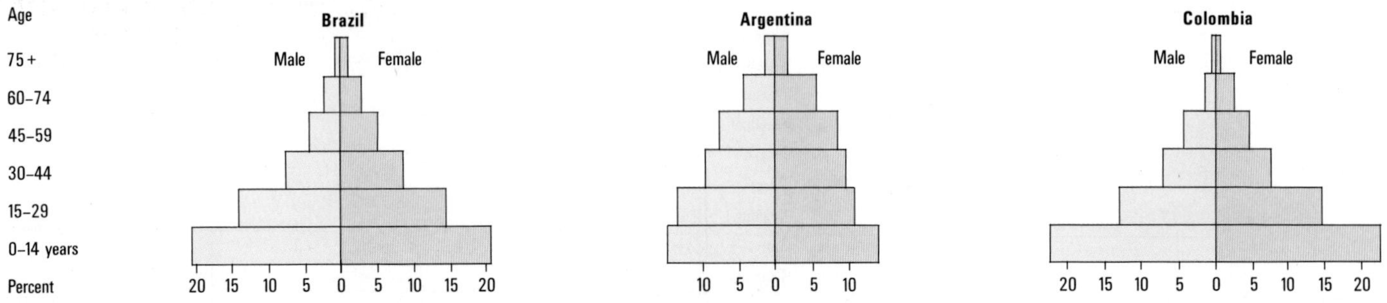

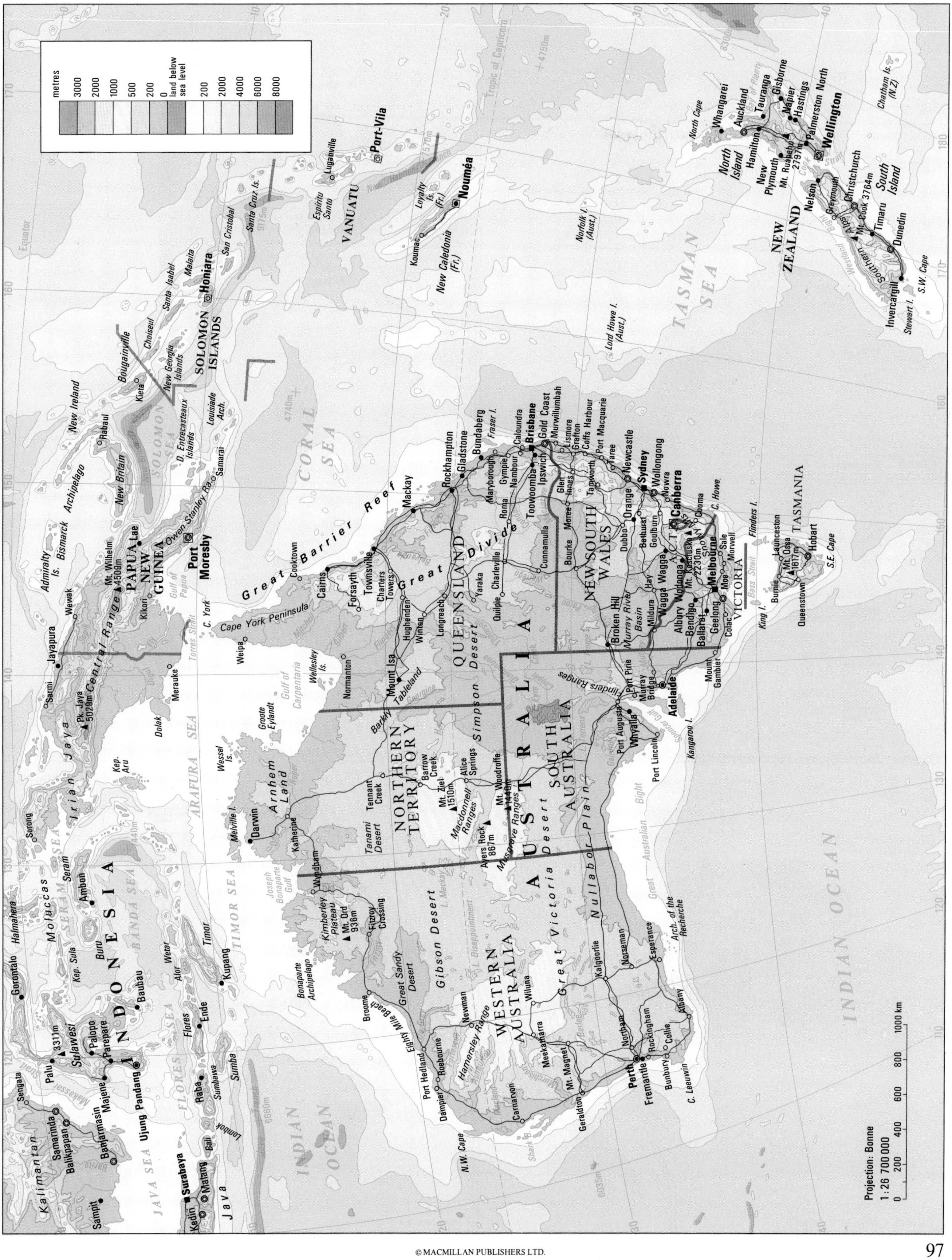
metres
3000
2000
1000
500
200
0
land below sea level
200
2000
4000
6000
8000
AUSTRALIA
WESTERN AUSTRALIA
NORTHERN TERRITORY
SOUTH AUSTRALIA
QUEENSLAND
NEW SOUTH WALES
VICTORIA
TASMANIA
A.C.T.
NEW ZEALAND
North Island
South Island
PAPUA NEW GUINEA
SOLOMON ISLANDS
VANUATU
New Caledonia (Fr.)
INDONESIA
Irian Jaya
Kalimantan
Sulawesi
Java
Moluccas
Timor
INDIAN OCEAN
TASMAN SEA
CORAL SEA
ARAFURA SEA
TIMOR SEA
BANDA SEA
SERAM SEA
FLORES SEA
JAVA SEA
SOLOMON SEA
Great Barrier Reef
Great Dividing Range
Cape York Peninsula
Gulf of Carpentaria
Arnhem Land
Great Sandy Desert
Gibson Desert
Great Victoria Desert
Simpson Desert
Tanami Desert
Nullarbor Plain
Flinders Ranges
Macdonnell Ranges
Musgrave Ranges
Hamersley Range
Kimberley Plateau
Barkly Tableland
Great Australian Bight
Bass Strait
Torres Strait
Ayers Rock 867m
Mt. Kosciusko 2230m
Mt. Wilhelm 4509m
Pk. Jaya 5029m
Mt. Cook 3764m
Mt. Ruapehu 2797m
Canberra
Sydney
Melbourne
Brisbane
Adelaide
Perth
Fremantle
Darwin
Hobart
Port Moresby
Honiara
Port-Vila
Nouméa
Wellington
Auckland
Christchurch
Dunedin
Alice Springs
Townsville
Cairns
Rockhampton
Mackay
Newcastle
Wollongong
Geelong
Launceston
Kalgoorlie
Broken Hill
Mount Isa
Katherine
Tennant Creek
Port Hedland
Broome
Wyndham
Albany
Esperance
Port Augusta
Port Lincoln
Whyalla
Jayapura
Ambon
Kupang
Ujung Pandang
Surabaya
Projection: Bonne
1:26 700 000
0 200 400 600 800 1000 km

AUSTRALASIA – CLIMATE

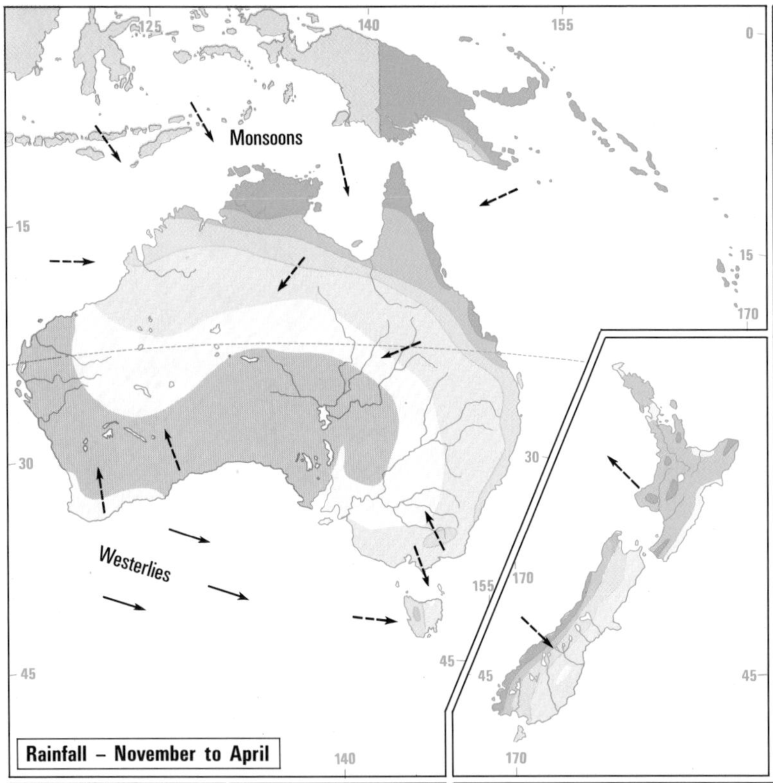

Rainfall – November to April

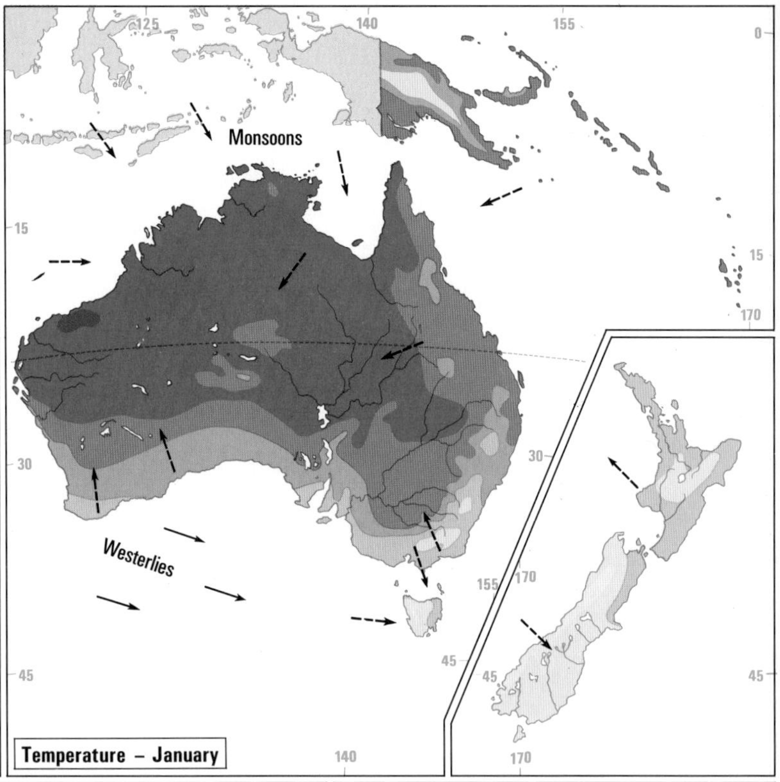

Temperature – January

Rainfall in millimetres

- 1000–1500
- 750–1000
- 500–750
- 250–500
- 125–250
- 0–125

Australia

0 400 800 1200 1600 kilometres

New Zealand

0 200 400 600 800 kilometres

Prevailing winds

- ——→ Blowing all year
- – – → November to April
- ········→ May to October

Surface temperature °C

32
28
24
20
16
12
8
4
0
−4

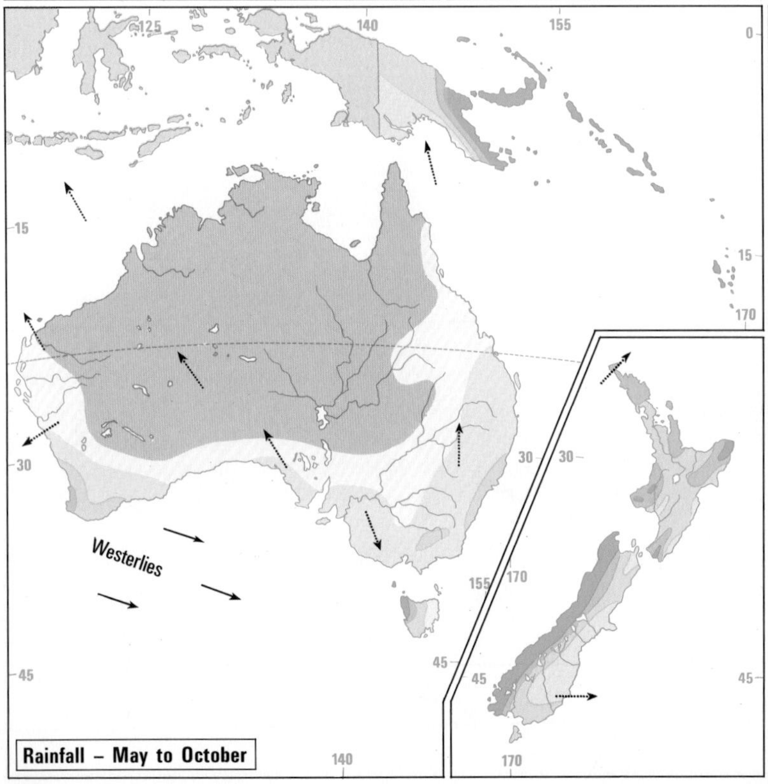

Rainfall – May to October

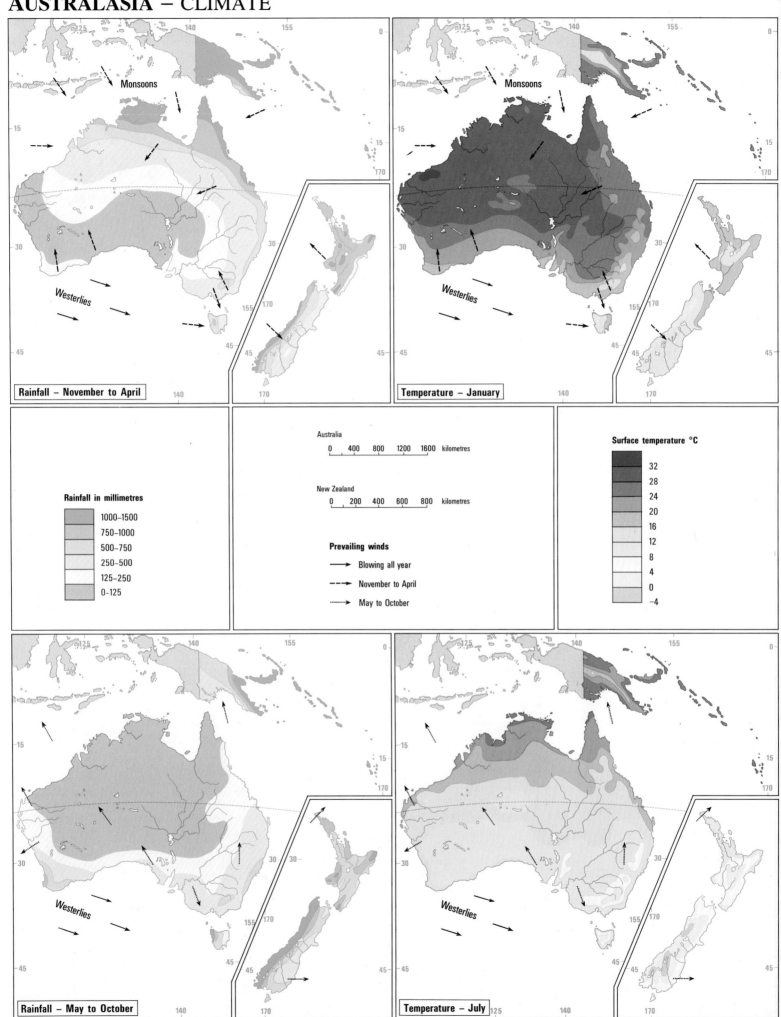

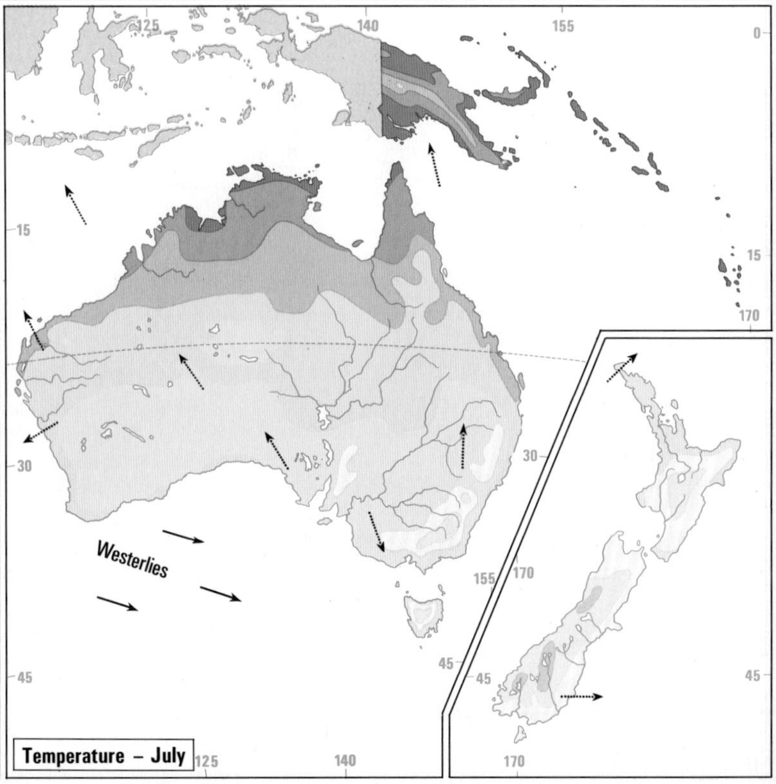

Temperature – July

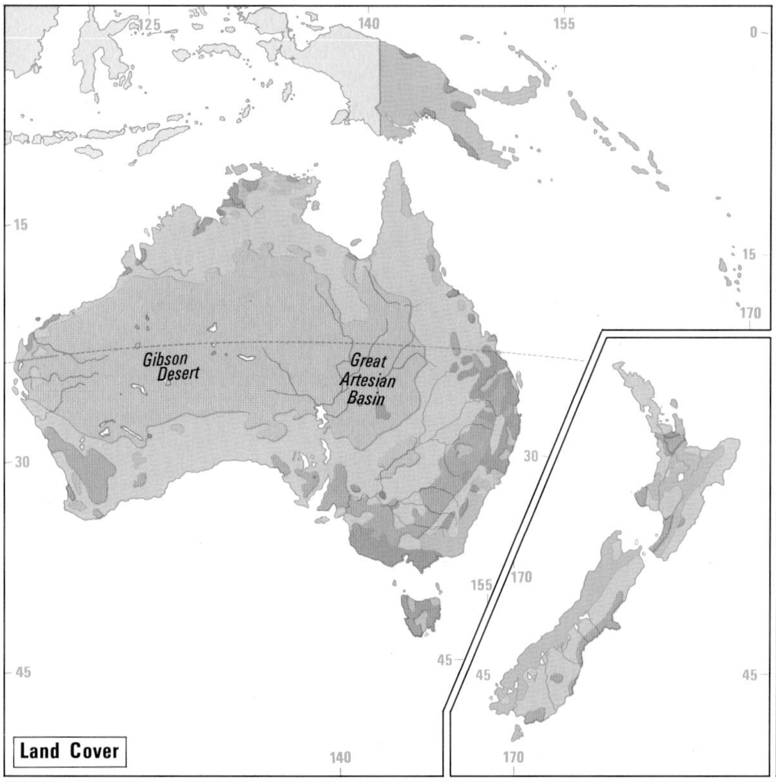

Land Cover

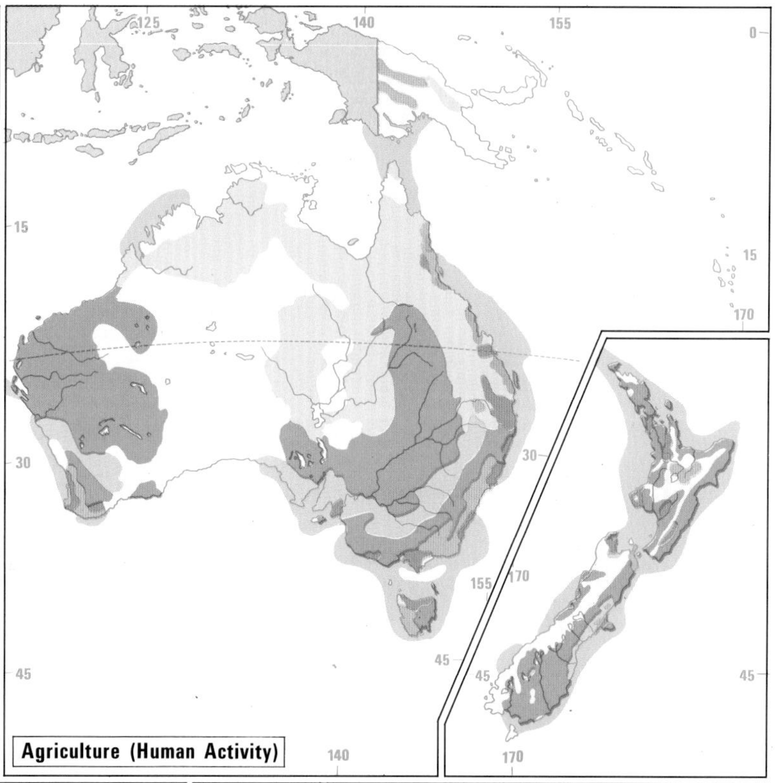

Agriculture (Human Activity)

Land Cover

- Cultivated land
- Cultivated land and grassland
- Cultivated land and woodland
- Forest and woodland
- Grassland
- Scrub and sparse grassland
- Swamp and marsh

Australia

0 400 800 1200 1600 km

New Zealand

0 200 400 600 800 km

Agriculture (Human Activity)

- Beef cattle
- Dairying
- Fishing
- Grain crops
- Sheep
- Sugar cane
- Timber production
- Unproductive

Minerals and Industry

- Oil
- Natural gas
- Coal
- Uranium
- Iron ore
- Steel
- Manganese
- Nickel
- Tungsten
- Asbestos
- Copper
- Bauxite
- Gold
- Lead and zinc
- Phosphates
- Potash
- Silver
- Sulphur
- Tin
- H.E.P. station
- Nuclear power station
- Major industrial centre
- Other industrial centre

Density (persons per sq. km)

- Over 100
- 25–100
- 3–25
- 1–3
- Under 1
- Cities over 1 million population

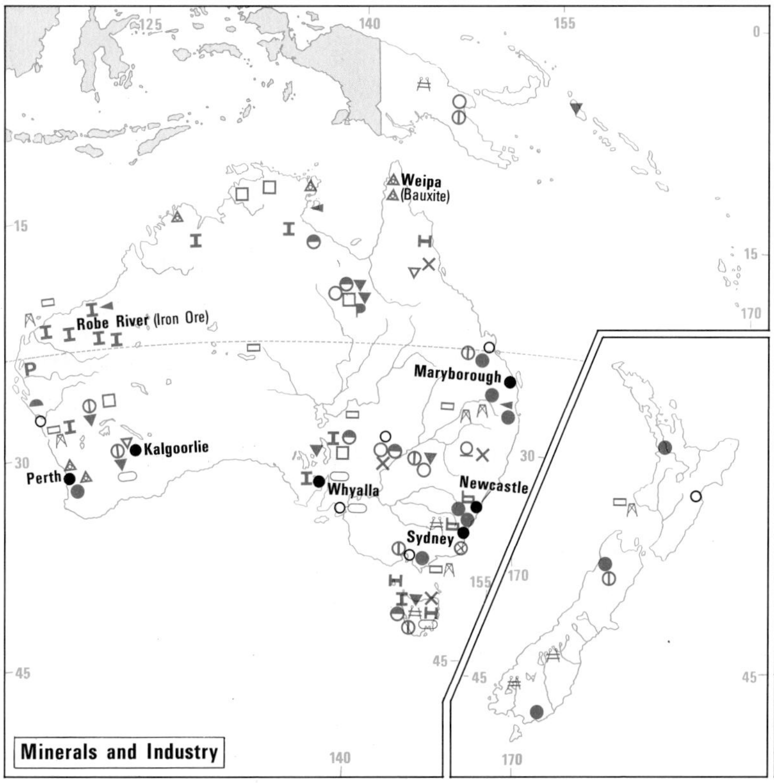

Minerals and Industry

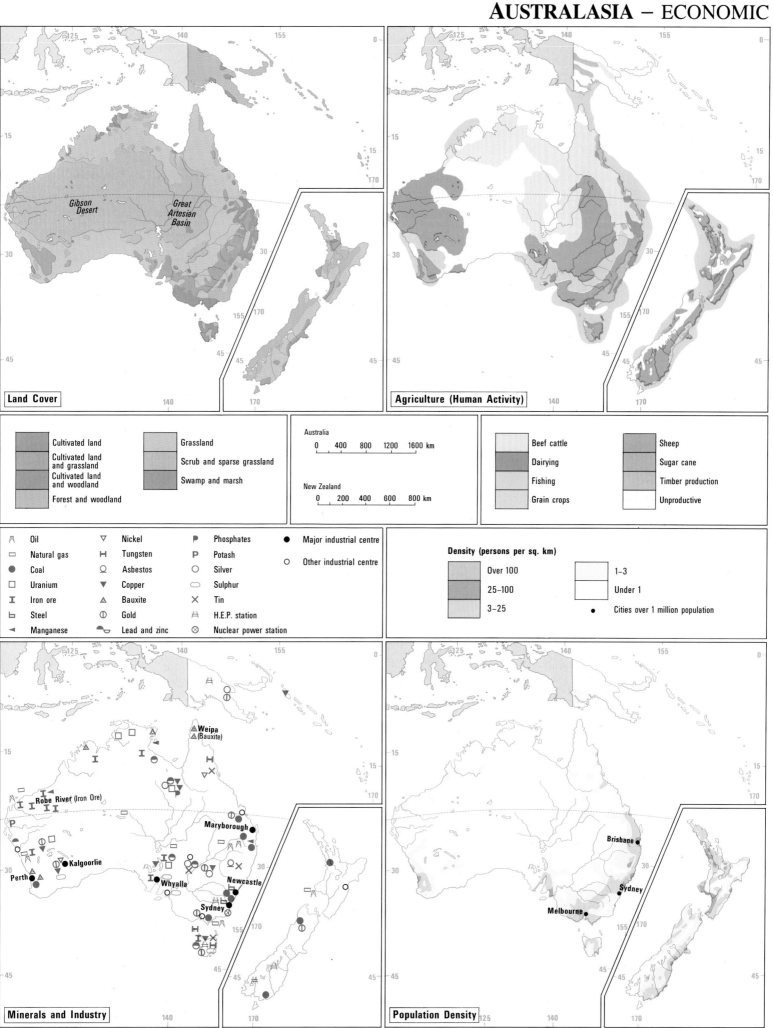

Population Density

POLAR REGIONS

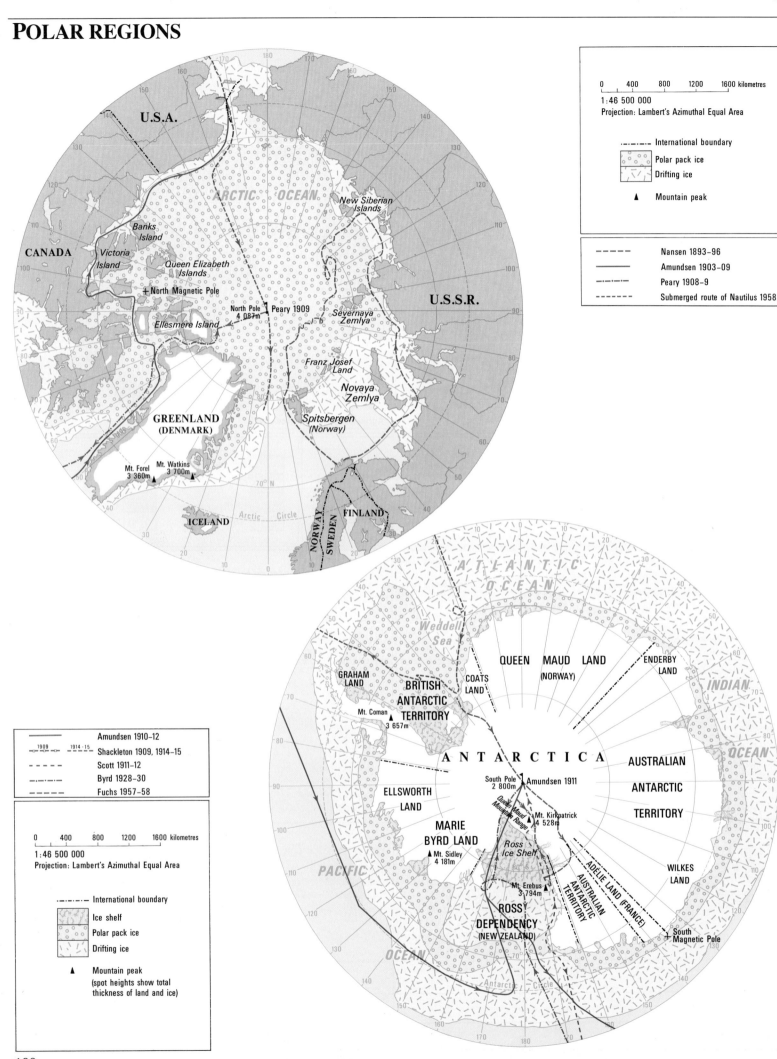

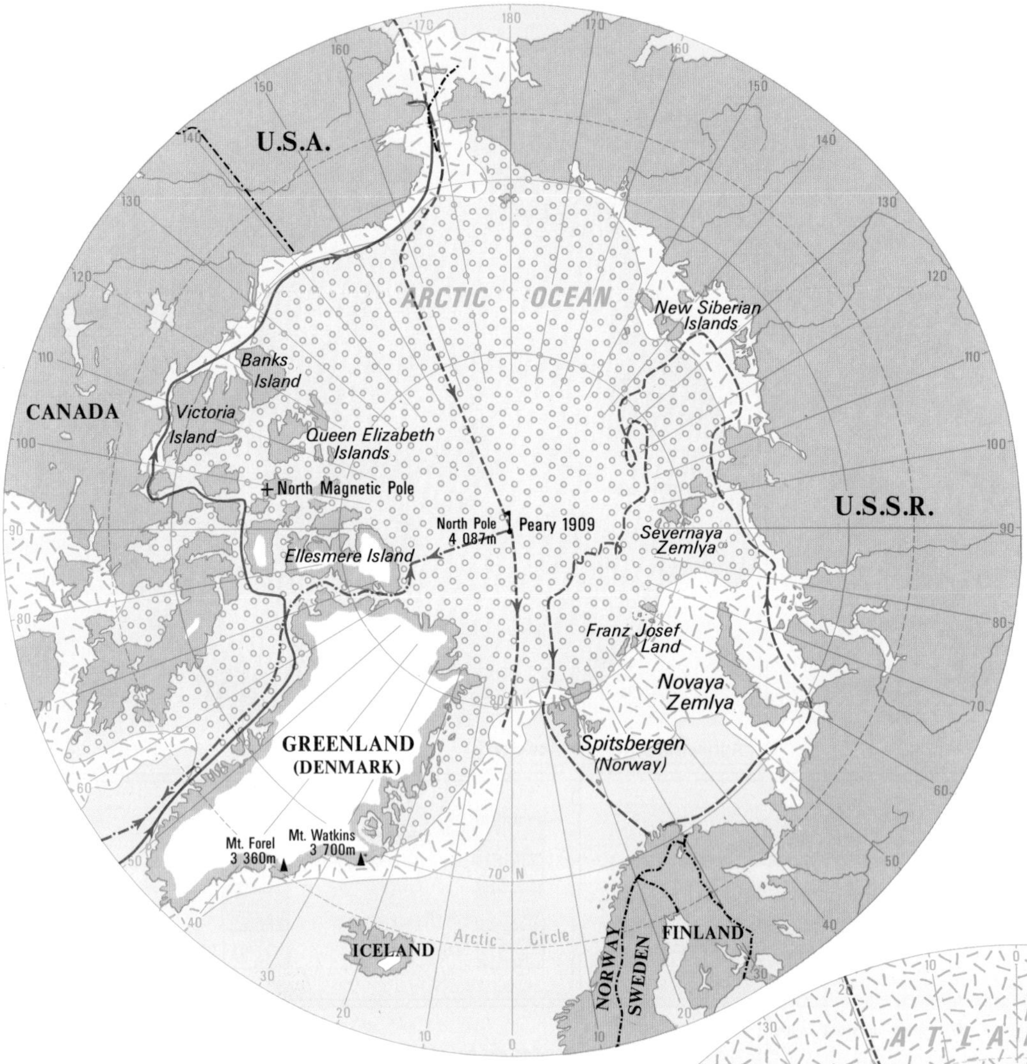

0 400 800 1200 1600 kilometres
1:46 500 000
Projection: Lambert's Azimuthal Equal Area

International boundary
Polar pack ice
Drifting ice
Mountain peak

Nansen 1893–96
Amundsen 1903–09
Peary 1908–9
Submerged route of Nautilus 1958

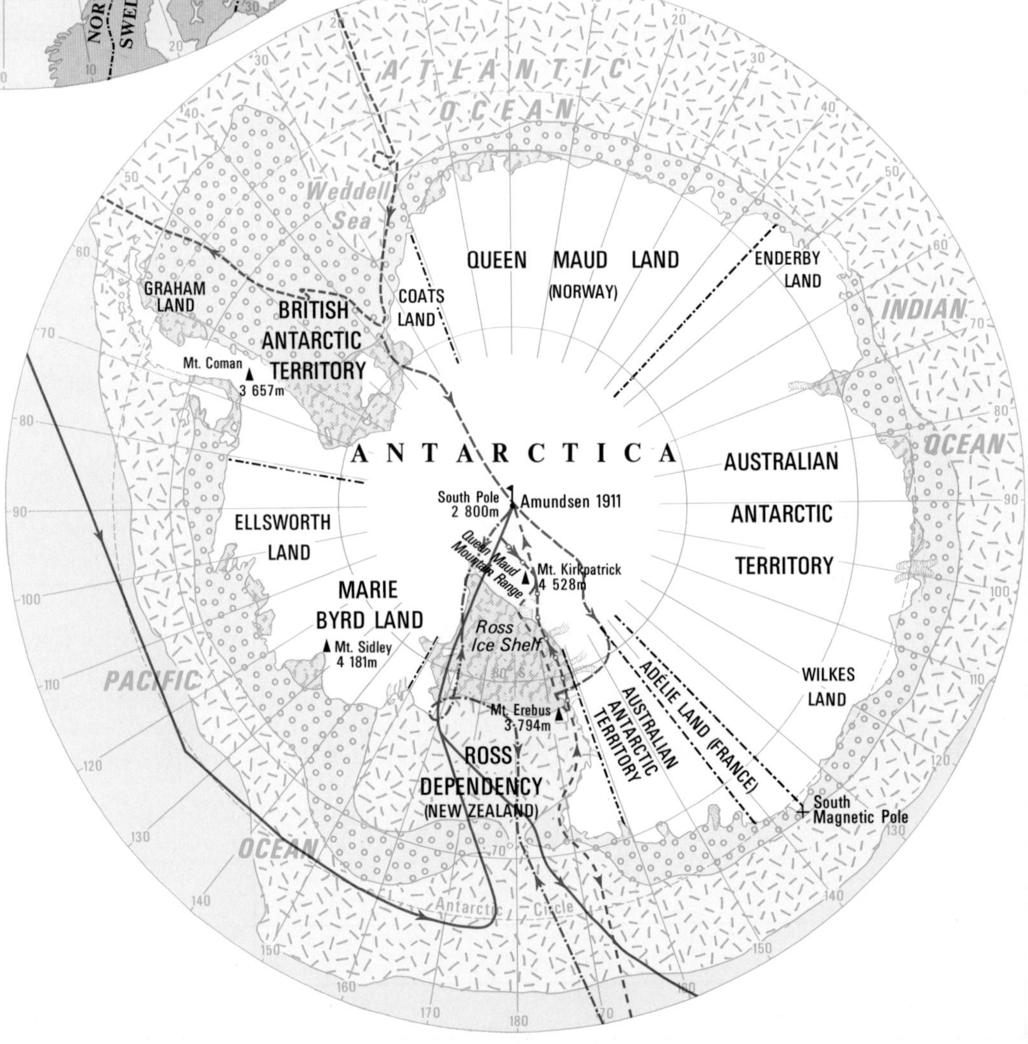

Amundsen 1910–12
Shackleton 1909, 1914–15
Scott 1911–12
Byrd 1928–30
Fuchs 1957–58

0 400 800 1200 1600 kilometres
1:46 500 000
Projection: Lambert's Azimuthal Equal Area

International boundary
Ice shelf
Polar pack ice
Drifting ice
Mountain peak
(spot heights show total thickness of land and ice)

Country	Area Sq. Km.	Total Population	Capital	Population of Capital
Africa				
Algeria	2 381 741	21 525 000	Algiers	1 722 000
Angola	1 246 700	8 339 000	Luanda	1 200 000
Benin	112 622	3 980 000	Porto Novo	208 000
Botswana	582 000	1 050 000	Gaborone	79 000
Burkina Faso	274 122	6 907 000	Ouagadougou	375 000
Burundi	27 834	4 537 000	Bujumbura	172 000
Cameroon	465 054	9 770 000	Yaoundé	436 000
Central African Republic	622 436	2 630 000	Bangui	387 000
Chad	1 284 000	5 079 000	N'Djaména	402 000
Comoros	1 862	374 000	Moroni	20 000
Congo	342 000	1 800 000	Brazzaville	422 000
Côte d'Ivoire	322 463	10 056 000	Yamoussoukro	70 000
Djibouti	23 000	297 000	Djibouti	200 000
Egypt	1 002 00	44 202 000	Cairo	14 000 000
Equatorial Guinea	28 051	304 000	Malabo	46 000
Ethiopia	1 221 900	42 169 000	Addis Ababa	1 412 000
Gabon	267 667	1 367 000	Libreville	350 000
Gambia, The	10 689	696 000	Banjul	45 000
Ghana	238 305	12 206 000	Accra	1 420 000
Guinea	245 857	5 715 000	Conakry	763 000
Guinea-Bissau	36 125	858 000	Bissau	109 000
Kenya	582 600	24 870 000	Nairobi	1 505 000
Lesotho	30 355	1 470 000	Maseru	45 000
Liberia	112 600	1 900 000	Monrovia	425 000
Libya	1 759 549	3 838 000	Tripoli	980 000
Madagascar	587 041	9 908 000	Antananarivo	600 000
Malaŵi	118 484	7 059 000	Lilongwe	187 000
Mali	1 240 142	7 915 000	Bamako	404 000
Mauritania	1 030 700	1 874 000	Nouakchott	150 000
Mauritius	2 040	1 000 000	Port Louis	149 000
Morocco	458 730	20 256 000	Rabat	893 000
Mozambique	799 380	13 527 000	Maputo	786 000
Namibia	824 269	1 033 000	Windhoek	111 000
Niger	1 186 408	6 475 000	Niamey	399 000
Nigeria	923 773	94 200 000	Lagos	1 477 000
Rwanda	26 338	6 030 000	Kigali	157 000
Senegal	196 192	6 540 000	Dakar	979 000
Seychelles	444	65 000	Victoria	15 000
Sierra Leone	73 326	3 354 000	Freetown	469 000
Somalia	637 657	5 270 000	Mogadishu	600 000
South Africa	1 221 037	31 586 000	Pretoria	739 000
Sudan	2 505 813	20 564 000	Khartoum	476 000
Swaziland	17 400	626 000	Mbabane	38 000
Tanzania	945 200	21 730 000	Dodoma	46 000
Togo	56 785	3 030 000	Lomé	366 000
Tunisia	164 150	6 966 000	Tunis	596 000
Uganda	236 000	13 990 000	Kampala	332 000
Western Sahara	252 120	164 000	El Aaiún	97 000
Zaire	2 344 885	34 250 000	Kinshasa	2 338 000
Zambia	752 614	6 650 000	Lusaka	538 000
Zimbabwe	390 308	8 174 000	Harare	656 000
Asia				
Afghanistan	652 090	17 672 000	Kabul	1 036 000
Bahrain	676	384 000	Manama	122 000
Bangladesh	143 999	99 235 000	Dhaka	3 440 000
Bhutan	46 500	1 286 000	Thimphu	20 000
Brunei	5 800	221 000	Bandar Seri Begawan	64 000
Burma	678 000	36 392 000	Rangoon	2 459 000
China	9 597 000	1 036 040 000	Beijing	9 500 000
Hong Kong (U.K.)	1 067	5 397 000	Victoria	1 184 000
India	3 166 829	748 000 000	New Delhi	273 000
Indonesia	1 919 400	173 000 000	Jakarta	7 636 000
Iran	1 648 000	45 190 000	Tehran	5 734 000
Iraq	434 924	15 400 000	Baghdad	3 190 000
Israel	20 770[1]	4 150 000	Jerusalem	429 000[2]
Japan	377 765	121 047 000	Tokyo	8 351 000
Jordan	97 740[3]	3 247 000	Amman	778 000
Kampuchea	181 035	7 200 000	Phnom Penh	600 000
Korea, North	122 098	19 630 000	Pyongyang	1 280 000
Korea, South	99 022	40 430 000	Seoul	9 501 000
Kuwait	17 818	1 709 000	Kuwait	168 000
Laos	236 800	3 585 000	Vientiane	210 000
Lebanon	10 452	3 500 000	Beirut	702 000
Malaysia	329 747	15 260 000	Kuala Lumpur	938 000
Maldives	298	200 000	Malé	42 000
Mongolia	1 565 000	1 866 000	Ulan Bator	479 000
Nepal	147 181	16 480 000	Kathmandu	235 000
Oman	272	1 500 000	Muscat	30 000
Pakistan	796 095	94 700 000	Islamabad	201 000
Philippines	300 000	54 450 000	Manila	1 630 000
Qatar	11 437	287 000	Doha	190 000
Saudi Arabia	2 200 000	12 400 000	Riyadh	667 000
Singapore	620	2 558 000	Singapore	2 500 000
Sri Lanka	65 609	15 599 000	Colombo	623 000
Syria	185 180	9 934 000	Damascus	1 251 000
Taiwan	36 179	19 135 000	T'ai-pei	2 450 000
Thailand	514 000	51 571 000	Bangkok	5 175 000
United Arab Emirates	92 100	1 622 000	Abu Dhabi	537 000

Country	Area Sq. Km.	Total Population	Capital	Population of Capital
Vietnam	329 566	58 300 000	Hanoi	2 647 000
Yemen Arab Rep.	195 000	7 162 000	Sana'a	427 000
Yemen, People's Dem. Rep.	287 682	2 500 000	Aden	264 000
Australasia				
Australia	7 682 300	15 452 000	Canberra	256 000
Fiji	18 333	691 000	Suva	74 000
New Caledonia (Fr)	18 576	150 000	Nouméa	60 000
New Zealand	268 704	3 291 000	Wellington	343 000
Papua New Guinea	462 840	3 330 000	Port Moresby	150 000
Solomon Islands	29 785	258 000	Honiara	24 000
Western Samoa	2 831	159 000	Apia	33 000
Central America & West Indies				
Bahamas	13 864	228 000	Nassau	135 000
Barbados	430	252 000	Bridgetown	7 000
Belize	22 963	162 000	Belmopan	3 000
Costa Rica	51 100	2 655 000	San José	245 000
Cuba	114 524	10 000 000	Havana	1 924 000
Dominican Rep.	48 442	6 588 000	Santo Domingo	1 313 000
El Salvador	21 393	5 480 000	San Salvador	452 000
Guatemala	108 889	8 335 000	Guatemala City	1 300 000
Haiti	27 750	5 272 000	Port-au-Prince	720 000
Honduras	112 088	4 092 000	Tegucigalpa	534 000
Jamaica	11 425	2 223 000	Kingston	101 000
Mexico	1 958 201	78 524 000	Mexico City	14 750 000
Nicaragua	148 000	2 908 000	Managua	615 000
Panama	78 046	2 179 000	Panama City	655 000
Puerto Rico	8 860	3 274 000	San Juan	435 000
Trinidad & Tobago	5 128	1 160 000	Port of Spain	66 000
Europe				
Albania	28 748	3 000 000	Tiranë	206 000
Andorra	465	43 000	Andorra la Vella	16 000
Austria	83 855	7 553 000	Vienna	1 531 000
Belgium	30 519	9 857 000	Brussels	982 000
Bulgaria	110 912	8 943 000	Sofia	1 094 000
Cyprus	9 251	657 000	Nicosia	161 000
Czechoslovakia	127 899	15 437 000	Prague	1 186 000
Denmark	43 075	5 111 000	Copenhagen	1 178 000
Finland	338 145	4 894 000	Helsinki	484 000
France	543 965	55 270 000	Paris	8 707 000
Germany, East	108 333	16 660 000	East Berlin	1 197 000
Germany, West	248 706	61 049 000	Bonn	292 000
Greece	131 957	9 970 000	Athens	886 000
Hungary	93 032	10 658 000	Budapest	2 072 000
Iceland	103 000	240 000	Reykjavik	89 000
Ireland, Rep. of	89 894	3 552 000	Dublin	915 000
Italy	301 268	57 080 000	Rome	2 827 000
Liechtenstein	160	27 000	Vaduz	5 000
Luxembourg	2 586	366 000	Luxembourg	76 000
Malta	246	322 000	Valletta	14 000
Monaco	2	27 000	-	-
Netherlands	33 963	14 484 000	Amsterdam	676 000
Norway	323 878	4 146 000	Oslo	447 000
Poland	312 683	37 063 000	Warsaw	1 649 000
Portugal	92 072	10 129 000	Lisbon	808 000
Romania	237 500	22 687 000	Bucharest	2 198 000
San Marino	61	22 000	San Marino	5 000
Spain	504 750	38 997 000	Madrid	4 727 000
Sweden	449 964	8 343 000	Stockholm	653 000
Switzerland	41 293	6 456 000	Bern	301 000
Turkey (Europe and Asia)	779 452	49 272 000	Ankara	3 463 000
United Kingdom	243 362	56 488 000	London	6 756 000
U.S.S.R (Europe and Asia)	22 400 000	276 300 000	Moscow	8 642 000
Yugoslavia	255 804	22 963 000	Belgrade	1 470 000
North America				
Canada	9 215 430	25 358 000	Ottawa	756 000
Greenland (Denmark)	2 175 600	53 000	Godthåb (Nuuk)	11 000
U.S.A. (including Alaska & Hawaii)	9 363 000	236 681 000	Washington D.C.	638 000
South America				
Argentina	2 778 815	30 097 000	Buenos Aires	9 969 000
Bolivia	1 098 581	6 082 000	La Paz	881 000
Brazil	8 511 965	135 564 000	Brasilia	1 177 000
Chile	757 626	12 074 000	Santiago	4 318 000
Colombia	1 141 748	28 217 000	Bogotá	3 968 000
Ecuador	270 670	9 052 000	Quito	1 110 000
French Guiana	83 533	78 000	Cayenne	38 000
Guyana	214 969	950 000	Georgetown	188 000
Paraguay	406 752	3 477 000	Asunción	708 000
Peru	1 285 216	19 700 000	Lima	5 258 000
Suriname	163 820	370 000	Paramaribo	151 000
Uruguay	186 296	2 991 000	Montevideo	1 362 000
Venezuela	912 050	16 054 000	Caracas	2 944 000

Notes

1. Area of Israel agreed in 1949, excluding areas taken under Israeli control since 1949.
2. Including East Jerusalem, annexed in 1949.
3. Including West Bank, currently occupied by Israel.

Population figures are the latest census figures or the latest official estimate where available.

WORLD – POLITICAL

Abbreviations

ALB.	Albania
Aust.	Australia
AUST.	Austria
BEL.	Belgium
CZECH.	Czechoslovakia
Den.	Denmark
DOM.REP.	Dominican Republic
Fr.	France
FR.GUIANA	French Guiana
HOND.	Honduras
HUNG.	Hungary
ISR.	Israel
LEB.	Lebanon
NETH.	Netherlands
N.Z.	New Zealand
Nor.	Norway
P.D.R.YEMEN	People's Democratic Republic of the Yemen
REP.	Republic
SW.	Switzerland
U.A.E.	United Arab Emirates
U.K.	United Kingdom
U.S.A.	United States of America
YEMEN A.R.	Yemen Arab Republic
ZIMB.	Zimbabwe

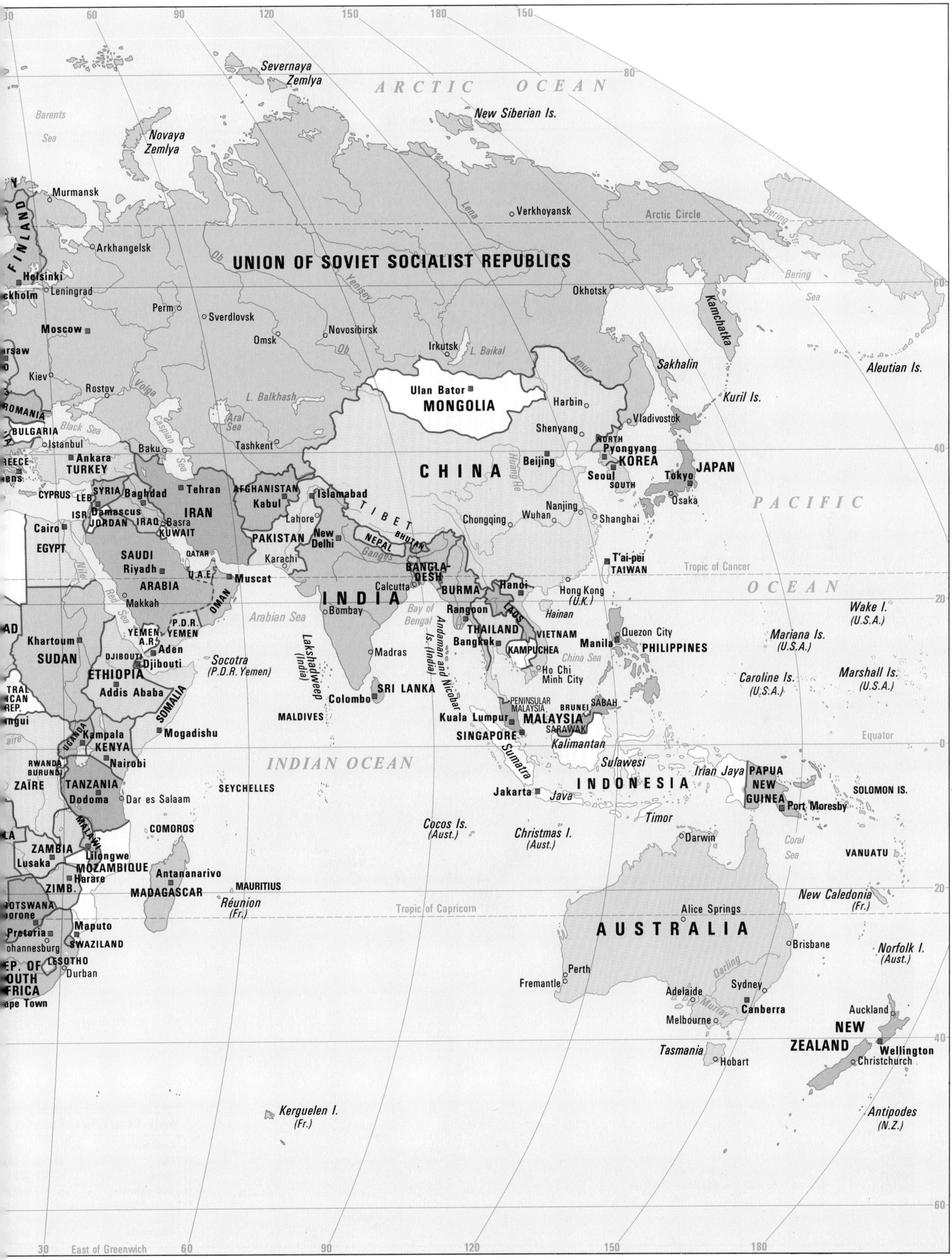
ARCTIC OCEAN
Severnaya Zemlya
New Siberian Is.
Barents Sea
Novaya Zemlya
Murmansk
FINLAND
Verkhoyansk
Arctic Circle
Bering Str.
Arkhangelsk
UNION OF SOVIET SOCIALIST REPUBLICS
Helsinki
Leningrad
Bering Sea
Okhotsk
Kamchatka
Perm
Sverdlovsk
Moscow
Novosibirsk
Omsk
Irkutsk
L. Baikal
Sakhalin
Aleutian Is.
Kiev
Rostov
Ulan Bator
MONGOLIA
Harbin
Kuril Is.
ROMANIA
L. Balkhash
Aral Sea
Vladivostok
BULGARIA
Black Sea
Shenyang
Istanbul
Baku
Tashkent
NORTH
Pyongyang
KOREA
Ankara
TURKEY
CHINA
Beijing
JAPAN
Seoul
SOUTH
Tokyo
CYPRUS
LEB.
SYRIA
Baghdad
Tehran
AFGHANISTAN
Kabul
Islamabad
Osaka
ISR.
Damascus
JORDAN
IRAQ
Basra
KUWAIT
IRAN
Lahore
TIBET
Nanjing
Shanghai
PACIFIC
Cairo
EGYPT
PAKISTAN
New Delhi
NEPAL
BHUTAN
Chongqing
Wuhan
SAUDI
Riyadh
ARABIA
QATAR
U.A.E.
Muscat
Karachi
BANGLA-DESH
T'ai-pei
TAIWAN
Tropic of Cancer
Calcutta
BURMA
Hanoi
Hong Kong (U.K.)
OCEAN
Makkah
OMAN
INDIA
Bombay
Arabian Sea
Bay of Bengal
Rangoon
LAOS
Hainan
Wake I. (U.S.A.)
P.D.R. YEMEN
YEMEN A.R.
Aden
Khartoum
SUDAN
THAILAND
Bangkok
VIETNAM
Manila
Quezon City
PHILIPPINES
Mariana Is. (U.S.A.)
DJIBOUTI
Djibouti
Socotra (P.D.R. Yemen)
Lakshadweep (India)
Madras
Andaman and Nicobar Is. (India)
KAMPUCHEA
China Sea
ETHIOPIA
Addis Ababa
Ho Chi Minh City
Caroline Is. (U.S.A.)
Marshall Is. (U.S.A.)
SOMALIA
SRI LANKA
Colombo
PENINSULAR MALAYSIA
BRUNEI
SABAH
MALDIVES
Kuala Lumpur
MALAYSIA
SARAWAK
UGANDA
Kampala
KENYA
Mogadishu
SINGAPORE
Kalimantan
Equator
Sumatra
Nairobi
INDIAN OCEAN
Sulawesi
RWANDA
BURUNDI
ZAÏRE
TANZANIA
Dodoma
Dar es Salaam
SEYCHELLES
Jakarta
Java
INDONESIA
Irian Jaya
PAPUA NEW GUINEA
Port Moresby
SOLOMON IS.
MALAWI
COMOROS
Cocos Is. (Aust.)
Christmas I. (Aust.)
Timor
ZAMBIA
Lusaka
Lilongwe
MOZAMBIQUE
Harare
ZIMB.
Antananarivo
MADAGASCAR
Darwin
Coral Sea
VANUATU
MAURITIUS
Réunion (Fr.)
Tropic of Capricorn
New Caledonia (Fr.)
Alice Springs
Maputo
Pretoria
Johannesburg
SWAZILAND
LESOTHO
Durban
AUSTRALIA
Brisbane
Norfolk I. (Aust.)
Perth
Fremantle
Darling
Sydney
Adelaide
Murray
Canberra
Melbourne
Auckland
NEW ZEALAND
Wellington
Christchurch
Tasmania
Hobart
Kerguelen I. (Fr.)
Antipodes (N.Z.)
East of Greenwich
30
60
90
120
150
180
80
40
20
0

WORLD – PHYSICAL AND OCEANS

ARCTIC OCEAN
Ellesmere Island
Queen Elizabeth Islands
Greenland
Banks I.
BEAUFORT SEA
Victoria Island
Baffin Island
Baffin Bay
Bering Strait
Brooks Range
Arctic Circle
Davis Strait
Denmark Strait
Iceland
NORWEGIAN SEA
Alaska Range
Mt. McKinley 6194 m
Mackenzie Mts.
Gt. Bear Lake
Gt. Slave Lake
Alaska Pen.
Gulf of Alaska
Hudson Bay
C. Farewell
British Isles
NORTH SEA
Churchill
Nelson
L. Winnipeg
Mt. Robson 3954 m
Rocky Mountains
Laurentian Plateau
Newfoundland
NORTH AMERICA
Cascade Ra.
L. Superior
L. Huron
L. Michigan
Missouri
Appalachian Mts.
C. Sable
Bay of Biscay
Mt. Blanc 4807 m
Alps
Pyrenees
Great Basin
Sierra Nevada
Mt. Whitney 4418 m
Mt. Elbert 4399 m
Colorado
Ohio
Mississippi
NORTH ATLANTIC OCEAN
North American Basin
Mid Atlantic Ridge
North East Atlantic Basin
Azores
Bermuda
Atlas Mts.
Mt. Toubkal 4167 m
Canary Is.
Canary Basin
Sierra Madre
Rio Grande
Gulf of Mexico
4023m
6493m
Tropic of Cancer
Bahama Is.
Milwaukee Depth 9200m
Hawaiian Is.
C. San Lucas
Citlaltepetl 5700 m
Greater Antilles
Puerto Rico Trench
Hoggar
Sahara
AFR
Cape Verde Is.
C. Vert
Cape Verde Basin
6390m
CARIBBEAN SEA
Lesser Antilles
Guatemala Trench
6669m
Fouta Djalon
Niger
L. Chad
Jos Plateau
Mt. Cameroon
PACIFIC OCEAN
G. of Panama
Llanos
Orinoco
Guiana Highlands
Galapagos Is.
Equator
Gulf of Guinea
5110m
Chimborazo 6267 m
Negro
Amazon
Madeira
C. São Roque
Selvas
SOUTH AMERICA
Ascension
East Pacific Ridge
Peru Basin
Peru-Chile Trench
Andes
Mato Grosso
Brazilian Highlands
Brazil Basin
Angola Basin
St. Helena
L. Titicaca
Atacama Desert
8055m
Tropic of Capricorn
Paraná
Paraguay
Easter I.
Aconcagua 6960m
Pampas
SOUTH ATLANTIC OCEAN
Walvis Ridge
Tristan da Cunha
Cape Basin
Patagonia
Argentine Basin
8245m
Falkland Is.
Tierra del Fuego
C. Horn
Magellan Strait
South Georgia
South Sandwich Is.
8265m
SCOTIA SEA
Pacific-Antartic Basin
South Shetland Is.
Antarctic Pen.
metres
6000
5000
4000
3000
2000
1000
500
200
0
land below sea level
2000
4000
6000
8000
Projection : Modified Winkel's
1:83 000 000 (Equatorial Scale)
0 1000 2000 3000 km
180
150
120
90
60
30
0
80
60
40
20

World-Geographical Statistics

Continents	
Asia	43 608 000 k
Africa	30 335 000 k
North America	25 349 000 k
South America	17 611 000 k
Europe	10 498 000 k
Australasia	8 923 000

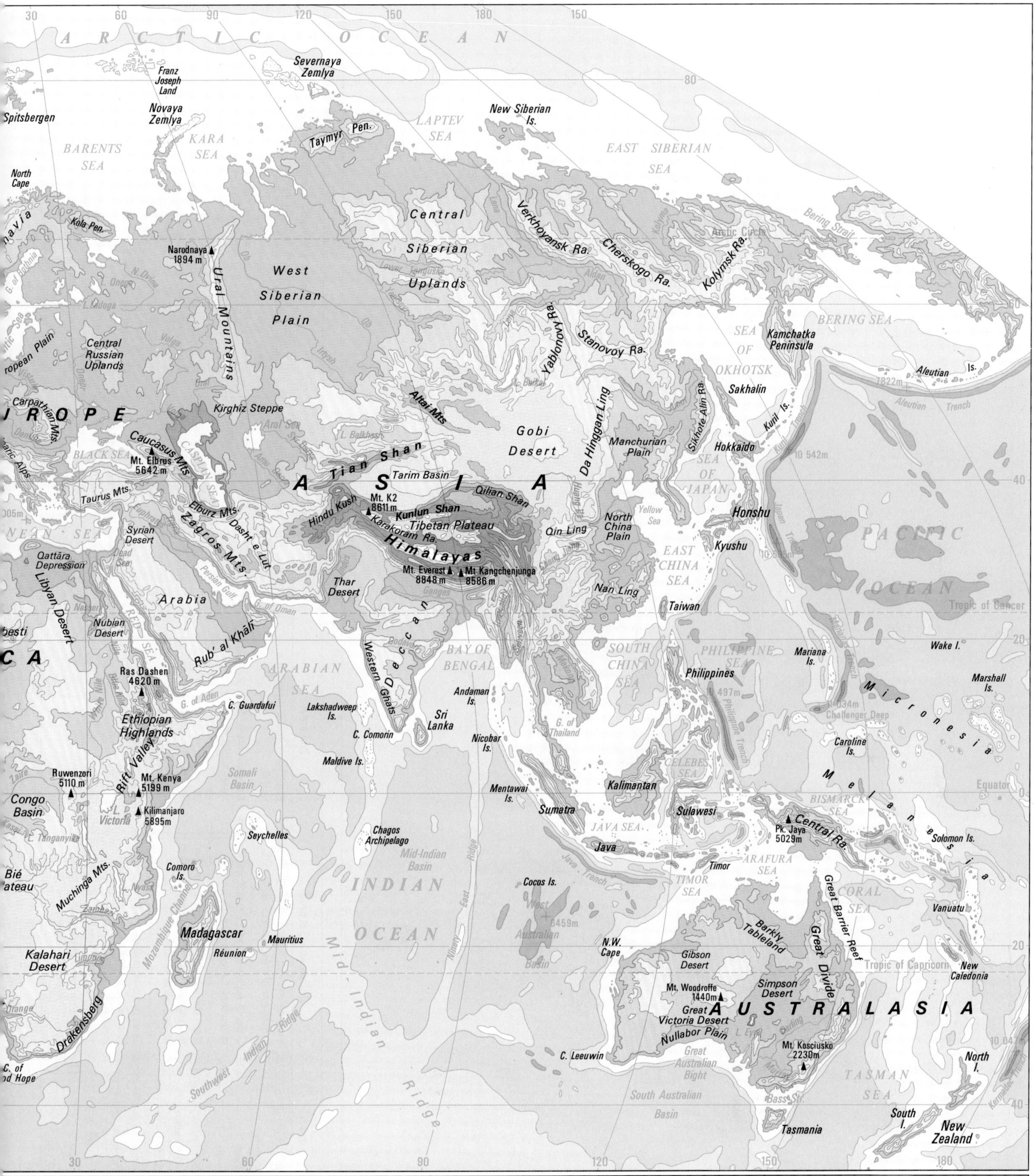

ceans and Seas	
acific Ocean	165 384 000 km²
tlantic Ocean	82 217 000 km²
ndian Ocean	73 481 000 km²
rctic Ocean	14 056 000 km²
editerranean Sea	2 505 000 km²
outh China Sea	2 318 000 km²

Highest Mountains	
Mt. Everest, Nepal/China	8 848 m
Mt. K2, Pakistan/China	8 611 m
Mt. Kangchenjunga, Nepal/China	8 586 m
Mt. Makalu, Nepal/China	8 482 m
Mt. Dhaulagiri, Nepal	8 222 m
Mt. Nanga Parbat, Pakistan	8 126 m

Deepest Sea Trenches	
Marianas Trench, Pac. Oc.	11 034 m
Tonga Trench, Pac. Oc.	10 882 m
Japan Trench, Pac. Oc.	10 595 m
Kuril Trench, Pac. Oc.	10 542 m
Philippine Trench, Pac. Oc.	10 497 m
Kermadec Trench, Pac. Oc.	10 047 m

Longest Rivers	
Nile, Africa	6 695 km
Amazon, South America	6 516 km
Mississippi-Missouri, N. America	6 019 km
Chang Jiang, Asia	5 470 km
Ob-Irtysh, Asia	5 410 km
Huang He, Asia	4 840 km

Largest Inland Seas and Lakes	
Caspian Sea, U.S.S.R./Iran	372 000 km²
Lake Superior, U.S.A./Canada	82 400 km²
Lake Victoria, East Africa	67 900 km²
Aral Sea, U.S.S.R.	66 500 km²
Lake Huron, U.S.A./Canada	59 600 km²
Lake Michigan, U.S.A.	58 020 km²

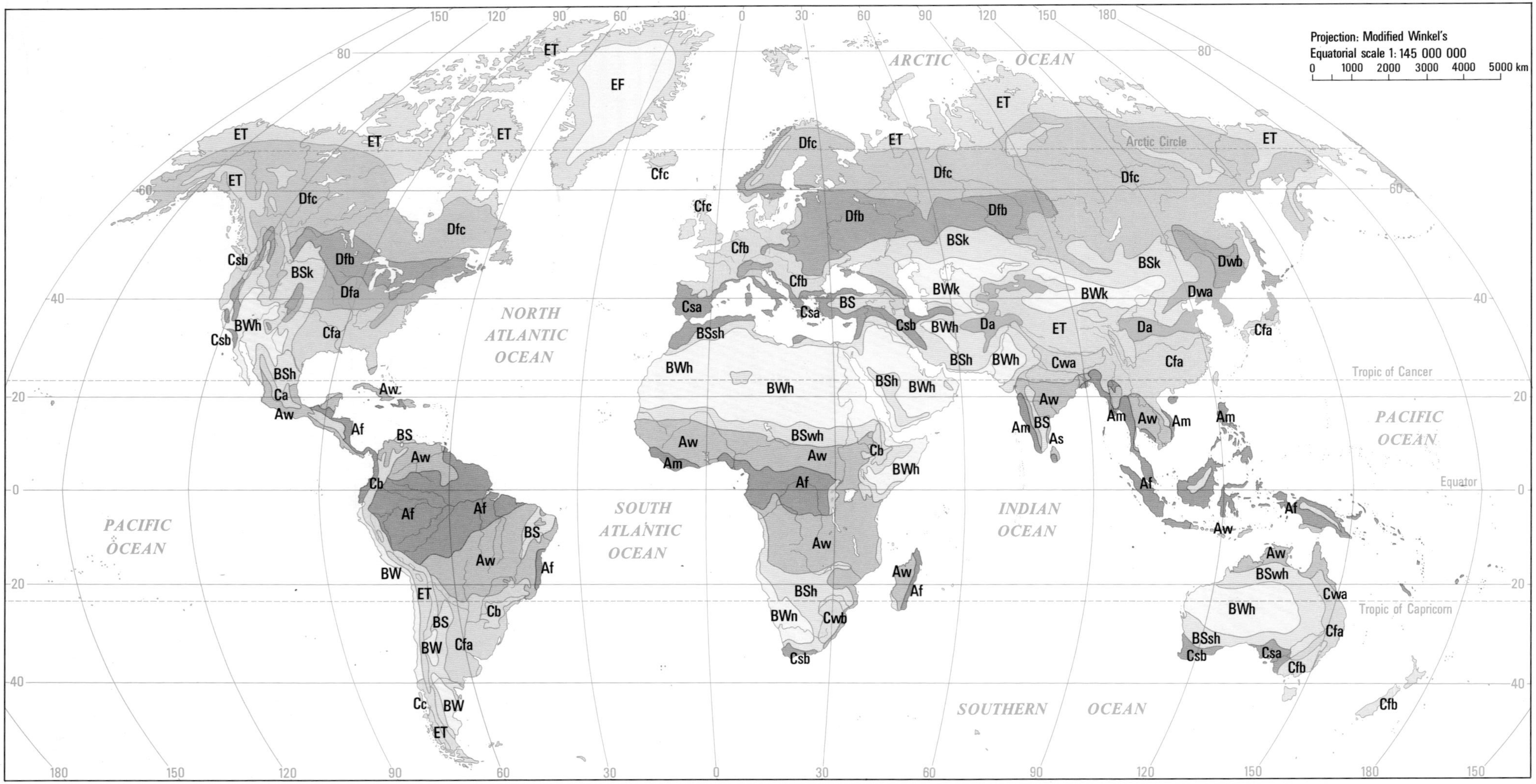

Climatic Regions		Definition - 1st order
A Tropical Rain Climates	Rain forest	Rainy climate with no defined winter. Mean temperature of coldest month above 18°C.
	Savanna	
B Arid Climates	Steppe BS	Limits of regions defined by rainfall and temperature formulae.
	Desert BW	
C Temperate Rain Climates	Mediterranean	Mean temperature of the coldest month in the range 18°C to -3°C. Warmest month in no case below 10°C.
	Humid subtropical	
	Maritime west coast	
D Cold Snow Climates	Continental warm summer	Mean temperature of the coldest month below -3°C. Warmest month in no case below 10°C.
	Continental cool summer	
	Subarctic	
E Polar	Tundra ET	Mean temperature of every month below 10°C.
	Ice cap EF	

2nd order	
f	Every month has at least 60 mm rainfall.
m	Monsoon climate. Short dry season. Heavy rainfall during the rest of the year.
s	Dry season in summer.
w	Dry season in winter.

3rd order	
a	Hot summer. Warmest month above 22°C.
b	Warm summer. Warmest month below 22°C. 4 – 12 months above 10°C.
c	Cool summer. Warmest month below 22°C. 1 – 3 months above 10°C.
d	Severe winter. Coldest month below -38°C.
h	Hot. All months above 18°C.
k	Cold winter. Yearly mean temperature below 18°C. Warmest month above 18°C.
n	Frequent fog.

Köppen's Climatic Classification

The above map shows the world divided into major climatic regions according to the Köppen system of classification. In this system each region has been defined according to fixed values of temperature and precipitation. The classification is characterised by a shorthand code of letters, each of which designates a major climatic group. The table on the left gives a brief definition of these regions. The five main groups (A, B, C, D and E) have then been subdivided — the 2nd order subdivision relating to the seasonal characteristics of precipitation, and the 3rd order to temperature.

As with all systems of classification, only the broadest climatic types can be recognised. In general, many of the values used by Köppen to establish the climatic boundaries coincide fairly closely with the limits of the main vegetation types. It is for this reason, and because of its relative simplicity that the Köppen system is widely used and accepted, especially for teaching purposes.

For the map on World – Vegetation see page 111

These graphs illustrate mean daily maximum and minimum temperatures and mean rainfall values for each month. The altitude for each station above sea level is given in metres. The colour behind each station relates to Köppen's climatic classification as defined on page 106.

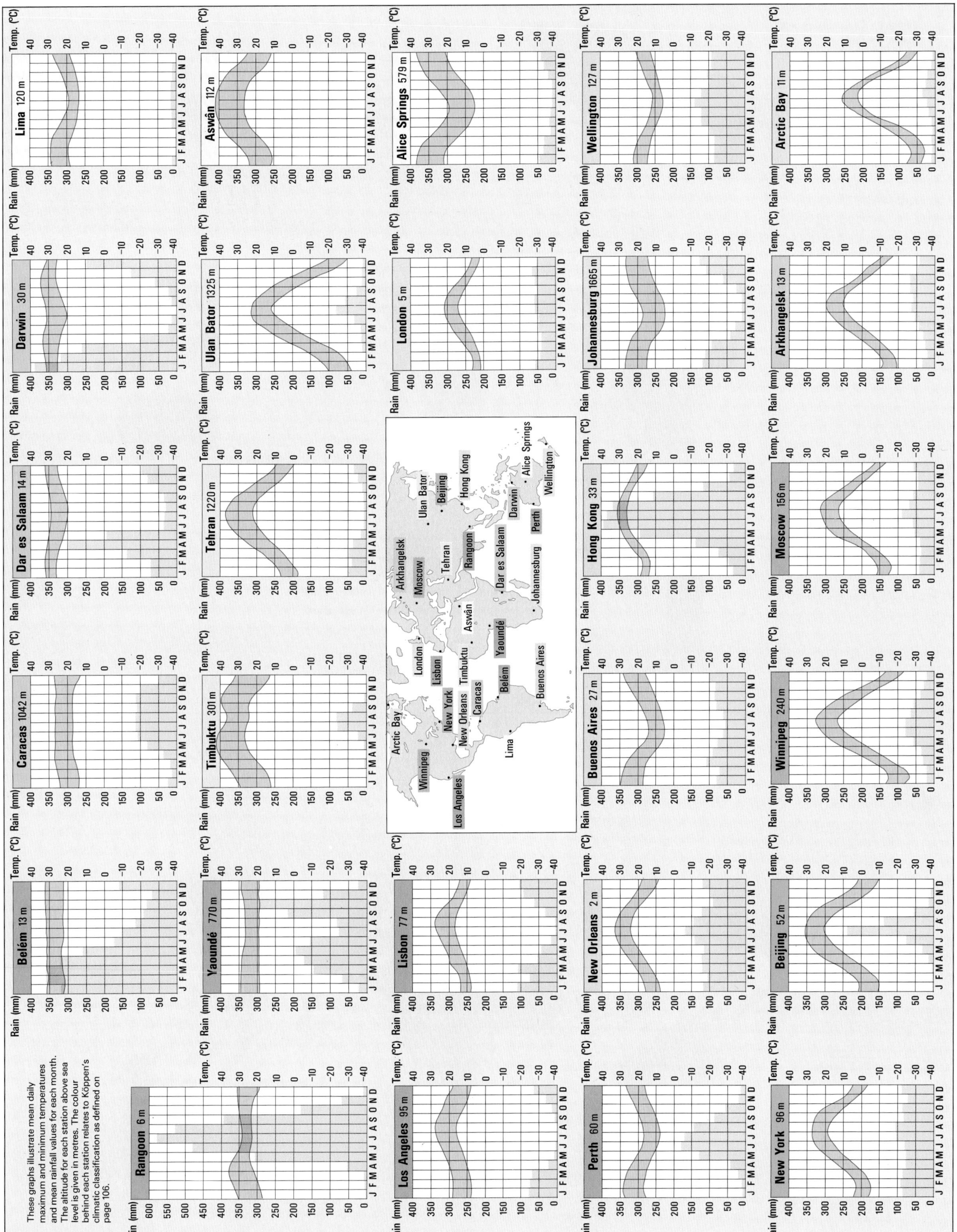

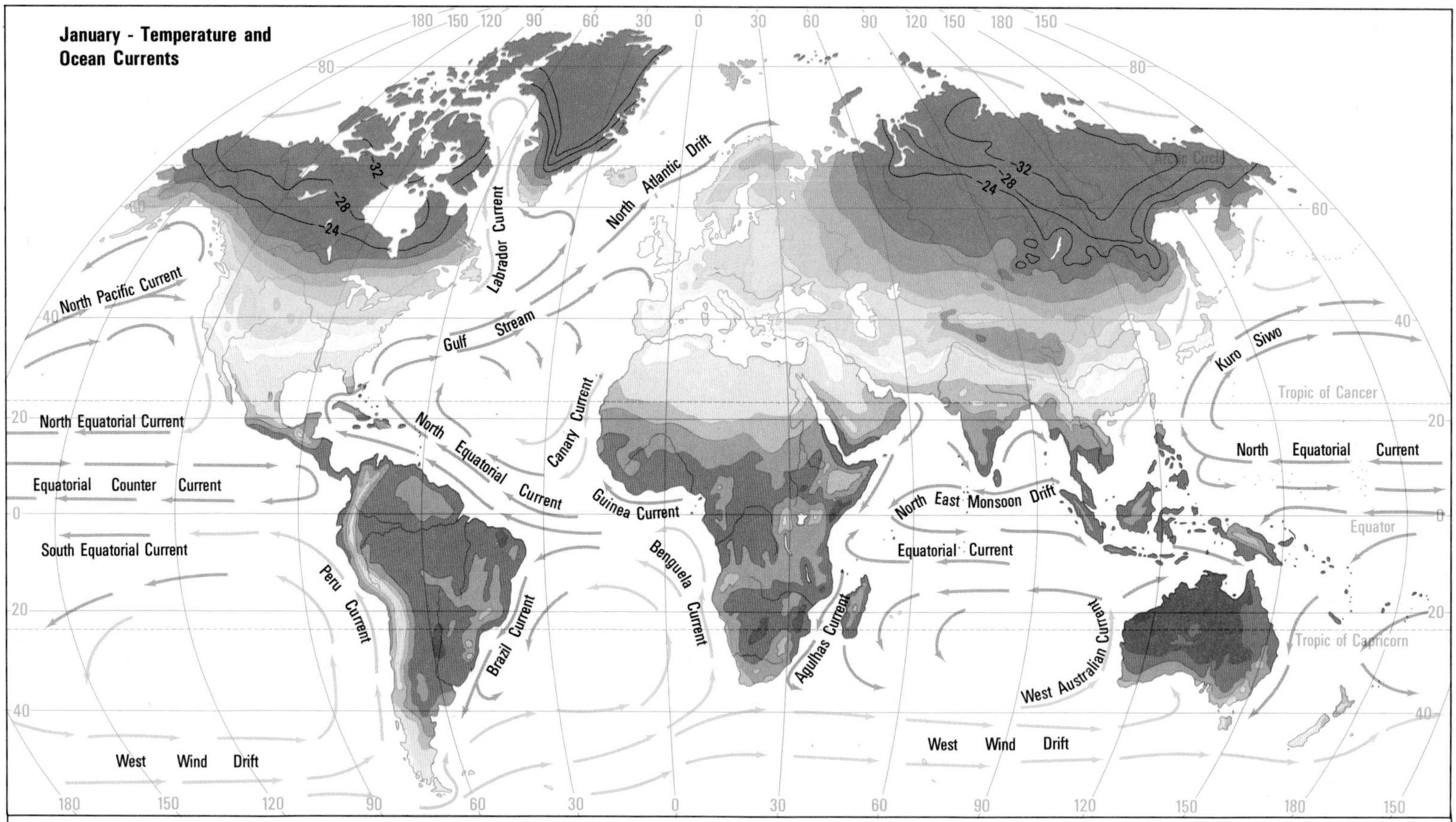

Surface temperature °C

36	8
32	4
28	0
24	−4
20	−8
16	−12
12	−16
	Below −20

Ocean Currents

Warm current

Cold current

Projection: Modified Winkel's

1:182 000 000

0 2000 4000 6000 km

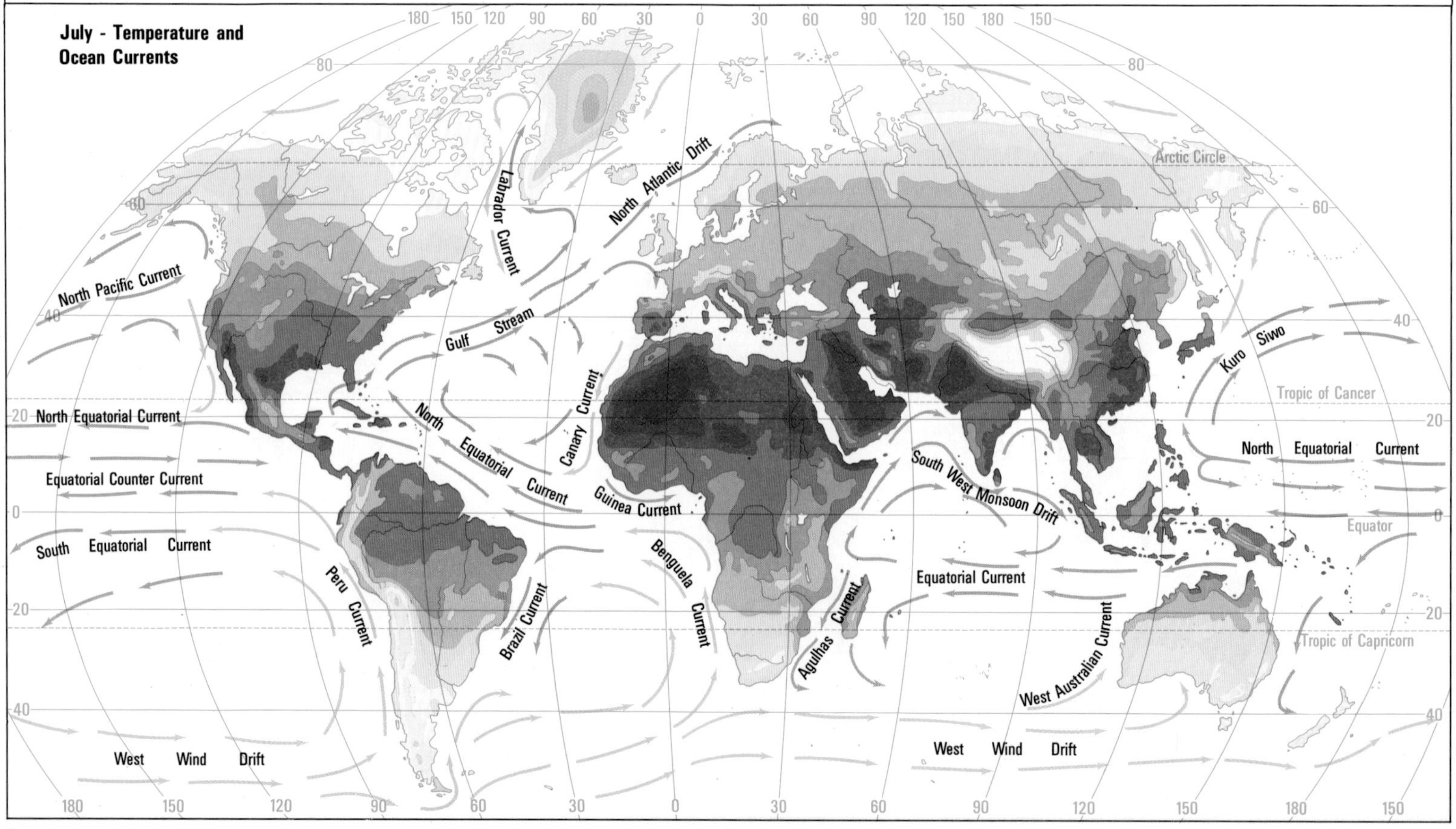

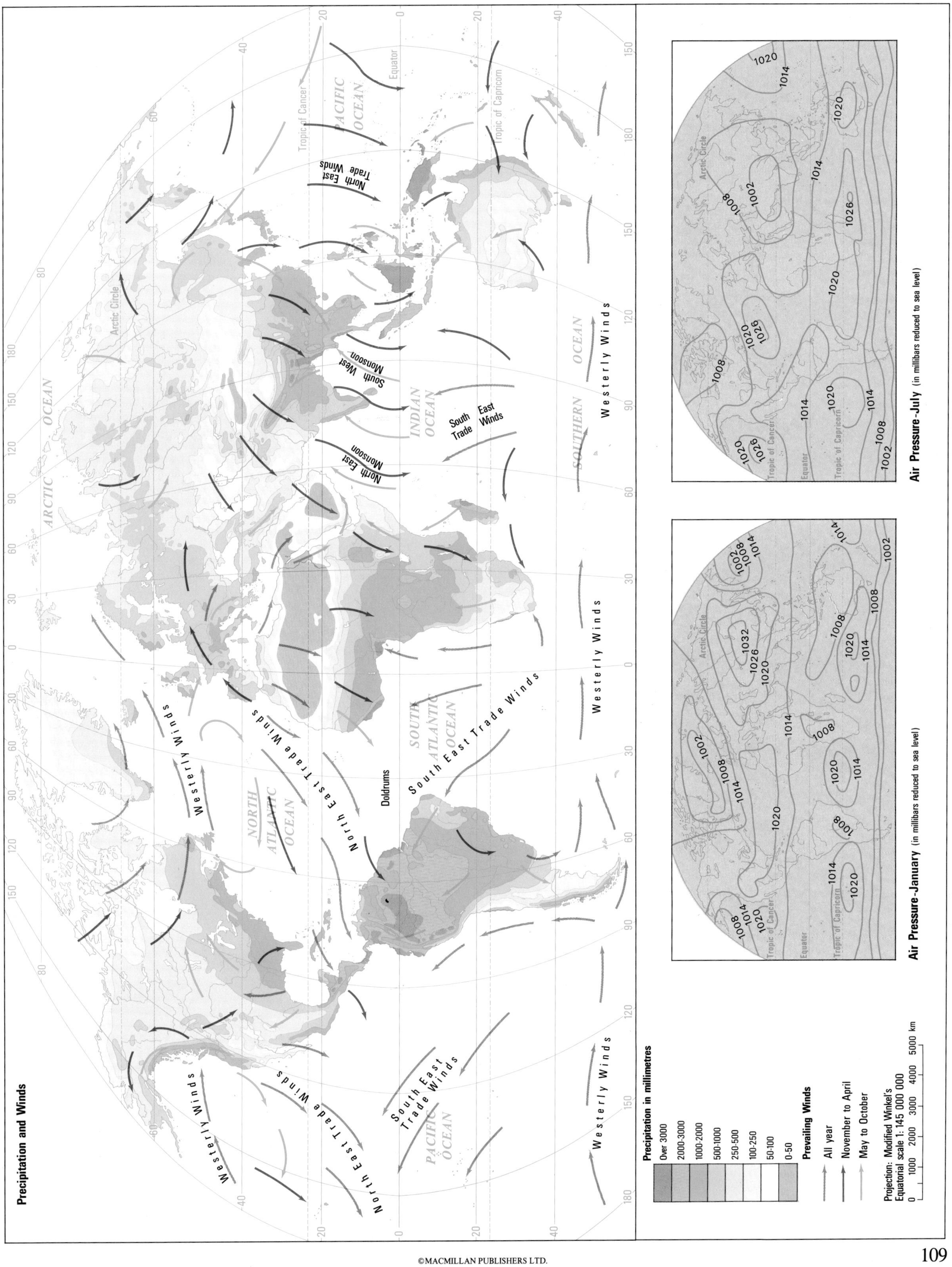
Precipitation and Winds
ARCTIC OCEAN
PACIFIC OCEAN
NORTH ATLANTIC OCEAN
SOUTH ATLANTIC OCEAN
INDIAN OCEAN
SOUTHERN OCEAN
Arctic Circle
Tropic of Cancer
Equator
Tropic of Capricorn
Westerly Winds
North East Trade Winds
South East Trade Winds
Doldrums
South West Monsoon
North East Monsoon
South East Trade Winds
Precipitation in millimetres
Over 3000
2000-3000
1000-2000
500-1000
250-500
100-250
50-100
0-50
Prevailing Winds
All year
November to April
May to October
Projection: Modified Winkel's
Equatorial scale 1: 145 000 000
0 1000 2000 3000 4000 5000 km
Air Pressure-January (in millibars reduced to sea level)
Air Pressure-July (in millibars reduced to sea level)

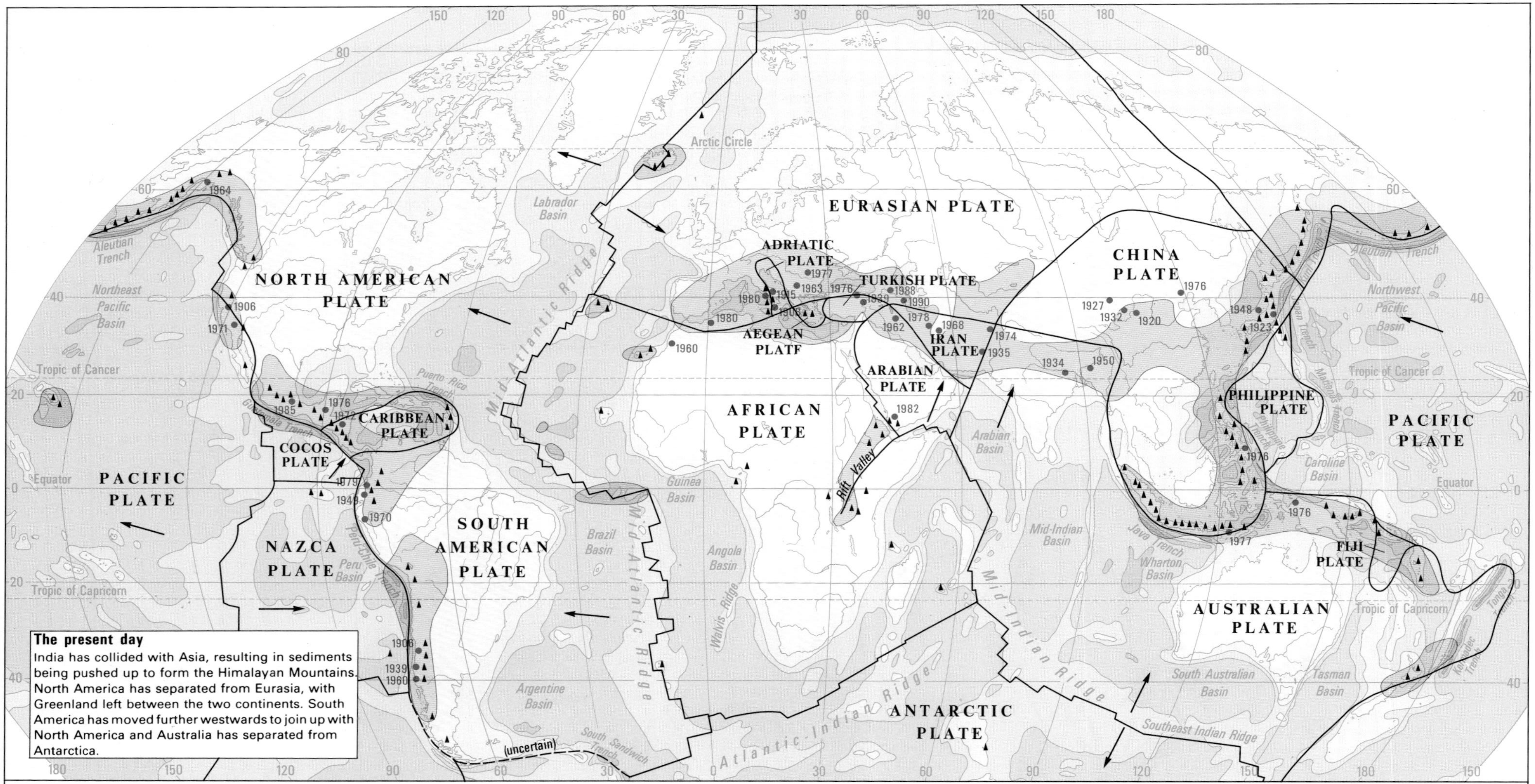

The present day
India has collided with Asia, resulting in sediments being pushed up to form the Himalayan Mountains. North America has separated from Eurasia, with Greenland left between the two continents. South America has moved further westwards to join up with North America and Australia has separated from Antarctica.

Evolution of the Continents

180 million years ago

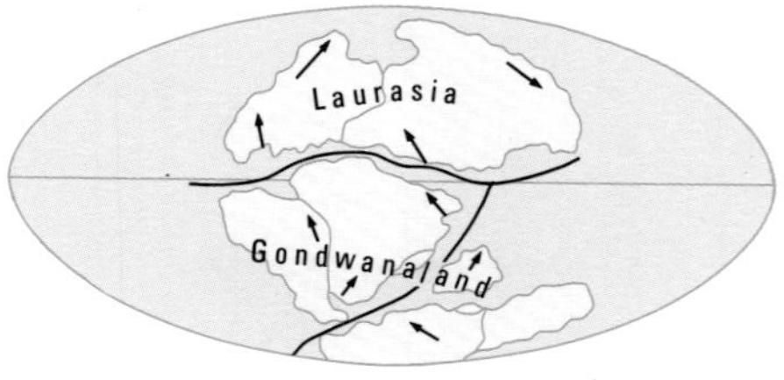

The original single landmass of Pangaea split into two continental blocks; Laurasia, which drifted northwards and Gondwanaland which began to break up. India separated, and the South American/African block moved away from Antarctica-Australia.

135 million years ago

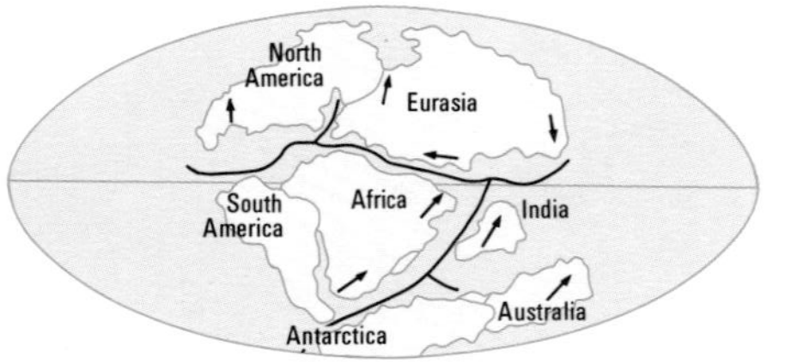

Gondwanaland and Laurasia continued to drift northwards and the Tethy's Sea between Africa and Eurasia started to close up in the east. The North Atlantic and Indian Oceans opened up further as the South Atlantic began to form. India continued to move north towards Asia.

65 million years ago

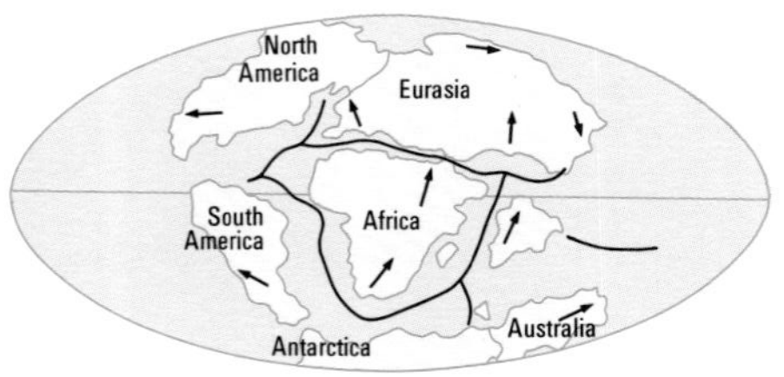

Madagascar broke from Africa while Australia remained connected to Antarctica. South America separated from Africa and, as it moved north and west, the South Atlantic Ocean opened up behind it. The Mediterranean Sea was now recognisable as the Tethy's Sea finally closed.

Tectonic plate boundary
Direction of plate movement
Major earthquake in the last 100 years
Recent active volcano
Areas liable to frequent earthquakes

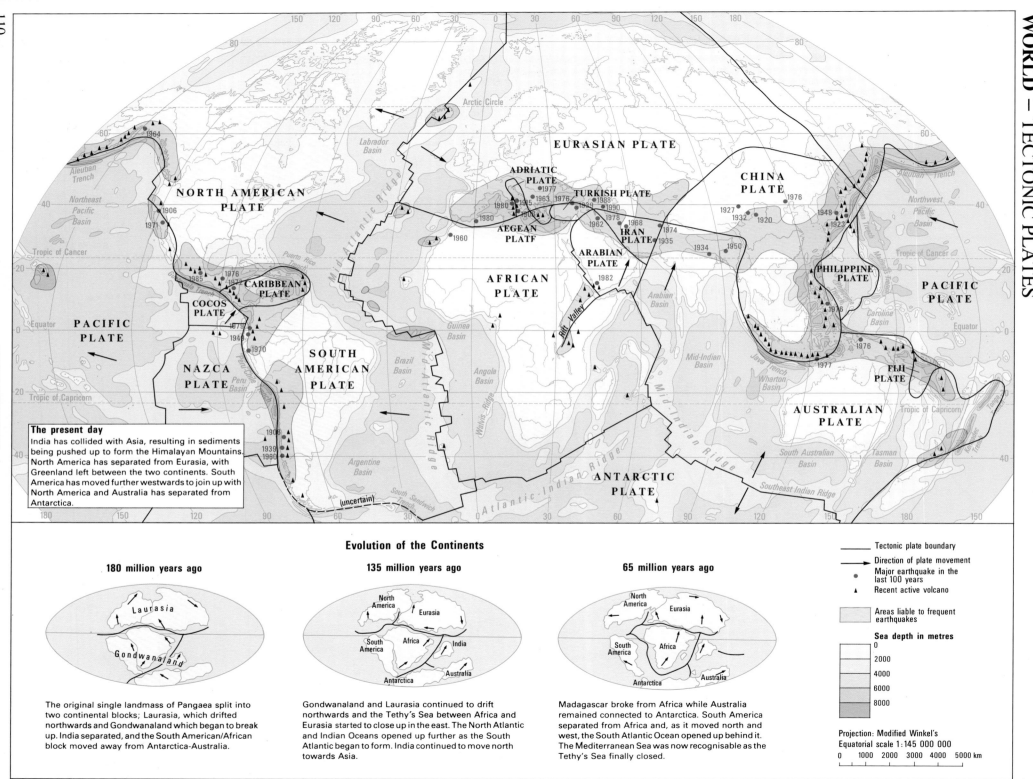

Sea depth in metres
0
2000
4000
6000
8000

Projection: Modified Winkel's
Equatorial scale 1:145 000 000
0 1000 2000 3000 4000 5000 km

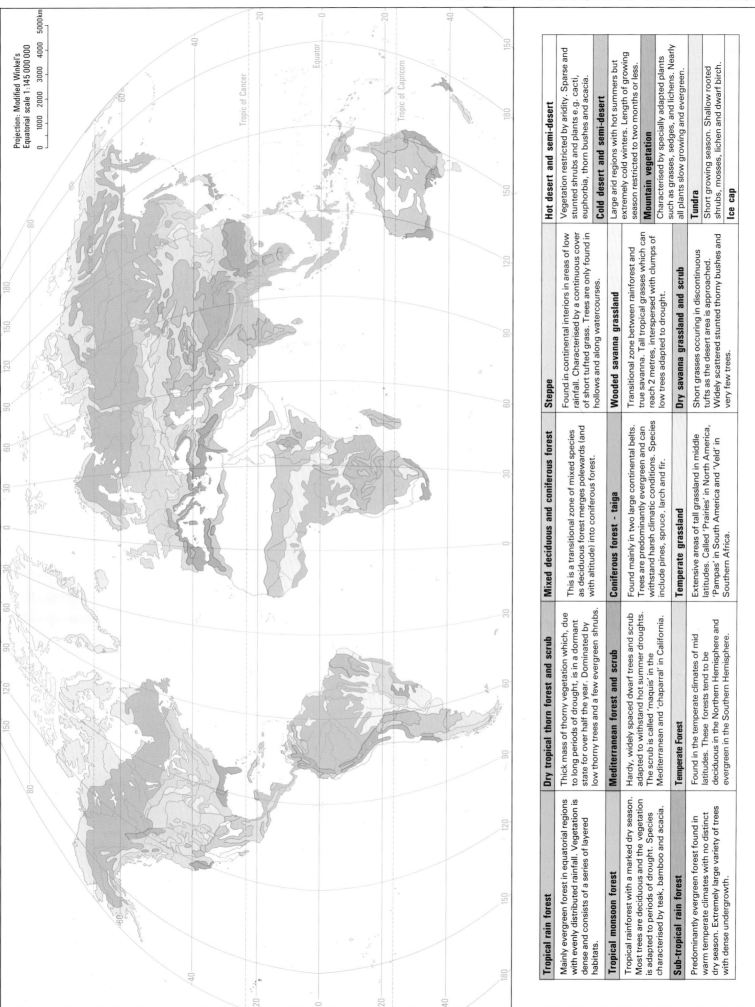

Tropical rain forest
Mainly evergreen forest in equatorial regions with evenly distributed rainfall. Vegetation is dense and consists of a series of layered habitats.

Tropical monsoon forest
Tropical rainforest with a marked dry season. Most trees are deciduous and the vegetation is adapted to periods of drought. Species characterised by teak, bamboo and acacia.

Sub-tropical rain forest
Predominantly evergreen forest found in warm temperate climates with no distinct dry season. Extremely large variety of trees with dense undergrowth.

Dry tropical thorn forest and scrub
Thick mass of thorny vegetation which, due to long periods of drought, is in a dormant state for over half the year. Dominated by low thorny trees and a few evergreen shrubs.

Mediterranean forest and scrub
Hardy, widely spaced dwarf trees and scrub adapted to withstand hot summer droughts. The scrub is called 'maquis' in the Mediterranean and 'chaparral' in California.

Temperate Forest
Found in the temperate climates of mid latitudes. These forests tend to be deciduous in the Northern Hemisphere and evergreen in the Southern Hemisphere.

Mixed deciduous and coniferous forest
This is a transitional zone of mixed species as deciduous forest merges polewards (and with altitude) into coniferous forest.

Coniferous forest - taiga
Found mainly in two large continental belts. Trees are predominantly evergreen and can withstand harsh climatic conditions. Species include pines, spruce, larch and fir.

Temperate grassland
Extensive areas of tall grassland in middle latitudes. Called 'Prairies' in North America, 'Pampas' in South America and 'Veld' in Southern Africa.

Steppe
Found in continental interiors in areas of low rainfall. Characterised by a continuous cover of short tufted grass. Trees are only found in hollows and along watercourses.

Wooded savanna grassland
Transitional zone between rainforest and true savanna. Tall tropical grasses which can reach 2 metres, interspersed with clumps of low trees adapted to drought.

Dry savanna grassland and scrub
Short grasses occuring in discontinuous tufts as the desert area is approached. Widely scattered stunted thorny bushes and very few trees.

Hot desert and semi-desert
Vegetation restricted by aridity. Sparse and stunted shrubs and plants e.g. cacti, euphorbia, thorn bushes and acacia.

Cold desert and semi-desert
Large arid regions with hot summers but extremely cold winters. Length of growing season restricted to two months or less.

Mountain vegetation
Characterised by specially adapted plants such as grasses, sedges, and lichens. Nearly all plants slow growing and evergreen.

Tundra
Short growing season. Shallow rooted shrubs, mosses, lichen and dwarf birch.

Ice cap

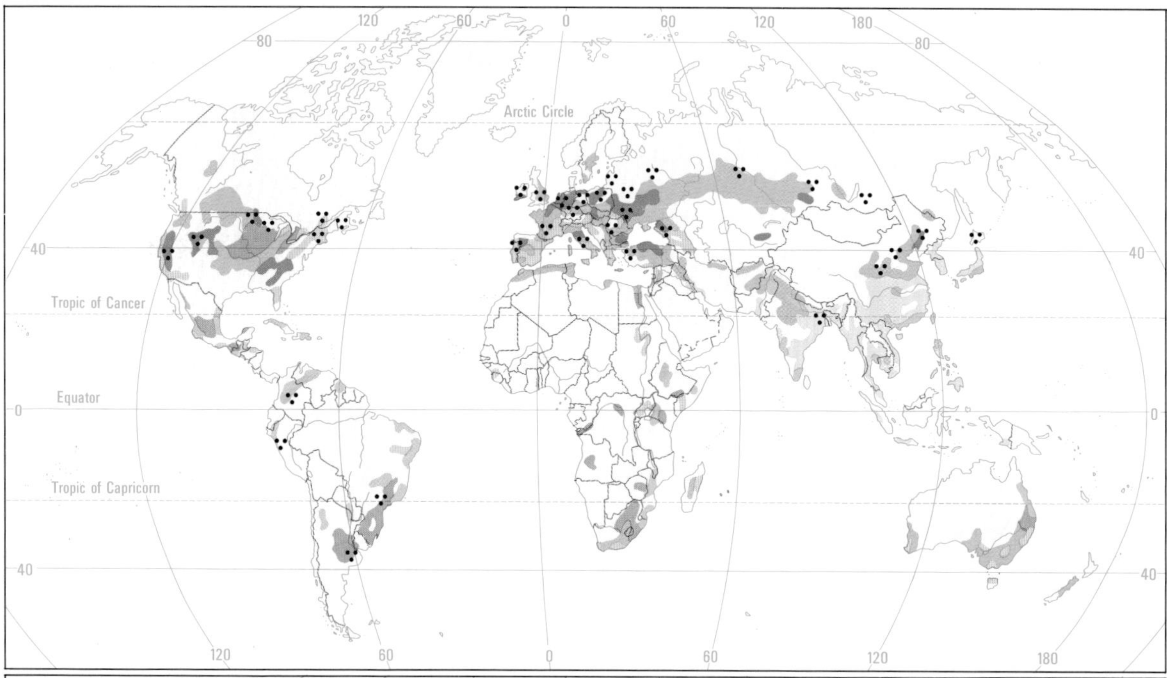

Selected Food Crops

- Wheat
- Maize
- Rice
- Sugar beet
- Sugar cane
- Fruit
- ∴ Potatoes

World Cereal Production 1985

Wheat Production
World total: 510 million tonnes

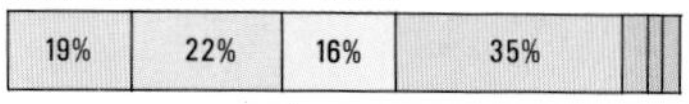

Maize Production
World total: 490 million tonnes

Rice Production
World total: 466 million tonnes

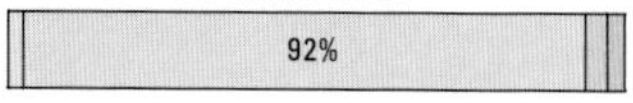

- North America
- Europe
- U S S R
- Asia
- South America
- Africa
- Australasia

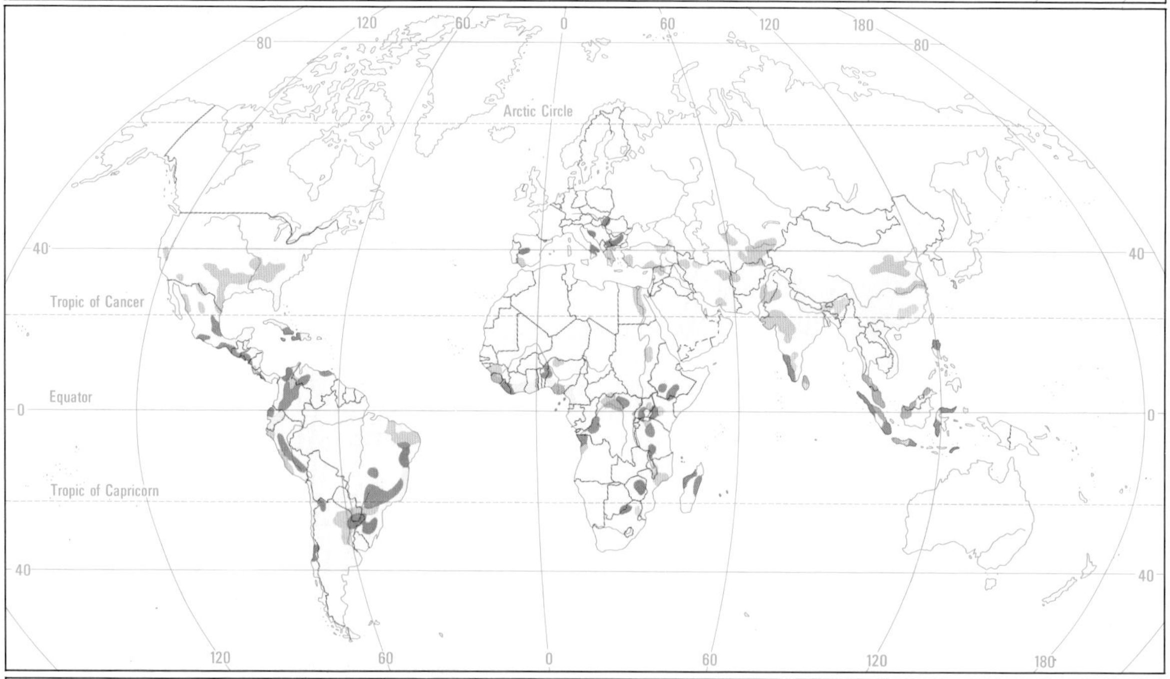

Selected Commercial Crops

- Cocoa
- Coffee
- Tea
- Cotton
- Tobacco
- Rubber

Coffee Production 1985
World total: 6 028 000 tonnes

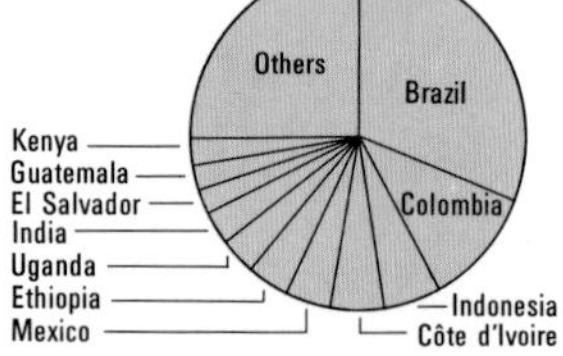

Tea Production 1985
World total: 2 333 000 tonnes

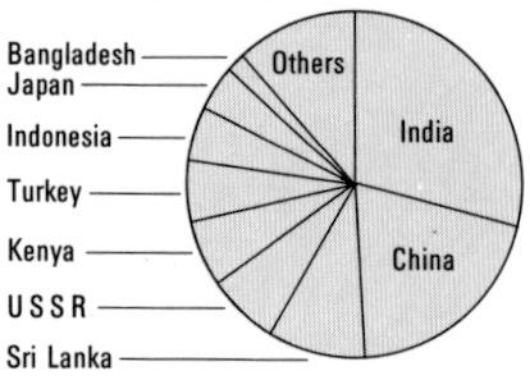

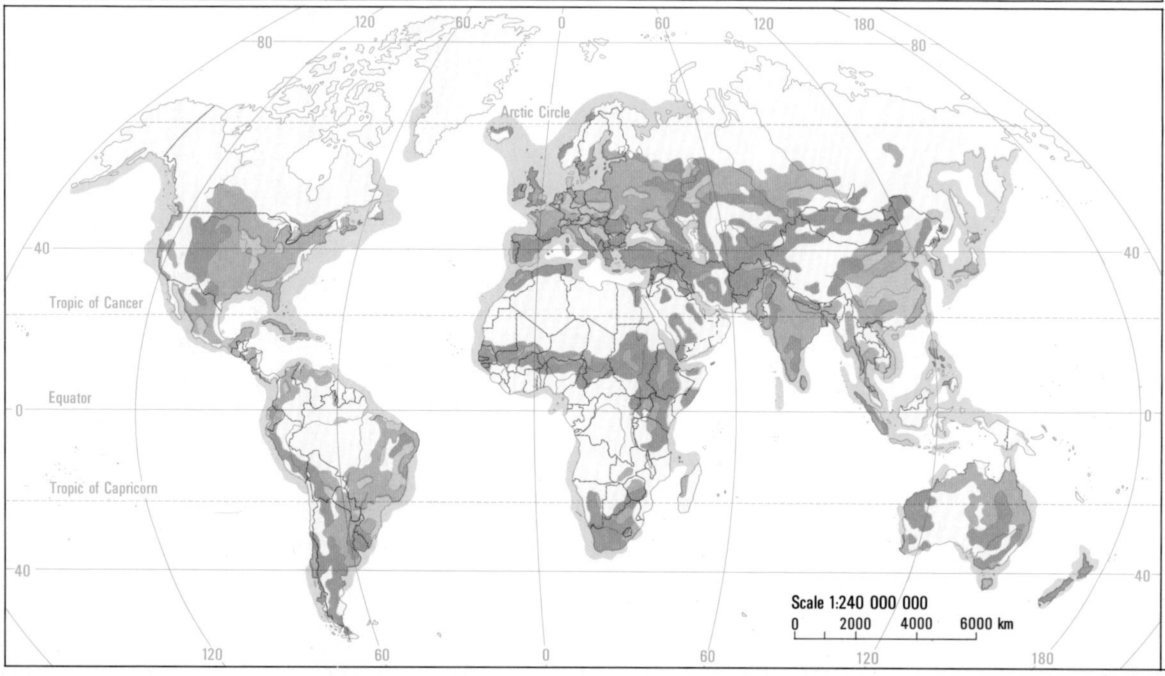

Livestock and Fishing

- Cattle
- Sheep
- Pigs
- Fishing

Cattle Stocks 1985
World total: 1 268 million head

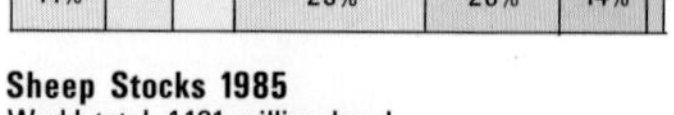

Sheep Stocks 1985
World total: 1121 million head

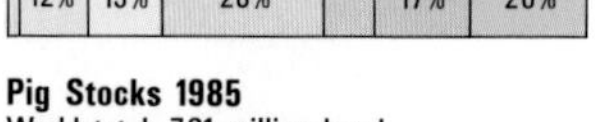

Pig Stocks 1985
World total: 791 million head

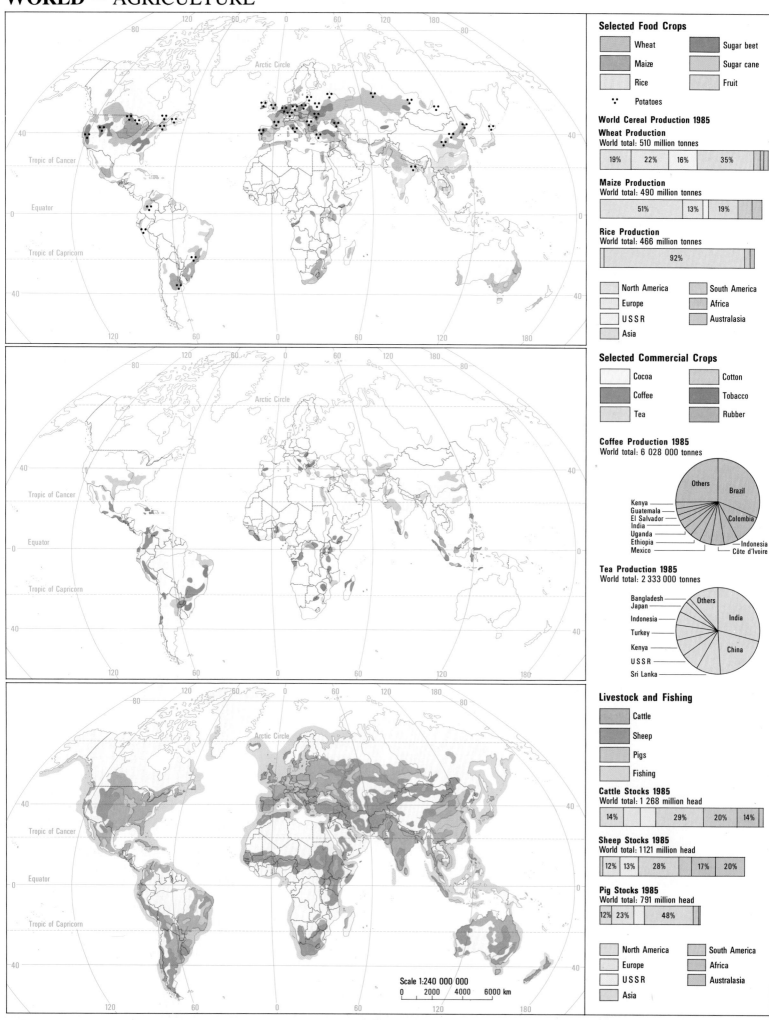

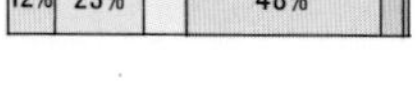

- North America
- Europe
- U S S R
- Asia
- South America
- Africa
- Australasia

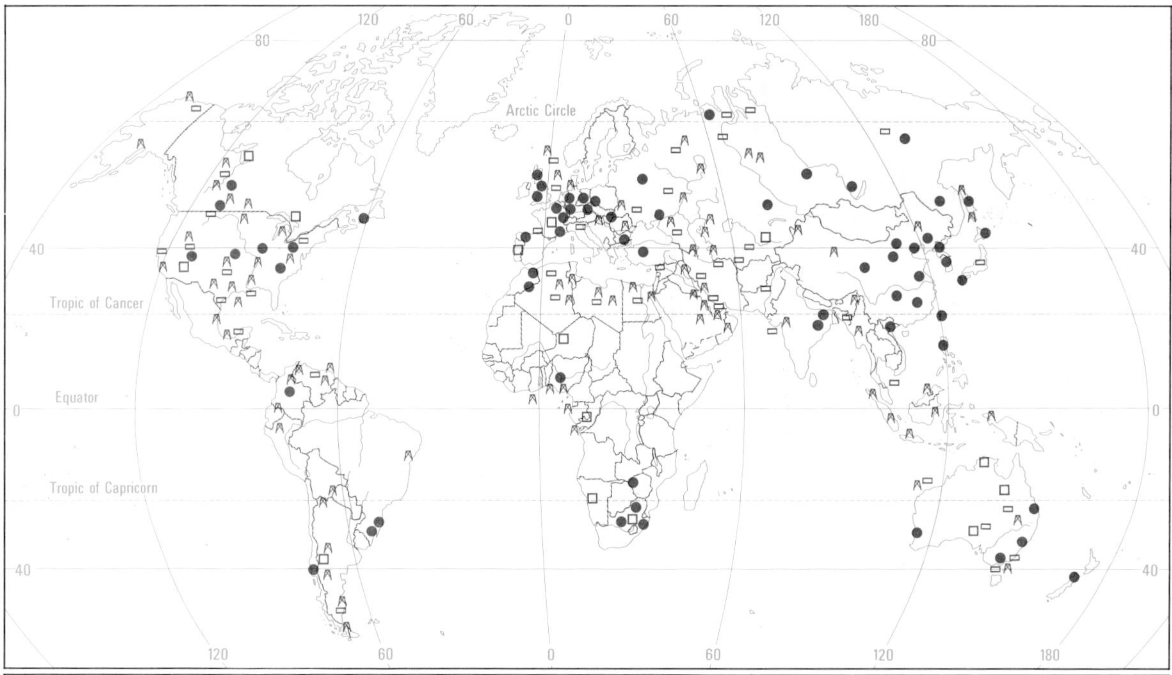

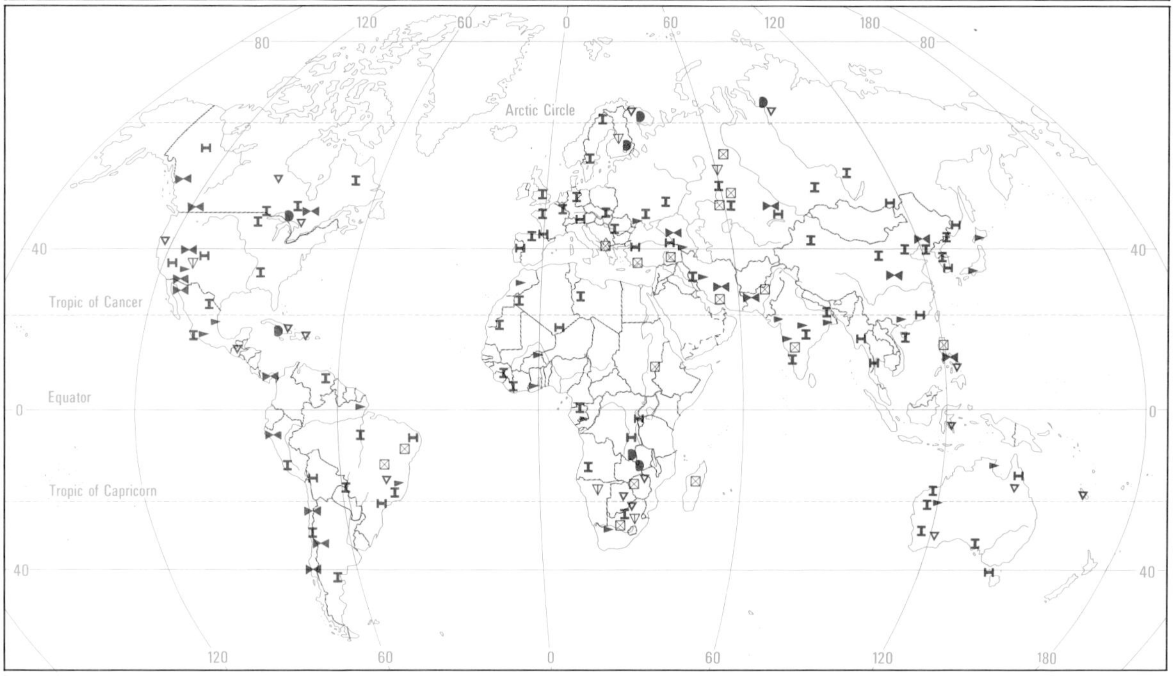

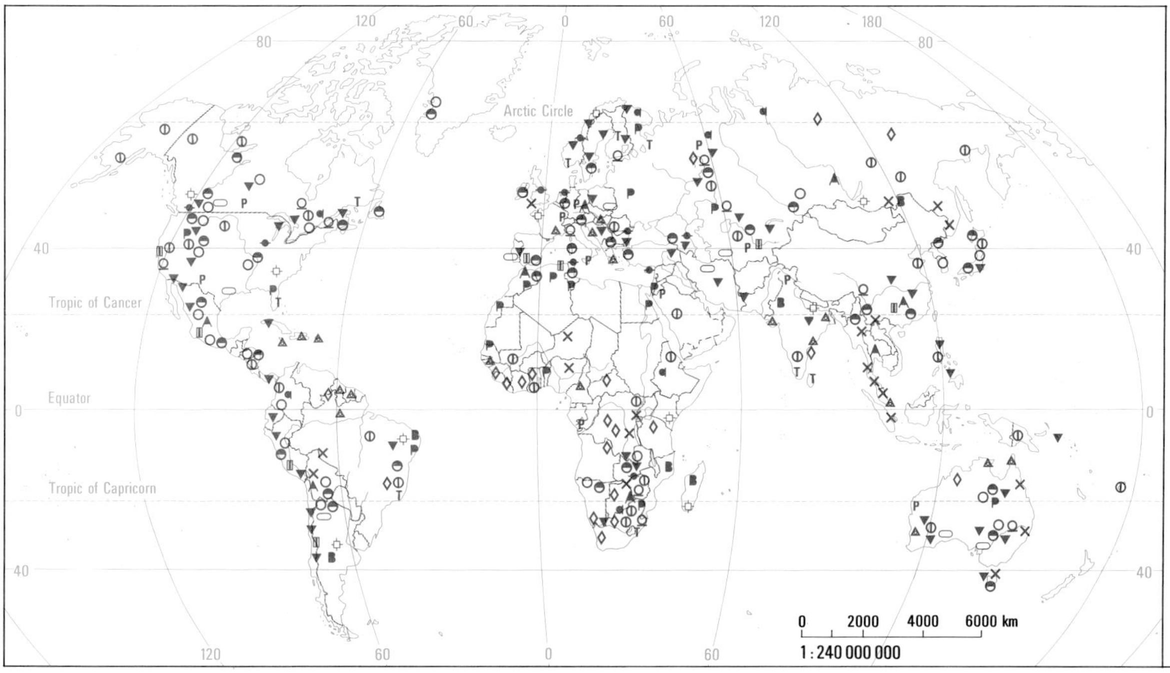

Fuel

- ● Coal
- Oil
- Gas
- □ Uranium

World Fuel Reserves at end 1985

Coal Reserves
World total : 954 486 million tonnes

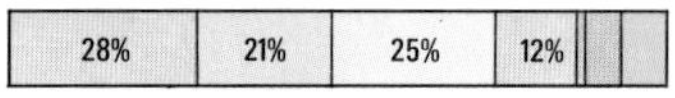

Oil Reserves (Proved)
World total : 95 800 million tonnes

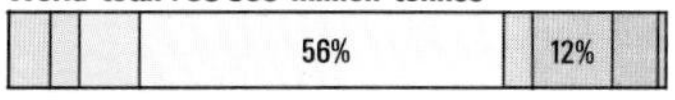

Gas Reserves (Proved)
World total : 98 trillion cubic metres

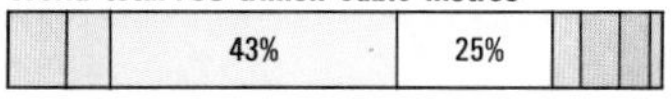

- North America
- Europe
- USSR
- Middle East
- Asia
- South America
- Africa
- Australasia

Iron Ore and Ferro-Alloys

- Chrome
- Cobalt
- Iron Ore
- Manganese
- Molybdenum
- Nickel
- Tungsten
- Vanadium

Iron Ore Production 1984
World total : 483 million tonnes

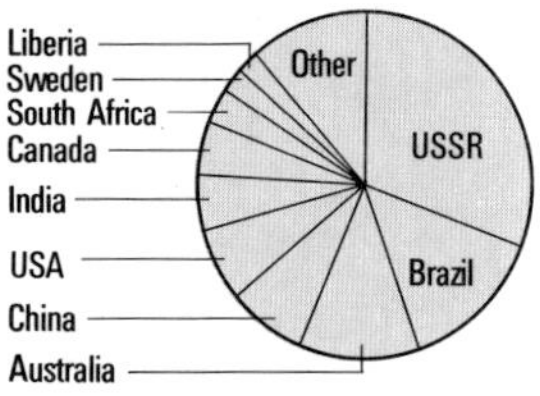

Chrome Production 1984
World total : 9 183 000 tonnes

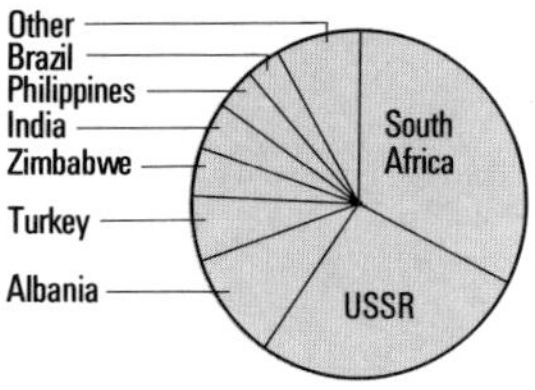

Other Minerals

- Antimony
- Asbestos
- Bauxite
- Beryllium
- Copper
- Diamonds
- Gold
- Lead and zinc
- Mercury
- Mica
- Phosphates
- Platinum
- Potash
- Pyrites
- Silver
- Sulphur
- Tin
- Titanium

Copper Production 1984
World total: 8 283 000 tonnes

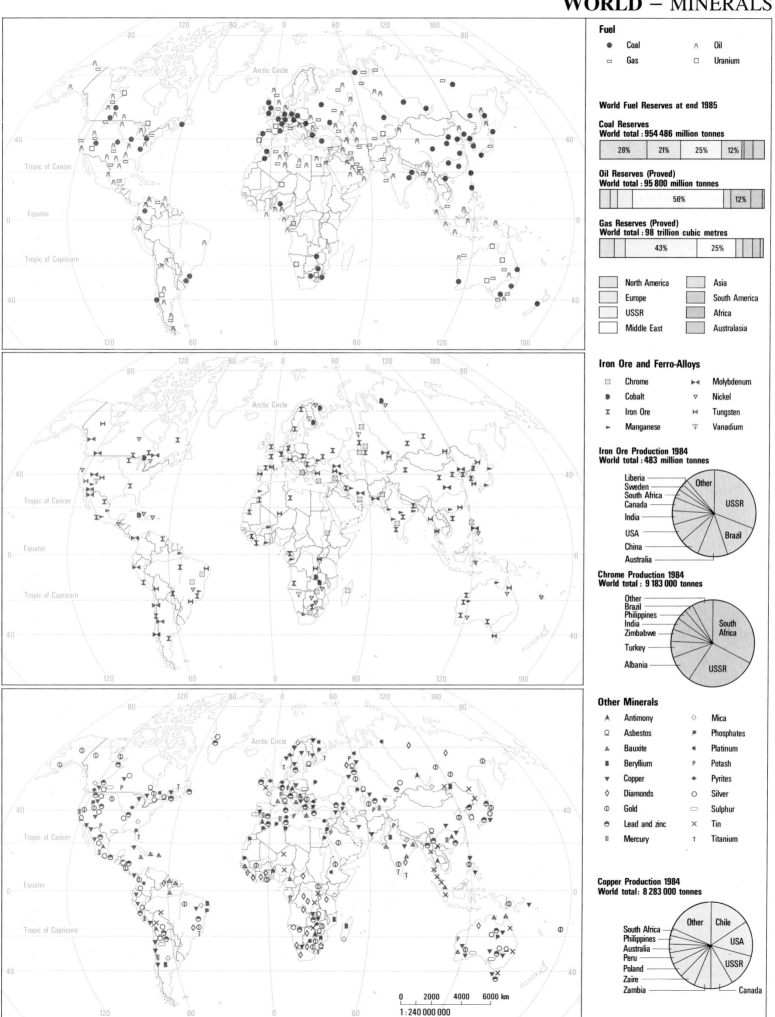

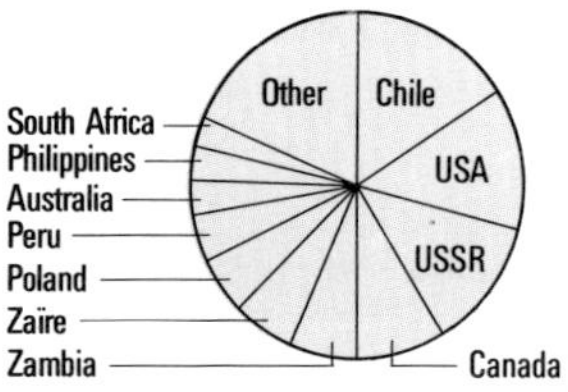

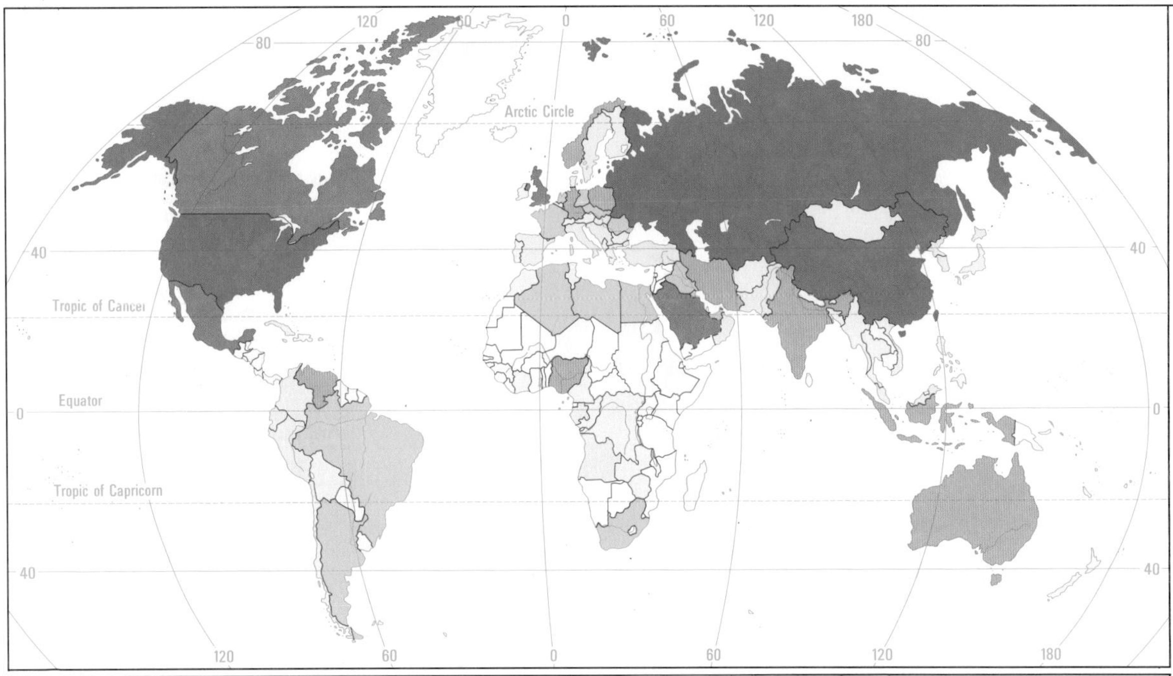

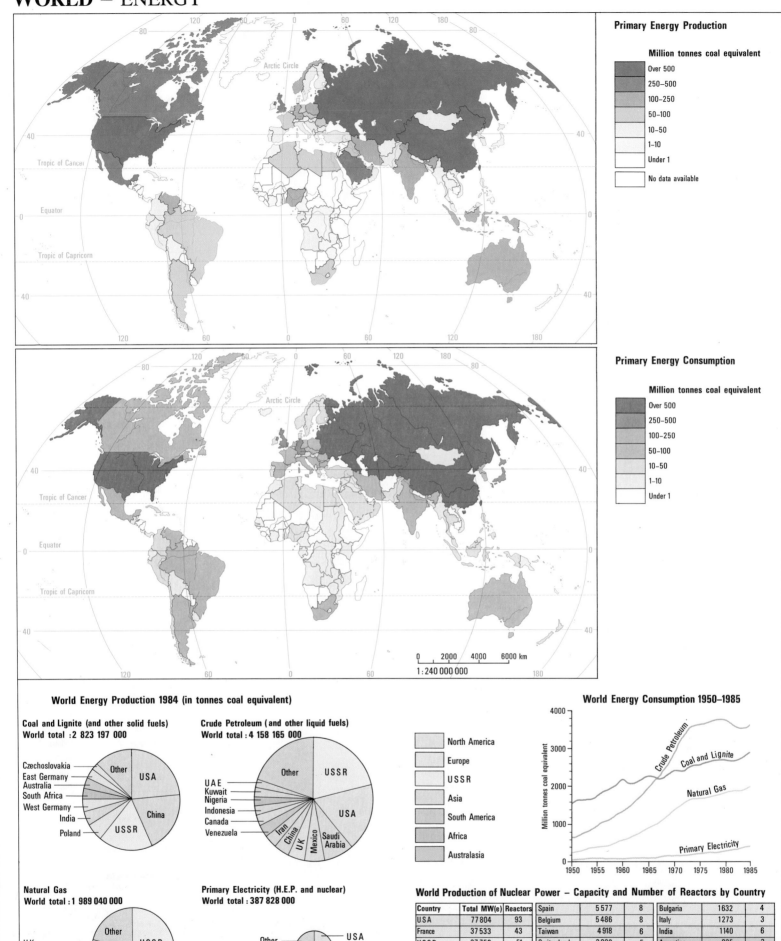

Primary Energy Production

Million tonnes coal equivalent

- Over 500
- 250–500
- 100–250
- 50–100
- 10–50
- 1–10
- Under 1
- No data available

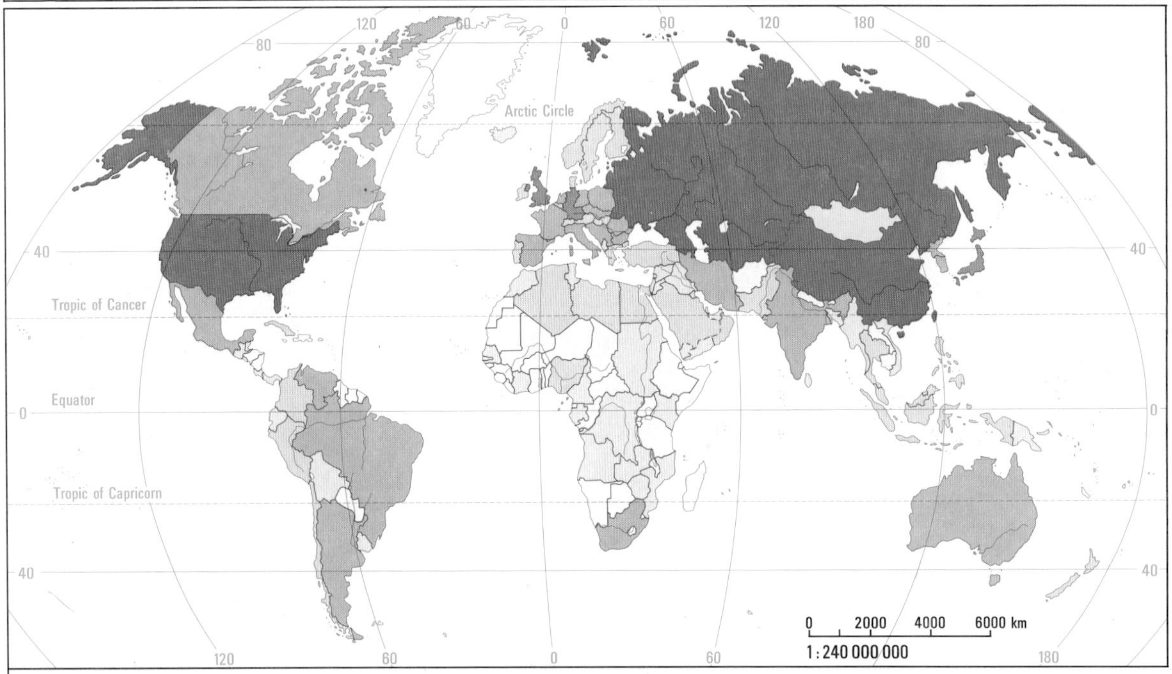

Primary Energy Consumption

Million tonnes coal equivalent

- Over 500
- 250–500
- 100–250
- 50–100
- 10–50
- 1–10
- Under 1

World Energy Production 1984 (in tonnes coal equivalent)

Coal and Lignite (and other solid fuels)
World total : 2 823 197 000

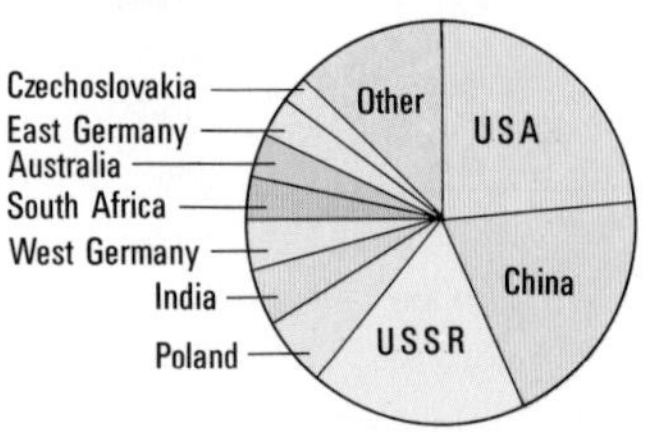

Crude Petroleum (and other liquid fuels)
World total : 4 158 165 000

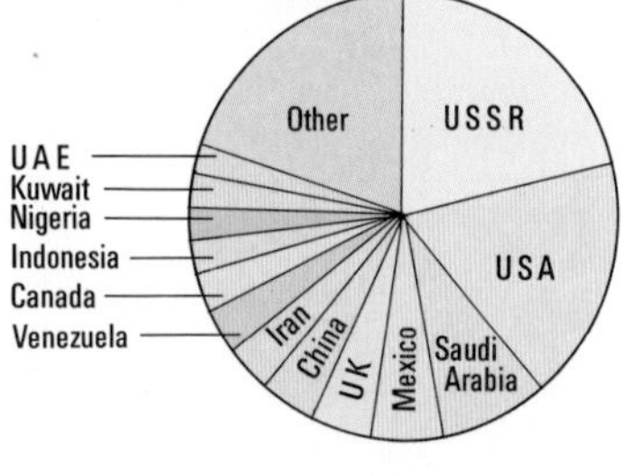

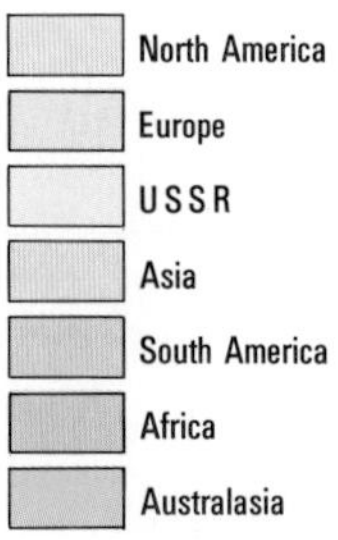

Natural Gas
World total : 1 989 040 000

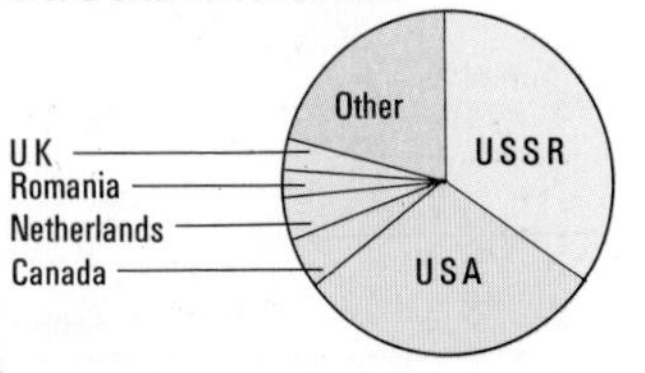

Primary Electricity (H.E.P. and nuclear)
World total : 387 828 000

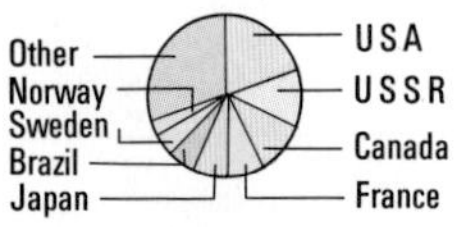

World Energy Consumption 1950–1985

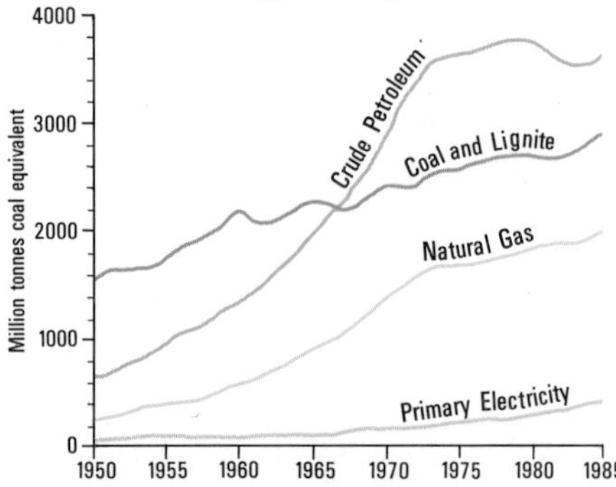

World Production of Nuclear Power – Capacity and Number of Reactors by Country

Country	Total MW(e)	Reactors
USA	77804	93
France	37533	43
USSR	27756	51
Japan	23665	33
West Germany	16413	19
UK	10120	38
Canada	9776	16
Sweden	9455	12
Spain	5577	8
Belgium	5486	8
Taiwan	4918	6
Switzerland	2882	5
South Korea	2720	4
Finland	2310	4
Czechoslovakia	1980	5
South Africa	1840	2
East Germany	1694	5
Bulgaria	1632	4
Italy	1273	3
India	1140	6
Argentina	935	2
Hungary	825	2
Yugoslavia	632	1
Brazil	626	1
Netherlands	508	2
Pakistan	125	1

1 MW (e) = 10^6 watts electrical

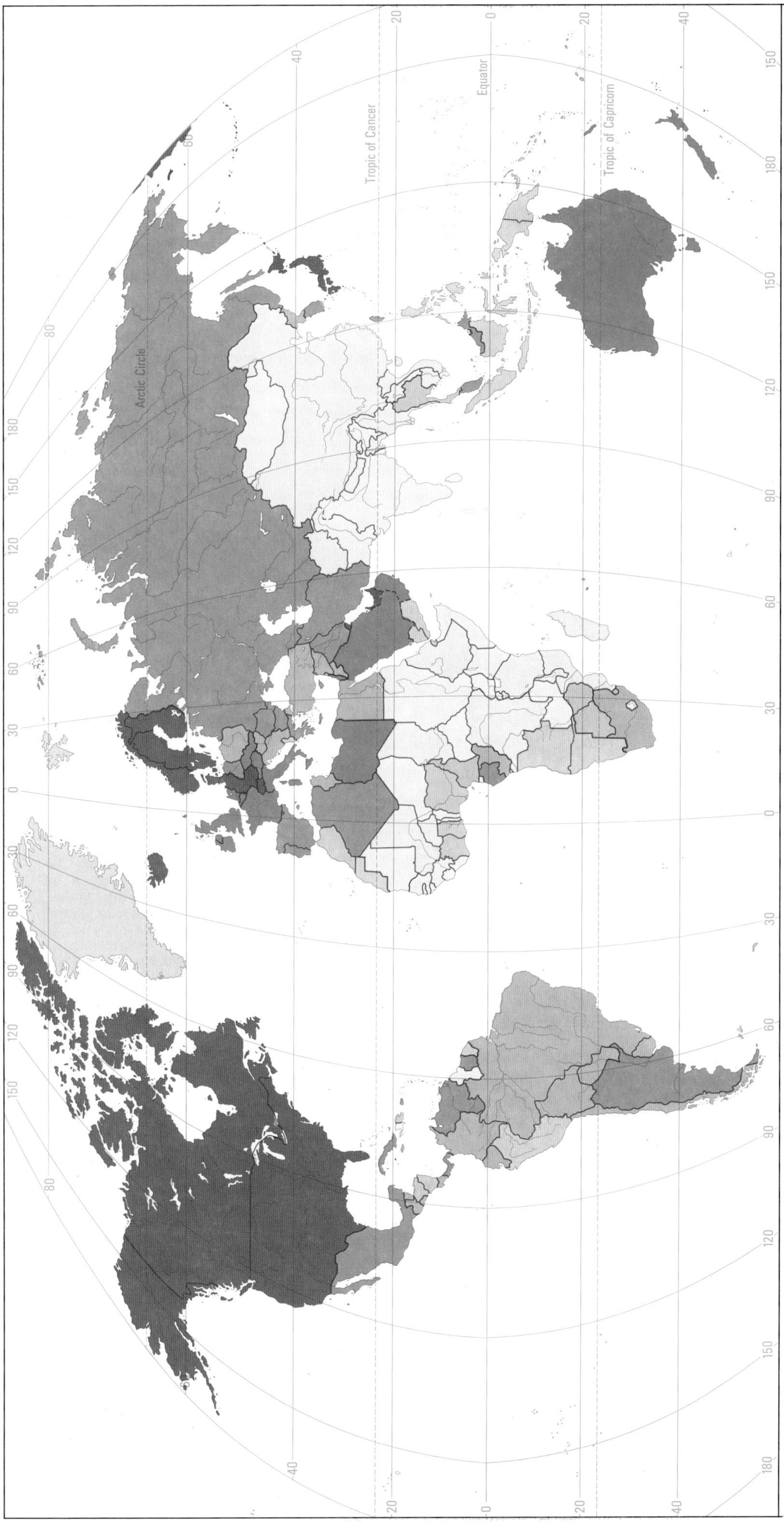
Arctic Circle
Tropic of Cancer
Equator
Tropic of Capricorn

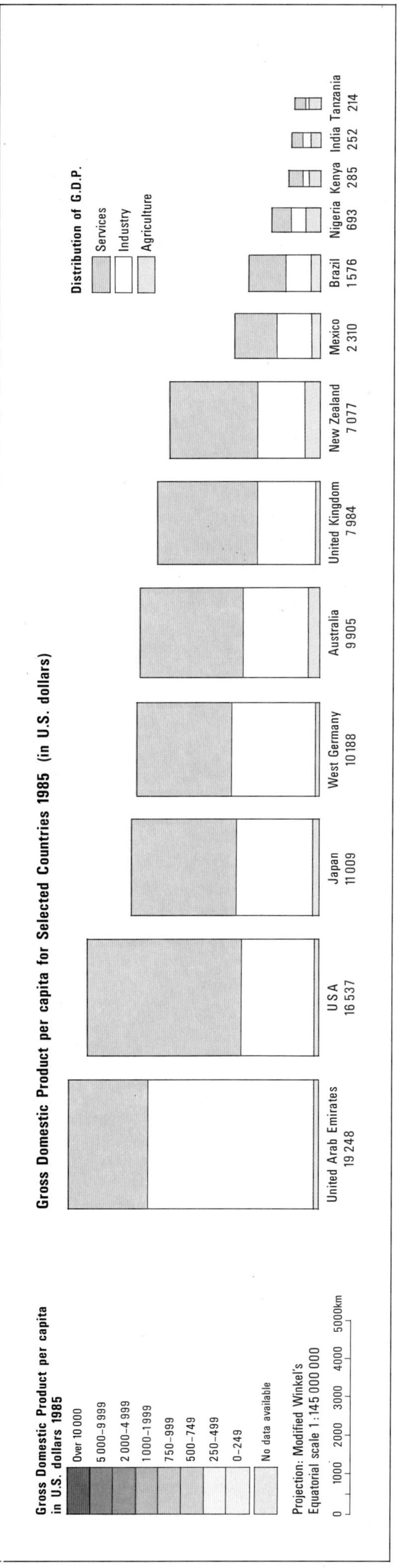
Gross Domestic Product per capita for Selected Countries 1985 (in U.S. dollars)
United Arab Emirates 19 248
USA 16 537
Japan 11 009
West Germany 10 188
Australia 9 905
United Kingdom 7 984
New Zealand 7 077
Mexico 2 310
Brazil 1 576
Nigeria 693
Kenya 285
India 252
Tanzania 214
Distribution of G.D.P.
Services
Industry
Agriculture
Gross Domestic Product per capita in U.S. dollars 1985
Over 10 000
5 000–9 999
2 000–4 999
1 000–1 999
750–999
500–749
250–499
0–249
No data available
Projection: Modified Winkel's
Equatorial scale 1:145 000 000
0 1000 2000 3000 4000 5000km

WORLD – SELECTED INTERNATIONAL ORGANISATIONS

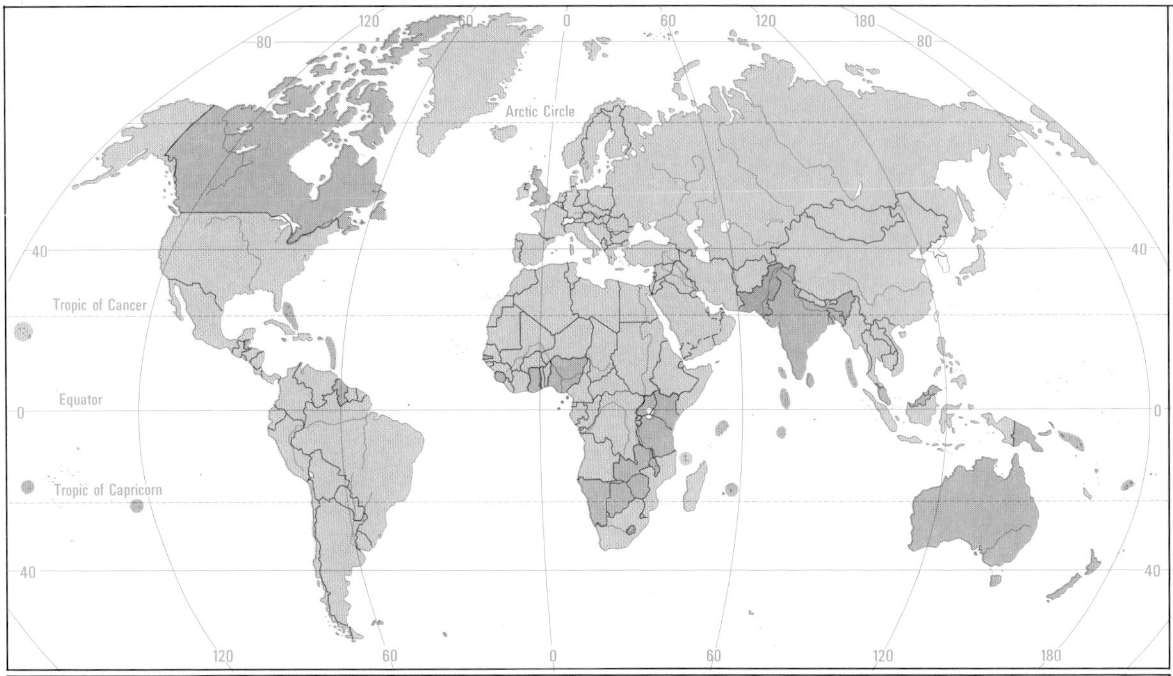

United Nations and The Commonwealth

- United Nations (UN)
- The Commonwealth and United Nations

On 24 October 1945 the UN organisation was established to maintain international peace and security, and to develop co-operation in solving economic, social, cultural and humanitarian problems. The main advisory body is the General Assembly whilst the Security Council bears the responsibility for maintaining international peace.

The Commonwealth is a voluntary association of sovereign independent states, founded in 1949. There are now 50 members comprising the United Kingdom and most of its former colonies and dependencies. The Secretariat, established in 1965, is the central co-ordinating body and has several divisions responsible for areas such as education, technology and health.

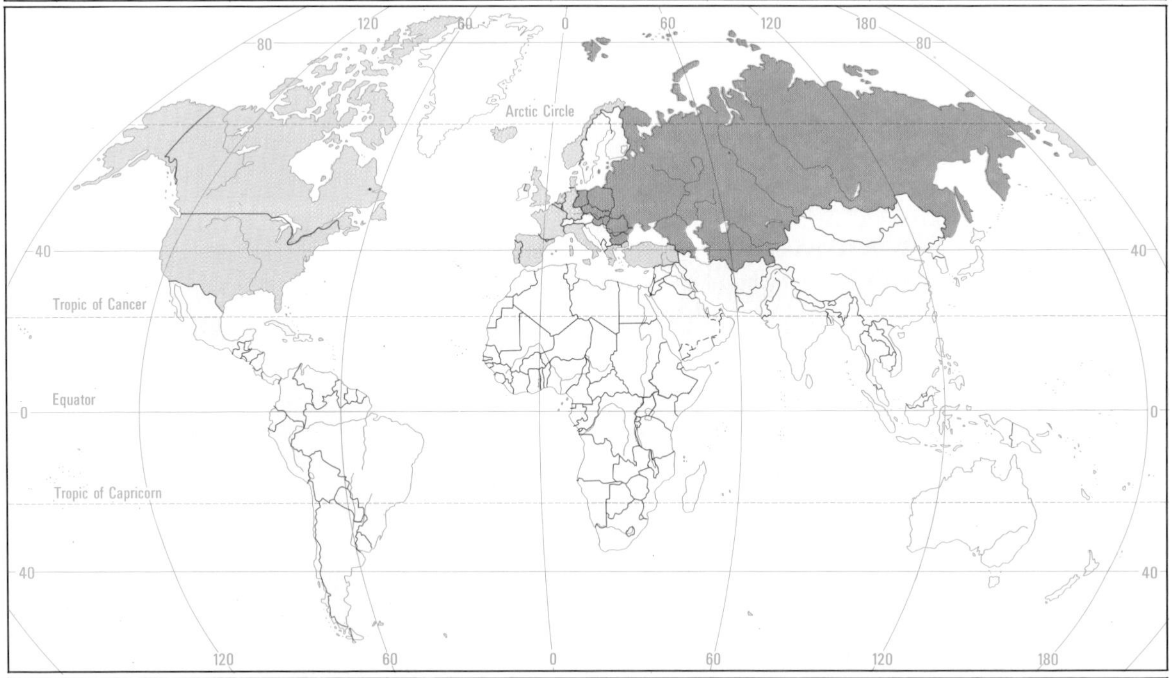

North Atlantic Treaty Organisation and The Warsaw Pact

- North Atlantic Treaty Organisation (NATO)
- The Warsaw Pact

NATO was founded in 1949 as a military alliance linking a group of European states with the USA and Canada. All 16 member states are bound to protect each other against armed attack. NATO also seeks to encourage economic and social co-operation amongst its member states. The Secretariat headquarters are in Brussels and the military headquarters near Mons, Belgium.

The Warsaw Treaty of Friendship, Co-operation and Mutual Assistance (The Warsaw Pact) was signed on 14 May 1955 by the USSR and its six Eastern European allies. It was formed as a military and defence organisation to act as a counterforce to NATO. The Soviet Army Marshal is the head of the Pact's forces and the headquarters are in Moscow.

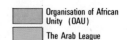

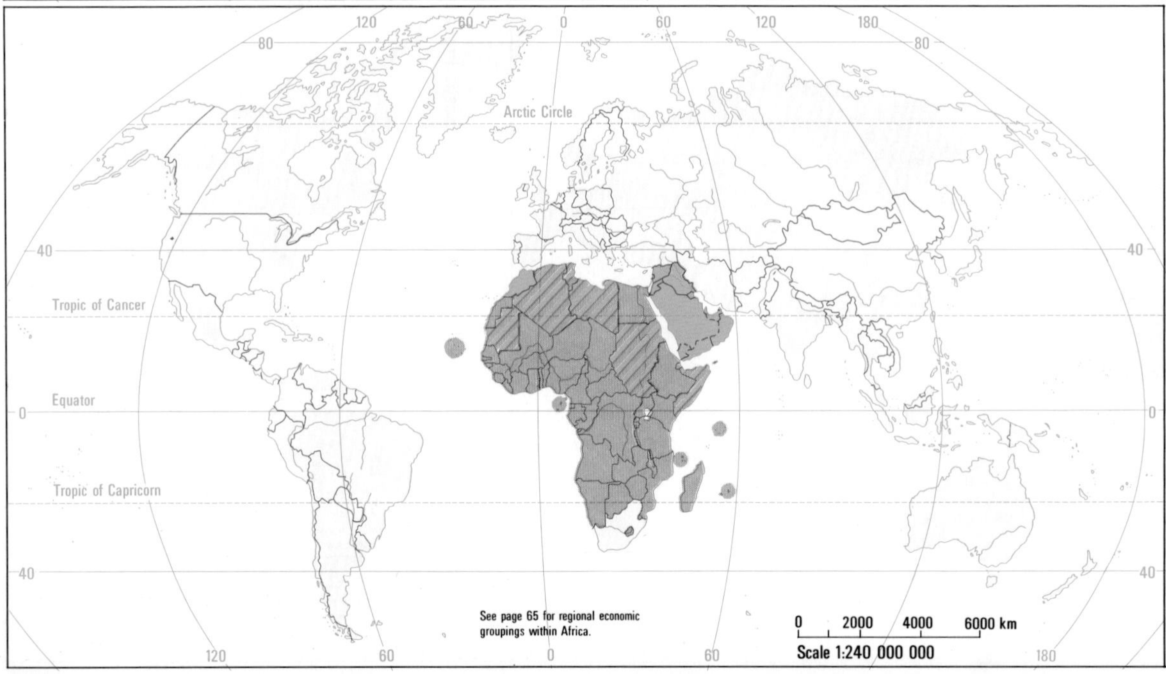

Organisation of African Unity and The Arab League

- Organisation of African Unity (OAU)
- The Arab League

The OAU was formed on 25 May 1963 in Addis Ababa by the independent states of Africa. Its main aims are to promote unity and solidarity between its members, and to eliminate colonialism in Africa. The Assembly of Heads of State and Government meets annually to co-ordinate policies on political, economic, cultural, health, scientific and defence issues.

The Arab League is a voluntary association of sovereign Arab states founded on 22 March 1945. Its main aim is to promote unity and co-operation. Following Egypt's suspension in March 1979, Tunis has been made the temporary headquarters of the League, its Secretariat and permanent committees.

Selected Regional Economic Organisations

- Andean Group
- Association of South East Asian Nations (ASEAN)
- Central American Common Market (CACM)
- Caribbean Community and Common Market (CARICOM)
- Organisation for Economic Co-operation and Development (OECD)

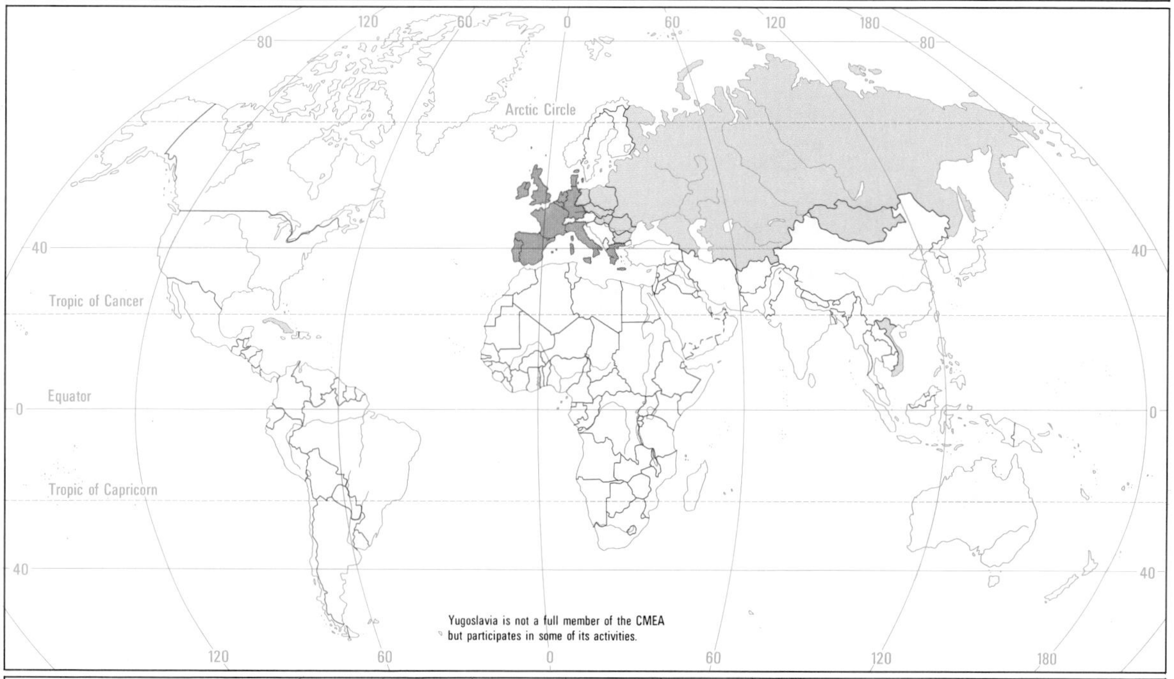

The Council for Mutual Economic Assistance and The European Community

- Council for Mutual Economic Assistance (CMEA or COMECON)
- European Community (EC or Common Market)

The CMEA was founded on 25 January 1949 to assist the economic development of its member states through the sharing of resources and the co-ordination of efforts. One of the Council's aims is to achieve greater economic and technical progress and in particular to promote a higher level of industrialisation.

The EC was founded on 1 January 1958 by the signing of the Treaty of Rome. The principal aims are to promote economic co-operation and common development of member states. This is achieved by the creation of a unified market without tariffs and a common external customs tariff. Free movement of labour, services and capital between members is also encouraged.

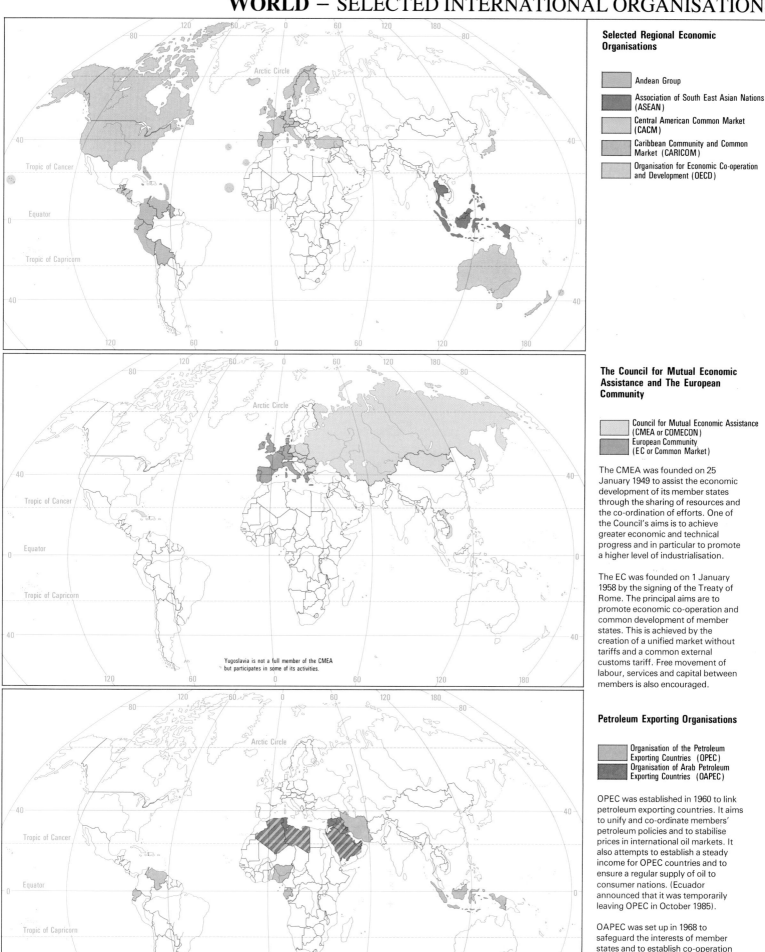

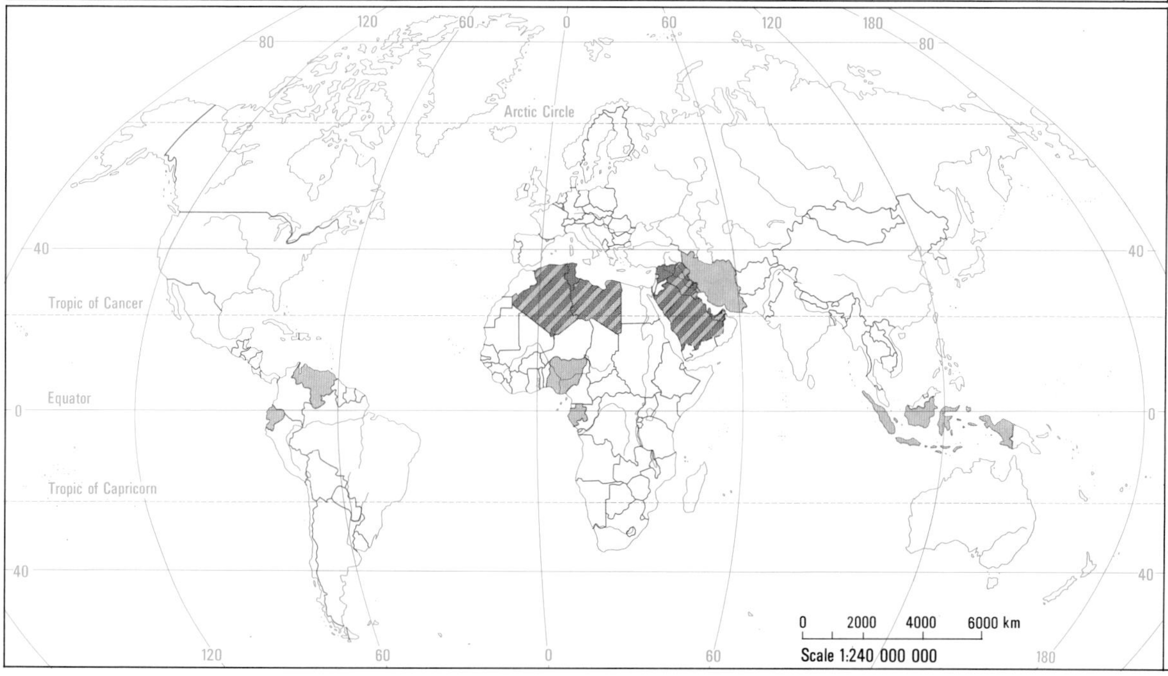

Petroleum Exporting Organisations

- Organisation of the Petroleum Exporting Countries (OPEC)
- Organisation of Arab Petroleum Exporting Countries (OAPEC)

OPEC was established in 1960 to link petroleum exporting countries. It aims to unify and co-ordinate members' petroleum policies and to stabilise prices in international oil markets. It also attempts to establish a steady income for OPEC countries and to ensure a regular supply of oil to consumer nations. (Ecuador announced that it was temporarily leaving OPEC in October 1985).

OAPEC was set up in 1968 to safeguard the interests of member states and to establish co-operation within the petroleum industry. It aims to co-ordinate several aspects of the industry including attempts to link petroleum research institutes within the Arab States. (Egypt's membership was suspended from 17 April 1979).

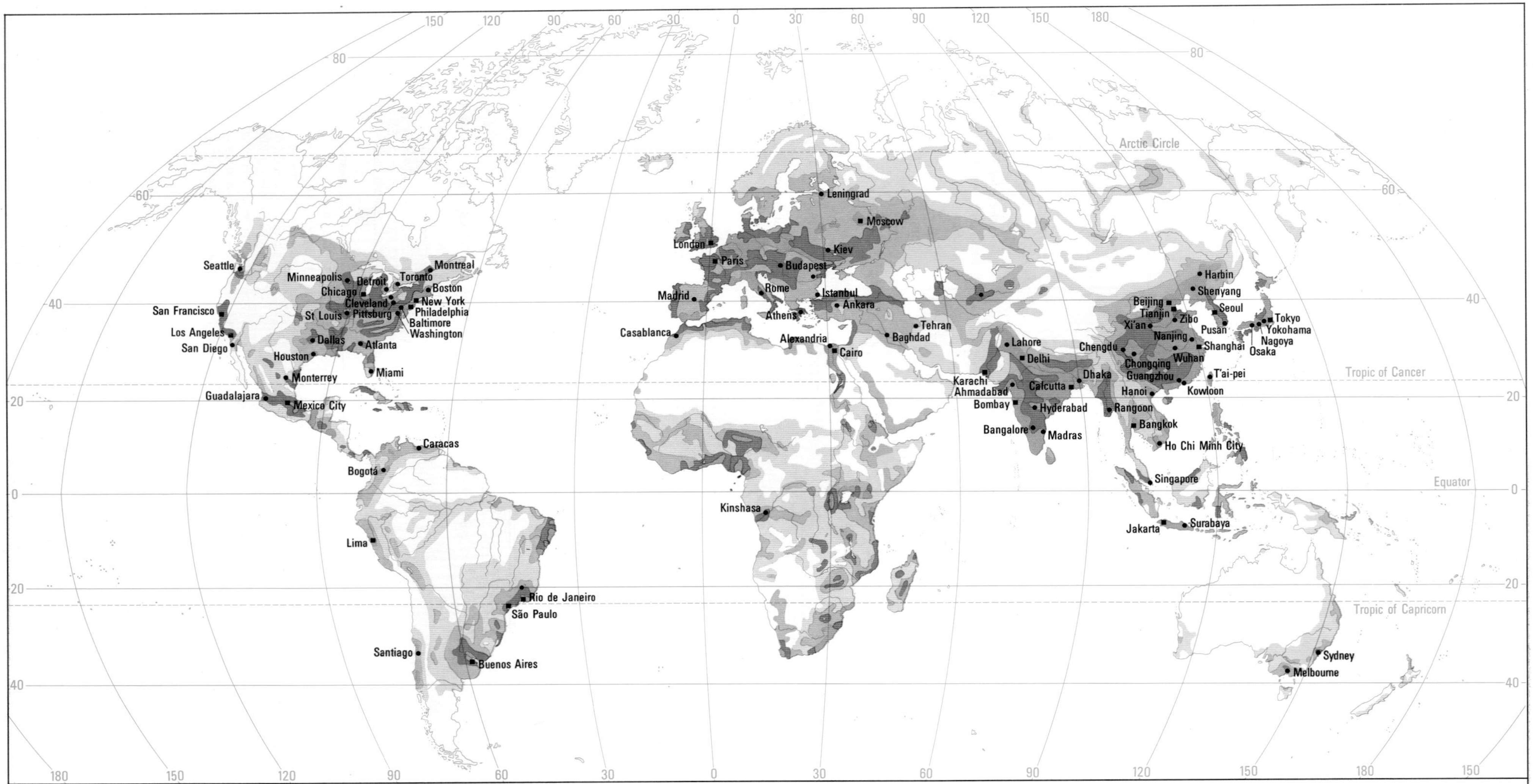

Population Density (persons per square km)

- Over 100
- 50–100
- 10–50
- 1–10
- Under 1

Population of Cities

- ■ Over 5 000 000
- • 2 000 000–5 000 000

Projection: Modified Winkel's
Equatorial scale 1: 145 000 000
0 1000 2000 3000 4000 5000km

Estimated Population of Selected Countries (in millions)

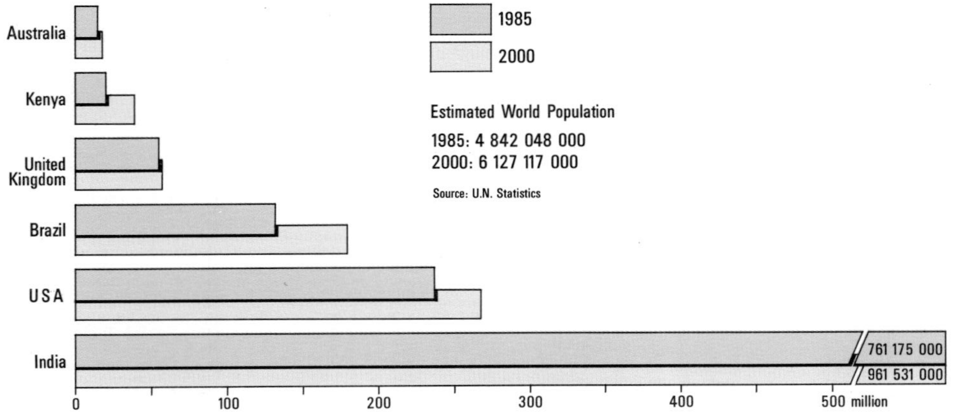

Estimated World Population
1985: 4 842 048 000
2000: 6 127 117 000

Source: U.N. Statistics

Urban Population as Percentage of Total Population

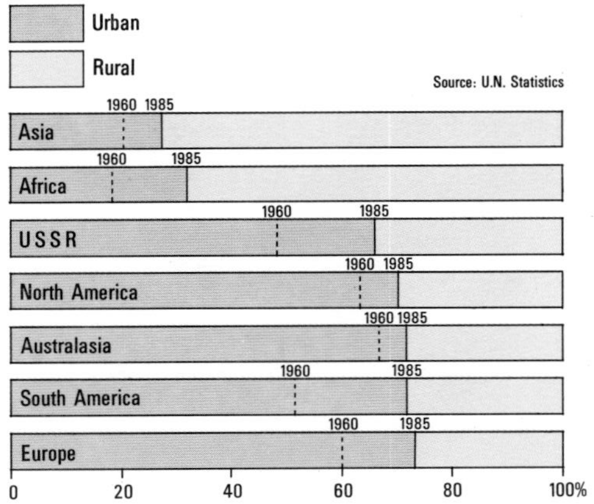

Abbreviations

A.C.T.	Australian Capital Territory	Pen.	Peninsula
Arch.	Archipelago	Pk.	Peak
B.	Bay	Plat.	Plateau
C.	Cape	Prov.	Province
Cent.	Central	Pt.	Point
Comm.	Commonwealth	R.	River
Dist.	District	Ra(s).	Range(s)
E.	East	Rep.	Republic
G.	Gulf	Res.	Reservoir
I(s).	Island(s)	S.	South
L.	Lake	SADR	Sahrawi Arab Democratic Republic
Mt(s).	Mountain(s)	Sd.	Sound
N.	North	Str.	Strait
NI.	Northern Ireland	UAE	United Arab Emirates
N.P.	National Park	UK	United Kingdom
N.R.	National Reserve	USA	United States of America
Nat. Sanct.	National Sanctuary	USSR	Union of Soviet Socialist Republics
Neth.	Netherlands	W.	West
PDRY	People's Democratic Republic of the Yemen	YAR	Yemen Arab Republic

INDEX

C

INDEX

INDEX

INDEX

N

O

T

INDEX